Green Energy and Technology

More information about this series at http://www.springer.com/series/8059

Nicolae Badea
Editor

Design for Micro-Combined Cooling, Heating and Power Systems

Stirling Engines and Renewable Power Systems

Springer

Editor
Nicolae Badea
"Dunarea de Jos" University of Galati
Galati
Romania

ISSN 1865-3529 ISSN 1865-3537 (electronic)
ISBN 978-1-4471-7186-7 ISBN 978-1-4471-6254-4 (eBook)
DOI 10.1007/978-1-4471-6254-4

Springer London Heidelberg New York Dordrecht

Printed on acid-free paper

Springer is part of Springer Science+Business Media (www.springer.com)

Acknowledgments

The authors acknowledge the financial support provided by EEA Grants Iceland, Lichtenstein, Norway, through Project RO 0054/2009 in achieving experimental trigeneration system.

Acknowledgments

Contents

Microgeneration Outlook

George Vlad Badea

Abstract This introductory chapter will blend both legal and technical aspects of microgeneration systems in order to acquaint the readers with the concept and roles of microgeneration systems, the perception of the European Union and the ways of promotion and development through policies and legal instruments. These notions are fundamental for readers and practitioners in the field of microgeneration systems since a variety of factors work in close connection and have a profound influence on the development of microgeneration systems. This chapter will make short explanatory remarks about the evolution (1) of the European Union and the energy sector in Europe in the transition to decentralised energy production and extensive use of microgeneration systems. Afterwards, the challenges (2) confronting the European energy sector are presented in order to understand the way problems are tackled by the European Union through policies (3) and legal instruments (4) to comprehend the use, promotion and trend for development of microgeneration systems (5).

1 Evolution

The past decades have witnessed important changes both in political and technological senses. Europe has evolved, people have evolved and brought along technological progress. Nonetheless, equally important challenges have arisen and Europe has to adapt to the new realities and find solutions in a reliable and sustainable way. In the energy sector, the reality is that the "existing energy systems need to be modernised" [1] in order to adapt to the economic, social and environmental contexts as Europe is struggling with "unprecedented challenges resulting from increased dependence on energy imports and scarce energy resources, and the need to limit climate change and to overcome the economic crisis" [2].

G.V. Badea (✉)
ICPE SA, Bucharest, Romania
e-mail: badea.george.vlad@gmail.com

N. Badea (ed.), *Design for Micro-Combined Cooling, Heating and Power Systems*,
Green Energy and Technology, DOI 10.1007/978-1-4471-6254-4_1

1.1 The European Union

Politically, "the European Union is a remarkable innovation in relations among states" [3] and both the Union and the energy sector have equally evolved. The European Union started in the 1950s with the European Economic Community (EEC), than the European Community (EC) in 1993 after the Treaty of Maastricht and becoming in 2009 after the Treaty of Lisbon, the European Union as we now know.

Legally speaking, there are more than 50 years since the entry in force of the first of the treaties that shaped the modern European Union. The current treaty in force since 1 December 2009 is the Treaty of Lisbon (the Treaty on the Functioning of the European Union—TFEU) being preceded by the Nice Treaty (2003), the Amsterdam Treaty (1999), the Maastricht Treaty (1993) and the Single European Act (1987) [4].

1.2 The European Energy Sector

It is important to know that whether founding or joining the European Union, the Member States (MS) freely undertook certain Treaty obligations which are fundamental for the proper development of the Union. And if, historically, the energy sector was an exclusive competence of the State, given its increasing importance, it has nowadays become a shared competence between the European Union and the MS which must be satisfied. Moreover, this delimitation of competences, either exclusive of shared between the European Union and the MS, is explicitly stated[1] in the Treaty.

As such, the energy sector is a shared competence[2] with a well established legal basis.[3] Accordingly, the European Union aims at ensuring the functioning of the

---

[1] Article 2, Para's 1 and 2 of TFEU

"1. When the Treaties confer on the Union *exclusive competence* in a specific area, only the Union may legislate and adopt legally binding acts, the MS being able to do so themselves only if so empowered by the Union or for the implementation of Union acts.
2. When the Treaties confer on the Union a *competence shared* with the MS in a specific area, the Union and the MS may legislate and adopt legally binding acts in that area. The MS shall exercise their competence to the extent that the Union has not exercised its competence. The MS shall again exercise their competence to the extent that the Union has decided to cease exercising its competence".

[2] Article 4, Para. 2 of TFEU
"1. Shared competence between the Union and the MS applies in the following principal areas:

(i) *energy*.

[3] Article 194, Para. 1 of TFEU

"1. In the context of the establishment and functioning of the internal market and with regard for the need to preserve and improve the environment, Union policy on energy shall aim, in a spirit of solidarity between MS, to:

energy market and the security of energy supply, promoting energy efficiency, energy savings, renewable energy sources (RES) and the interconnection of energy networks. On the other hand, the MS have a relative independence in determining the way in which their energy resources are explored and the free choice of energy sources.

1.3 Traditional Grids Versus Smart Grids

Technically, the mature European energy system that has "provided the vital links between electricity producers and consumers with great success for many decades" [5] is in fact adapting to the current realities (economical, environmental, social, technological, etc.). The European Union has started a transition from the traditional, centralised way of producing energy (Centralised Energy Production—CEP) from fossil fuels and nuclear-based power systems to a modern, decentralised way of producing energy (Decentralised Energy Production—DEP) from small-scale generation from RES, using low-carbon solutions such as the microgeneration systems. This, in turn, implies a shift in energy consumer's activity, from the traditional passive consumers to modern active consumers which become themselves producers [6]. To have a visual image of the above-mentioned, the traditional electricity grid in a simple depiction goes from production of electricity in power plants, transmission of electricity through high-voltage lines and distribution to consumers through low-voltage lines as presented in the following Fig. 1.

On the other hand, the new grids, commonly known as Smart Grids, are "intelligent energy supply systems" [8] that in Europe are being defined as "electricity networks that can intelligently integrate the behaviour and actions of all users connected to it—generators, consumers and those that do both—in order to efficiently deliver sustainable, economic and secure electricity supplies" [9]. This entails that Smart Grid covers the entire electricity chain from production to consumption, with bidirectional flows of both energy (import and export of energy, easy grid access) and information (real time interactions with electricity market), as shown in the next Fig. 2.

(Footnote 3 continued)

(a) ensure the functioning of the energy market;

(b) ensure security of energy supply in the Union;

(c) promote energy efficiency and energy saving and the development of new and renewable forms of energy; and

(d) promote the interconnection of energy networks".

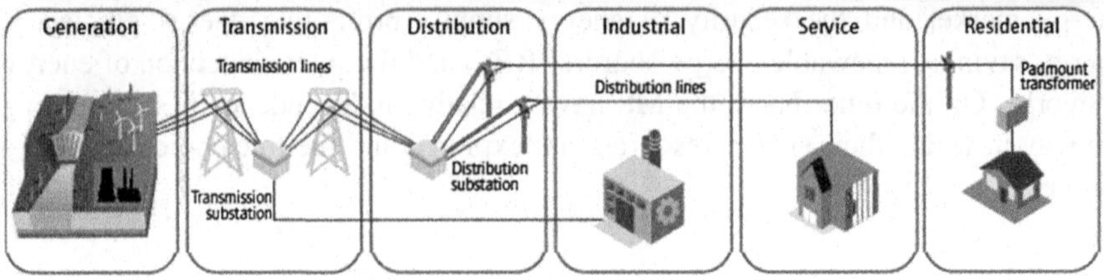

Fig. 1 Traditional electricity grid [7]

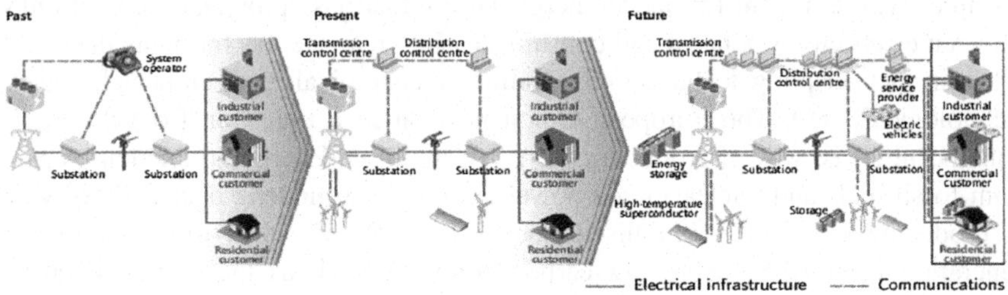

Fig. 2 A Smart Grid [10]

1.4 Microgeneration Systems

In the context of moving towards Smart Grids, new low-carbon solutions such as microgeneration systems are essential. The successful attainment of Smart Grids is largely dependent on the large-scale implementation of microgeneration systems by residential consumers and small and medium-sized enterprises (SMEs) [11].

Microgeneration systems are a form of "decentralised energy generation" [12] used for the "small-scale generation of heat and electric power by individuals, small businesses and communities to meet their own needs, as alternatives or supplements to traditional centralised grid connected power" [13]. Within a Smart Grid, the electricity produced by the microgeneration systems can be used not only for consumers needs but can also be delivered to the grid in exchange for certain revenues.

Microgeneration systems use as intake either fossil fuels, RES or a combination of both in order to generate heat, electricity or a cumulus of both heat and electricity. Considering the technologies used to obtain the heat or/and electricity, microgeneration systems are divided in three categories:

- Microheat based on heat pumps (air, water and ground source), biomass and solar thermal;
- Microelectricity based on solar PV, microwind turbines and microhydro;
- Micro Combined Heat and Power (micro-CHP) or Cogeneration systems based on the internal combustion engines, Stirling engines and Fuel cells.

In order to fully understand the role and increasing importance of microgeneration systems within the actual European context, a critical look into the challenges confronting the European energy sector is strongly needed. Consequently, the next section will present the current challenges in the energy sector in view of facilitating the interpretation of corresponding European policies and legislation.

2 Challenges in the European Energy Sector

"Energy is essential for Europe to function" [14], it is vital for industry, for transport, for households and businesses alike. It is present in every aspect of life. Derived from the Greek "Enérgeia", it meant being active or being at work [15]. Likewise, "the word energy incidentally equates with the Greek word for *challenge*" [16]. In this line of thought, the European energy sector has an entire array of challenges and barriers to cope with: from environmental issues to ensuring the security of supply for the European Union, from the liberalisation of the energy market to social, economic, financial and technological matters. Last but not least, there are political and legal shortcomings.

The European energy sector has to mitigate *environmental concerns* on climate changes resulted from greenhouse gas (GHG) emissions, ensure the *security of supply* by limiting the external vulnerability on energy imports in the context of increasing energy demands in the "global race for energy sources" [17], develop the infrastructure and connections for the European energy networks to avoid possible energy crisis or energy cut-offs. At the same time, the European Union has to develop a fully liberalised *Internal Energy Market*, in which fair competition is stimulated, investments are booming, innovation is fostered through research and development (R&D) of emerging technologies, in order for the Union to safeguard its place as a powerful global energy market.

There is also a *social dimension* implying that citizens should be able to take full benefit of energy liberalisation in terms of reduced energy prices, encouraging energy savings, boosting job creation and increased welfare on the overall [18]. The *economic and financial* [19] drawbacks, as well as the *technological* barriers have a bi-faceted nature. Firstly, they consist of insufficient investments in the development of low-carbon technologies (including microgeneration systems) in order to obtain highly efficient equipments at relatively affordable prices in a competitive market. Secondly, starting from the costs of equipments, the drawbacks refer to the inadequate incentives given by the European Union and its MS for the future owners of equipments (households, SMEs) leading thus to a slow adoption of RES technologies.

The *political* [20] and *legal* [21] barriers refer to the fact that, as it will be presented at a later stage, on the one hand the European Energy policies are in place, but the actual implementation is a lengthy process. On the other hand, the current legal framework still presents barriers and inconsistencies for the smooth shift to the new way of producing energy, Distributed Energy Production using the full potential of RES.

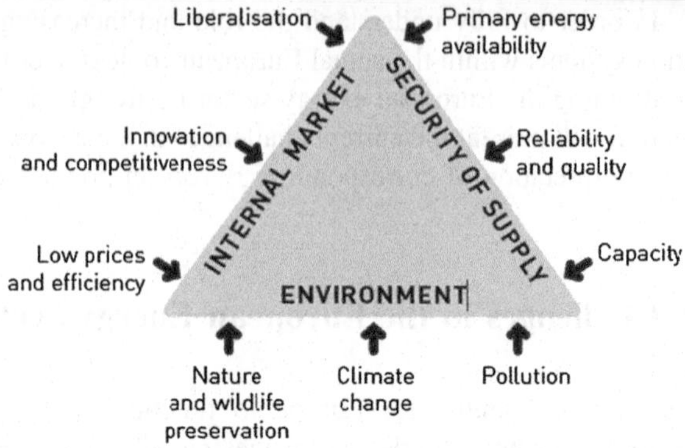

Fig. 3 Fundamental challenges [22]

2.1 *Connecting the Dots*

Connecting the dots between all the pointed challenges and in spite of all the differences among the states in the European Union, all of them have to solve three main problems, shown in Fig. 3:

- *Environment/Sustainable development*: limiting the environmental impact of energy production, transport and use [23].
- *Energy security/Security of supply*: ensuring the reliability and continuity of energy supply.
- *Internal market/Competition*: a highly competitive internal energy market with reduced energy prices for consumers (households, businesses).

A holistic approach is helpful to see how all these challenges present in the European energy sector interlink.

2.1.1 Environment/Sustainable Development

Environmental concerns are linked to the notion of sustainable development, having a well-established tradition both internationally [24][4] and at European level [25],[5] with the purpose of "promoting well-being of citizens now and in the future"

[4] The increase in environmental movements in the 1950s brought the concern for "sustainable development". However, it was not until 1987, when the United Nations released the Brundtland Report (Report of the World Commission on Environment and Development: Our Common Future) that the notion of "sustainable development" was firstly framed ("development which meets the needs of the present without compromising the ability of future generations to meet their own needs").

[5] Since 1985, the ECJ sought the importance of environmental protection in Procureur de la République v. Association de Défense des Bruleurs d'Huiles Usagées (See ECJ Case C-240/83). Sustainable developments was first enshrined in the Maastricht Treaty (1992) and reinforced in the Amsterdam Treaty (1997).

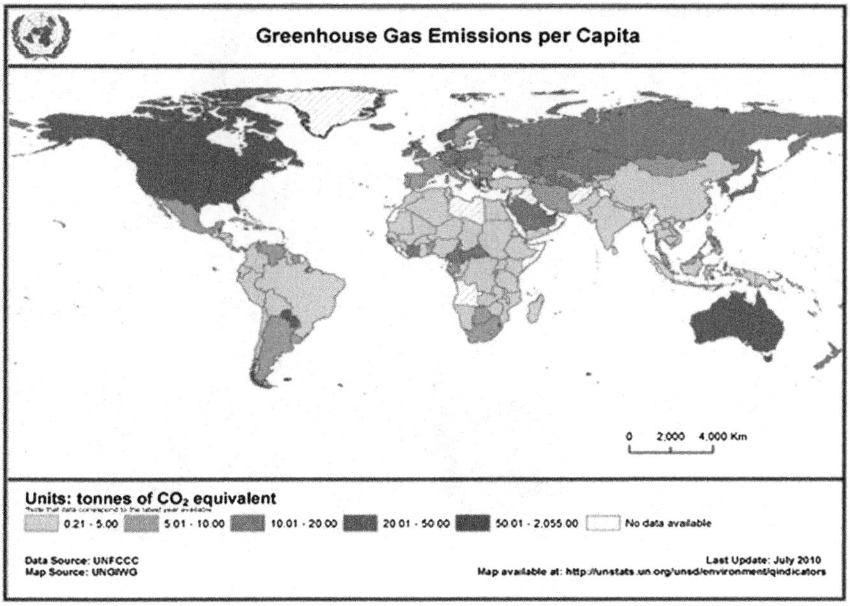

Fig. 4 UN map of GHG emissions [28]

[26]. However, the awareness and care for sustainable development have increased in the past years due to climate changes and air pollution. The leading polluters (Fig. 4) are the GHG with their main component carbon dioxide (CO_2).

Mitigating greenhouse gases emissions is an important challenge as they are at the root of climate changes and most air pollution [27].

In a global context, the European Union contributes (Fig. 5) with around 11 % of the total GHG emissions each year [29].

These emissions are shared among the MS and come from different sectors (Fig. 6). The need to reduce GHG emissions is obvious for the improvement of the environment contributing to sustainable development.

Moreover, such decreases are binding through both internationals agreements (Kyoto Protocol for instance) and European policies and legislation. An option to successfully reduce GHG emissions is exploiting the full potential of RES. In this sense, European efforts are being made to increase the share of RES in the energy production (Fig. 7).

Nonetheless, the share of RES in the energy consumption is halved (Fig. 8) and the energy mix differ across the MS (Fig. 9).

Between the energy production and energy consumption, a large share of energy is used by the households and services sectors, as noticed from Fig. 10.

The potential for these sectors to reduce GHG emissions is not yet fully tapped, though significant improvements are in progress (Fig. 11). Consequently, in these sectors *microgeneration systems* can be/are applied and can have a fundamental position to help decrease the GHG emissions.

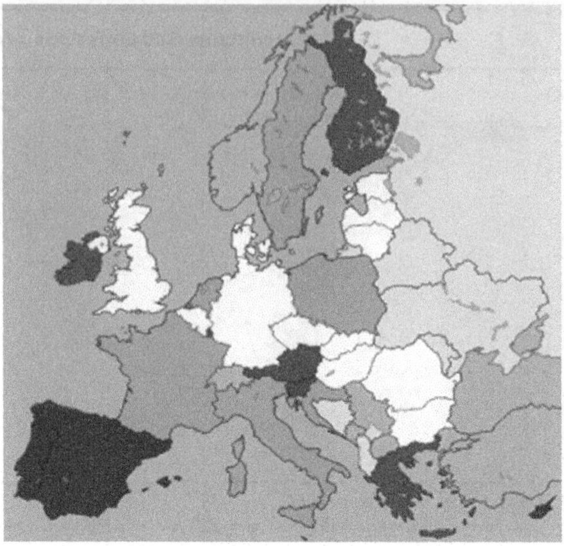

Fig. 5 GHG emissions in the MS [30]

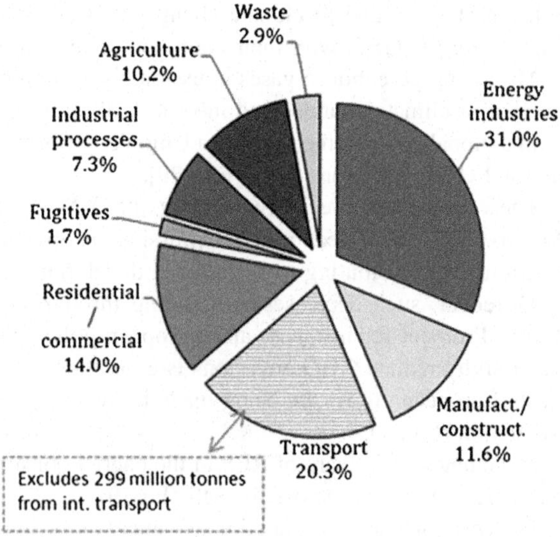

Fig. 6 EU GHG emissions by sector in 2011 [31]

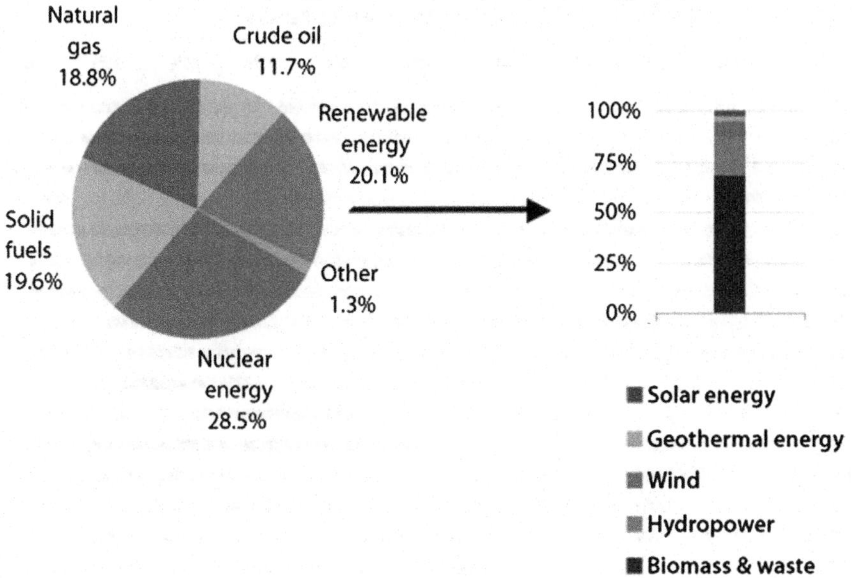

Fig. 7 EU primary energy productions in 2010 [32]

Fig. 8 EU energy consumption in 2011 [33]

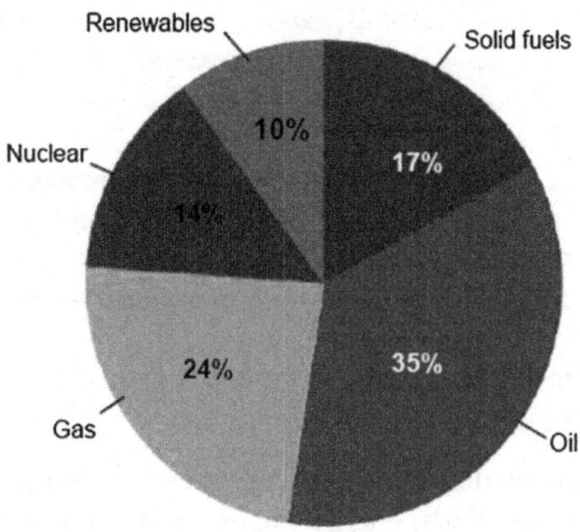

2.1.2 Energy Security/Security of Supply

With these environmental concerns in mind, the focus is on ensuring the *security of supply*, "Europe's biggest energy goal" [35]. In the context of lessening energy resources and constantly increasing energy demands, Europe is in a global race for energy sources [17].

In the European Union, the energy demand has been slowly but steadily increasing (Fig. 12). At least in the electricity sector, the increase is of 1.5 % each year [37]. Moreover, since the primary energy sources are insufficient to cover the

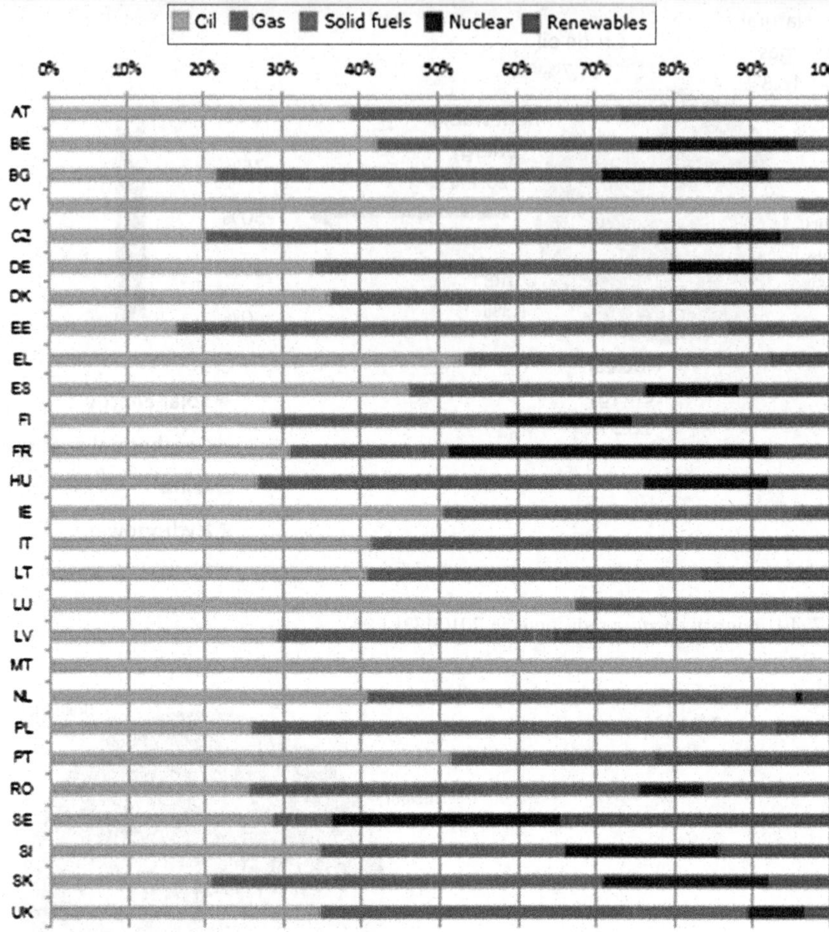

Fig. 9 EU MS energy mix in 2011 [34]

energy demand, the energy sector in Europe is also dependent on energy imports. If in the 1980s, energy imports were around 40 %, at the end of the 1990s, imports reached 45.1 % [38] to amount in 2010, 54.1 % [39] (with a peak value of 56.3 % in 2008). To this picture, under a "business as usual" scenario, projections estimate an increase up to 65 % by 2030 [40]. In addition, given this vulnerability on energy imports and the inadequate development of energy infrastructure and interconnections (as decentralised energy systems cannot be accommodated yet), an energy crisis could have a detrimental effect on the MS in terms of both political and economic risks.

Per a contrario, security of supply (or *energy security*) by definition implies the "uninterrupted availability of energy sources at an affordable price" [41] precisely to avoid such vulnerabilities and prospective crises or cut-offs. Therefore, energy

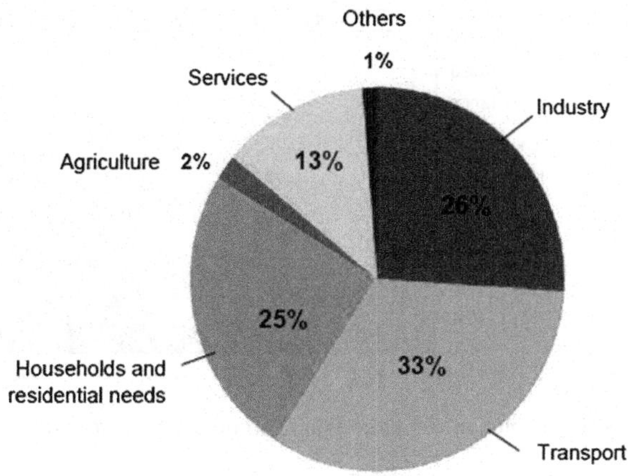

Fig. 10 EU final energy consumption in 2011 [33]

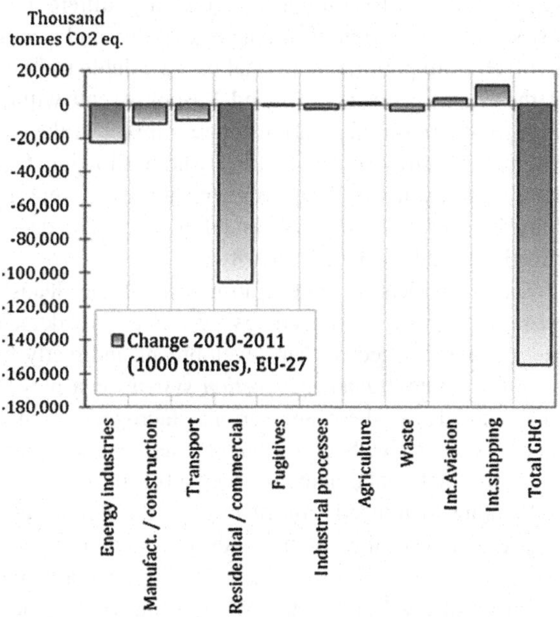

Fig. 11 EU GHG reductions 2010–2011 [31]

security entails on the one hand, long-term security of resource availability and on the other hand, short-term security related to supply reliability [42]. In order to provide energy that is readily available and a reliable source of power, there are several *key aspects* that need to be developed. Among them, microgeneration systems can have a determining role.

To begin with, relating to production of energy, the European Union must have a well-balanced energy supply system with diversified production technologies. It is also called fuel mix or *energy mix* (see Fig. 9 above) and comprises a "combination

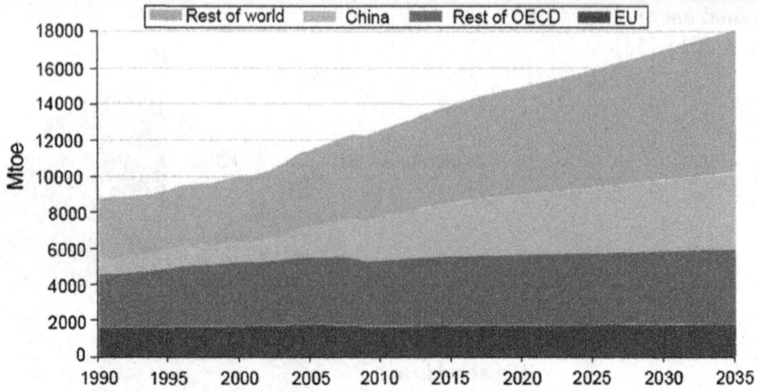

Fig. 12 Evolution and projection of world energy demand [36]

of sources used to generate energy at any given time and place" [43]. Every Member State of the European Union has a different energy mix as it depends on a series of factors varying from the availability and sizes of state resources to all costs involved from production to final use, available technologies and the characteristics of the energy demand (Demand Response), all within clear policy and legal settings. Furthermore, the various energy mixes in the MS can have clear advantages within the European Union as a whole allowing flexibility in meeting Member State's energy needs by maximising the use of energy resources, ensuring continuity in energy supply and enabling a certain degree of energy independence, directly related to energy security.

The diversification of generation technologies leads also to an easy and swift *fuel substation* as traditional energy resources are decreasing significantly and new energy sources need further development, indirectly helping energy security.

In this context, *microgeneration systems* can play their part, if spread and used wide enough, as there are various microgeneration technologies using different energy sources (see Sect. 4) that can help enhance the energy mix. Moreover, they are a sustainable alternative to fuel substitution.

Relating to transmission of energy, the key aspect in ensuring the security of supply and flexibility of the energy system is having proper *infrastructure* and *interconnections* to easily relocate readily available energy. This implies that both at European level and in the MS, starting from generation to final use, the infrastructure and interconnections in and between the MS must be adequately developed (transmission lines and interconnection systems) to ensure easy access to energy. This also applies to users of *microgeneration systems* as they need suitable grid access to export (excess) energy.

Another key aspect to ensure the security of supply is improving the *quality of energy*. It can also be called *high efficiency* in both producing and transmitting energy and can only be achieved through significant *investments*. The level of investment required should be consistent from both the European Union and the MS. With regards to high-efficiency transmission, this could imply transition to DEP (see Sect. 3). This shift also covers *microgeneration systems* as they are at the basis of DEP. The investments in the production of energy refer to improving generation technologies to become highly efficient in producing more energy with fewer resources, having also low-carbon emissions. Among these new highly efficient technologies, *microgeneration systems* are included, as the latest technologies on microgeneration use only RES as input and produce a considerable amount of energy (for instance micro-CHP produce cooling, heat and electricity). Furthermore, all these new technologies directly help improve the environment with their low-carbon emissions.

The last issue to be considered in ensuring the security of supply and connected to high efficiency regards *energy savings*. If the above arguments had the context of increasing energy demands, this key point emphasises how energy can also be saved. Using high-efficiency technologies and improved *energy management* energy savings can be obtained through *energy efficiency*. This means using less energy or recovering energy losses to cater for the same needs. To this end, a great potential for energy savings is in the household and commercial sectors as buildings across Europe use around 40 % of the total final energy consumption (see Fig. 11), being the largest end-use sector. In these sectors, the majority of energy is used by residential buildings as they cover around 76 % of the total building floor area [8], with 65 % single family houses [44]. Keeping in mind this large untapped potential for energy savings, the large-scale integration of *microgeneration systems* in these sectors can have a determining role in both ensuring the security of supply and helping mitigate the climate changes.

Internal Energy Market

Furthermore, having as starting point, the direct and ancillary effects that installing microgeneration systems in buildings can have, the final point of discussion in this sub-chapter is the *Internal energy market*. The internal energy market implies the interconnection of regional and national markets in the short run and the interconnection of an EU wide energy market in the long run. There are three decisive factors that determine the development of the internal energy market in the European energy sector: *liberalisation* (unbundling), *competition* and *innovation and investments*.

The entire idea of having a single internal energy market in the European Union began with the *liberalisation* process. It is a lengthy and ongoing process that started at the end of the 1990s and went through several steps[6] facing strong opposition along the way,[7] from the biggest energy companies[8] to the national regulatory cultures [45] and the incomplete implementation and enforcement of EU energy legislation by EU MS [46]. The rationale behind the liberalisation process is

[6] Conerstones: 1990s, 2003, 2006, 2009 and lastly 2012.

[7] The energy markets were dominated by national monopolies that wanted to preserve their status quo. See also Sect. 2.

[8] For instance in France, where the markets were dominated by giants such as EDF and GDF.

to unbundle all the links in the energy chain (see Fig. 1) and considerable efforts have been made in this sense.[9] The end result is to have a fully liberalised internal energy market for both gas and electricity.

The effects of market liberalisation are also closely connected with stimulation of fair and dynamic *competition* within the entire European Union. This is a strategic instrument entailing easy access to the market, giving consumers free choices for energy suppliers and creating a level playing field for new comers into the market. Consequently, the market becomes flexible, ensuring consumers' needs and responding to challenges, accessible, granting easy connection to all network users, reliable, assuring and improving the security and quality of supply and economic, providing best value through innovation, efficient energy management and level playing field for competition and regulation [47].

In order to ensure the efficient implementation of the internal energy market, *innovation* and *investments* is the key. The market must continuously meet the requirements of the vibrant business environment in "guaranteeing high security, quality and economic efficiency" [48]. Accordingly, investments in innovation (R&D) are fundamental not only to achieve the fully liberalised internal energy market but also to secure the energy and help mitigate climate changes.

Moreover, all these interlinked components of the internal energy market have a multi-layer effect. The fully liberalised highly efficient and competitive internal energy market makes energy a universal service for all consumers who should enjoy full benefits. Having a liberalised and competitive market put pressure on prices, linked to the definition for security of supply—energy at affordable prices. Combined with energy management, it leads to energy savings in terms of economic value (cutting the energy bill). Adding investments and innovation, it creates new and easy market entries for a large plethora of energy players, increases revenues and boosts job creation along the process.

The impact of an efficient internal energy market stretches over *microgeneration systems* too, covering the entire link from production to use. In this sense, promotion and investments in R&D increase the business opportunities for technological producers, while increased competition lowers the prices of equipments and leverages their efficiency. Combined with appropriate incentives, the market is accessible to all energy producers, "especially the smallest and those investing in renewable forms of energy" [49], high-efficiency local generation with zero or low-carbon emissions.

Lastly, finding optimum solutions to all these challenges is not an easy task. Nonetheless, their successful overcoming will ultimately lead to increased welfare of European citizens and strengthen the position of the European Union as a global energy leader. Together with the right legal instruments and enforcement measures, there might be a way of solving.

---

[9] The unbundling process started from unbundling Transmission System Operators (TSOs), Distribution System Operators (DSOs) to nowadays unbundling consumers, the last link in the energy chain and engaging them in the internal energy market. This is also the reason for previously starting directly from the downstream market of microgeneration systems.

3 Policy Overview

In order to overcome all the above-mentioned challenges and barriers facing the European energy sector, the European Union has started developing an "ambitious, competitive and long term—and to the benefit of all Europeans" [39] energy policy with the aim of moving "towards competitive, sustainable and secure energy throughout Europe" [50]. The main action directions are the following:

- European Energy Policy (Energy policy for Europe, Market-based instruments, Energy technologies, Financial instruments)
- Internal Energy Market (The market in gas and in electricity, Trans-European energy networks, Infrastructure, Security of supply, Public procurement, Taxation)
- Energy Efficiency (Energy efficiency of products, buildings and services)
- Renewable Energy (Electricity, Heating and Cooling, Biofuels)
- Nuclear Energy (Euratom, Research and technology, Safety, Waste)
- Security of Supply, external dimension and enlargement (Security of supply, External relations, European Energy Charter, Treaty establishing the Energy Community, Enlargement).

For the purpose of achieving a "competitive, reliable and sustainable energy sector" [51], the European Union has taken concrete policy steps [52]: adoption and implementation of a short-term Europe 2020 policy (The 2020 climate and energy *package*) and recent adoption of a 2030 framework policy (The 2030 *framework* for climate and energy policies), all under a long-term overarching Roadmap 2050 (*Roadmap* for moving to a low-carbon economy in 2050). Figure 13 provides a visual description on how the policies complement and support each other.

Starting from Europe 2020, the subsequent policies rely on previous results, reinforce and upgrade each other, constantly moving one step further. To keep consistency in reaching the objectives, the Europe 2020 policy has been created and is implemented to efficiently reach short-term goals. At the same time, ultimate goals have been set by Roadmap 2050, with intermediate results in the 2030 framework.

3.1 Europe 2020

Since 2007 [53], the European Union has elaborated an energy and climate change policy to tackle the deficiencies in the increasingly important energy sector. This policy has been incorporated into Europe 2020 [54], a comprehensive strategy for the period 2010–2020 aiming to deliver smart, sustainable and inclusive grown.[10]

Europe 2020 covers economic, social and energy issues promoting three mutually reinforcing priorities: development of an economy based on knowledge

[10] See http://ec.europa.eu/europe2020/index_en.htm.

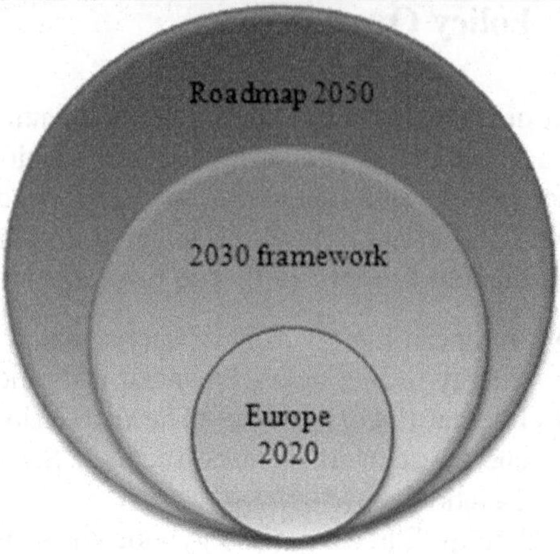

Fig. 13 Energy policy outlooks

and innovation (smart growth), promotion of an efficient, greener and more competitive economy (sustainable growth) and enhancement of high-employment economy with social and territorial cohesion (inclusive growth). Following the Lisbon Strategy (2000–2010), it has five headlines, among which the *20/20/20 climate and energy targets* [55] establishing clear key objectives to be reached by 2020:

- At least 20 % reduction of GHG emissions compared to the 1990 levels (30 % in right conditions);
- At least increase to 20 % in the share of EU energy consumption produced from RES;
- At least 20 % reduction in primary energy use by 20 % improvement of EU energy efficiency.

Moreover, the energy strategy encompasses five priority areas accompanied by supporting flagship initiatives to enhance and reinforce them:

- Achieving an energy efficient Europe;
- Building a truly pan-European integrated energy market;
- Empowering consumers and achieving the highest level of safety and security;
- Extending Europe's leadership in energy technology and innovation;
- Strengthening the external dimension of the EU energy market.

In order to develop the priority areas in ensuring the accomplishment of the key policy objectives, binding measures have been implemented through legislation. To this end, the climate and energy package adopted in 2009[11] comes to help mitigate climate changes (GHG emissions) and increase the share of RES. Furthermore, to improve EU's energy efficiency, a flagship initiative for Resource efficient Europe

[11] The climate and energy package: EU Emissions Trading Directive (EU ETS 2009/29/EC), Effort Sharing Decision (non-ETS 406/2009/EC), Carbon capture and geological storage (CCS 2009/31/EC) and Renewable Energy Sources Directive (RES 2009/28/EC).

[56] and an Energy Efficiency Plan [57] have been developed in 2011 to smoothen the transition to Smart grids [58] and in 2012, the Energy Efficiency Directive [59] has been adopted as enforcement measure.

There are several key elements in the EU energy policy related to *microgeneration systems*. To start with, the priority areas of an energy efficient Europe and empowerment of consumers to help ensure the security of supply cover directly microgeneration systems. To the same end, the EU places a big energy saving potential from buildings, again in direct relation to microgeneration systems. Furthermore, by increasing the use of RES, microgeneration systems are also covered and by promoting energy efficiency within the energy producers and end-users, the widespread deployment of microgeneration systems is accelerated. Taking into account that another priority area is the development of energy technology and innovation and the fact that investments in R&D are encouraged through both the Energy 2020 policy and a separate Europe 2020 headline (3 % of the EU's GDP should be invested in R&D), all these measures boost the renewables industry encouraging technological advancements in developing highly efficient microgeneration systems.

3.2 Roadmap 2050

At the same time with the development of short-term objectives within Europe 2020, the long-term goal has also been drafted [60] in 2011: reduction of GHG emissions by at least 80 % (95 % in right conditions) below 1990 levels by 2050 (as shown in Fig. 14). The result is the Energy Roadmap 2050 [61], a guideline to a prosperous, low-carbon European Union. In order to achieve the decarbonisation objective and in line with energy security and competitiveness goals, feasible and cost-efficient pathways are searched to make "the European economy more climate-friendly and less energy-consuming" [62].[12]

The road to a low-carbon economy by 2050 is also closely connected to *microgeneration systems* as clean technologies are essential for the European Union to cut most of its GHG emissions. Moreover, energy efficiency is regarded as a key driver in this transition. Consequently, innovation and investments in clean technologies and low or zero-carbon energy are needed and encouraged (innovation and green growth) as many such technologies exist today but need to be further developed. This implies a "much greater need for renewable sources of energy, energy-efficient building materials, hybrid and electric cars, 'smart grid' equipment, low-carbon power generation and carbon capture and storage technologies" [62]. In addition, locally produced energy from RES significantly increases, developing

[12] See http://ec.europa.eu/clima/policies/roadmap/index_en.htm, www.roadmap2050.eu and http://ec.europa.eu/energy/energy2020/roadmap/index_en.htm.

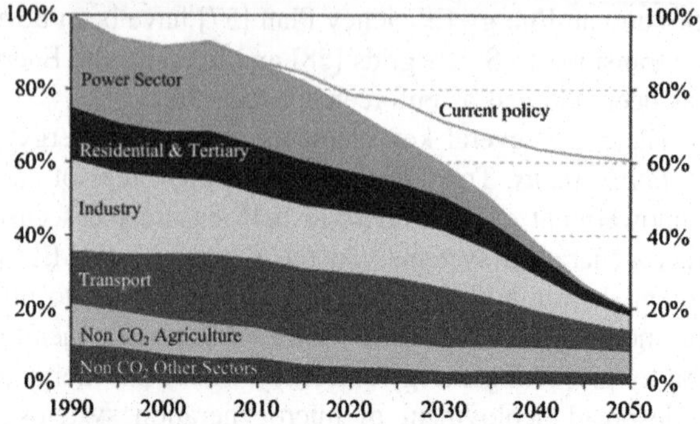

Fig. 14 EU GHG emission trend to 2050 [63]

DEP. All these converge to mitigating climate changes and ensuring energy security in a highly competitive European Union.

The Roadmap proposes also intermediate milestones on the pathway to at least 80 % GHG emissions reductions, 40 % by 2030 and 60 % by 2040. Naturally, the next consistent step of the EU energy policy is the creation of a framework with clear targets for 2030.

3.3 Framework

The 2030 framework for climate and energy policies, started in spring 2013 [64], has been presented on the 22 January 2014 [65] to continue the efforts in moving towards a competitive, secure and sustainable energy system in the European Union. Based on the Europe 2020 climate and energy package (Sect. 3.1) and taking into account the Roadmap 2050 (Sect. 3.2), this framework is to be debated in spring 2014.[13]

The aim of this policy framework is to continue to mitigate the challenges in the European energy sector as presented in Chap. "Decentralized Poly-generation of Energy: Basic Concepts" in order to "build a competitive and secure energy system that ensures affordable energy for all consumers, increases the security of the EU's energy supplies, reduces our dependence on energy imports and creates new opportunities for growth and jobs" [66].

It draws few concise key objectives on GHG emissions and RES: reduction of GHG emissions by at least 40 % compared to the 1990 levels; increase to at least 27 % in the share of EU energy consumption produced from RES. But it also leaves room for improvements without mentioning clear objectives in continuing the development of energy efficiency, reforming the EU ETS system by establishing a market stability reserve, key set indicators for progress assessment and proposal for a new governance framework based on national plans for competitive, secure and sustainable energy.

[13] See http://ec.europa.eu/clima/policies/2030/index_en.htm.

Microgeneration systems are also engaged as RES has a crucial position in the shift to a competitive, secure and sustainable economy and energy efficiency is seen as an essential contributor to the European climate and energy policies, with massive investments strongly encouraged in low-carbon generation. Nonetheless, it is still work in progress as on the 5 February 2014, the European Parliament criticised the framework advancing legally binding targets on CO_2 emissions, renewables and energy efficiency: 40 % reduction in GHG emissions, 30 % share of RES and 40 % improvement in energy efficiency.

4 Regulatory Framework

Taking into account the three main challenges existing in the European energy sector addressed by the European Union through specific "pillar" actions in the policy measures (namely, competitiveness, security of supply and sustainable development), the concrete solving solutions are being given and implemented through a legislative framework. The legal framework for microgeneration systems has met a constant evolution, starting from a pillar specific approach in which different legislative acts are correlated to the latest integrated approach in which the legislative act addresses all pillars.

The legal advances of microgeneration systems have started in 2004 with the CHP Directive [67] intended as general framework for the promotion of high-efficiency cogeneration addressing security of supply through energy efficiency. Later in 2006, the Annex III of ESD Directive [68] improved the security of supply through end-use efficiency and energy services. In 2009, microgeneration systems were covered in the RES Directive [69] that addressed the issue of sustainability by promoting the use of energy from renewable sources, touching also on security of supply. Competition issues such as the implications in the internal energy market were first tackled by Directive 2003/54/EC establishing common rules for the internal market in electricity which was repealed in 2009 by the adoption of the third internal energy market package among which the IEM Directive [70] introducing common rules for both internal gas and electricity markets. In 2010, the EPBD Directive [71] in Article 1, Paragraph 1 (b) tackled both the security of supply using energy efficiency and sustainability in the energy performance of buildings. The latest regulatory development is the EED Directive [59] on energy efficiency repealing the CHP and ESD Directives. It uses an integrated approach to address all the three issues of competitiveness, security of supply and sustainable development by placing energy efficiency in a central position.

Given this brief European regulatory evolution for microgeneration systems, currently in force are the IEM, RES, EPBD and EED Directives. In the following, a critical legal analysis of these relevant European Directives is necessary to understand the relation between microgeneration systems, users and MS in helping increase competition and ensuring security of supply, while leading to sustainable development at the same time.

4.1 IEM Directive

One of the primary objectives of the European Union in the European energy sector is the creation of genuine internal energy market with the purpose of "giving European consumers a choice between different companies supplying electricity at reasonable prices, and of making the market accessible for all suppliers, especially the smallest and those investing in renewable forms of energy" [72].

The Internal Electricity Market Directive within the Third Energy Package created the framework for competition by setting common rules for the internal market for electricity "with a view to improving and integrating competitive electricity markets in the Community".[14] The directive established the rules associated with "the organisation and functioning of the electricity sector, open access to the market, the criteria and procedures applicable to calls for tenders and the granting of authorizations and the operation of systems. It also lays down universal service obligations and the rights of electricity consumers and clarifies competition requirements",[15] all to "deliver real choice for all consumers of the European Union… so as to achieve efficiency gains, competitive prices, and higher standards of service and to contribute to security of supply and sustainability".[16]

For all consumers of energy (covering users of microgeneration systems), the Directive created market access by placing consumer interests "at the heart of this Directive".[17] An appropriate level of information and transparency is ensured with an overseeing role to the National Regulatory Authorities (NRAs) to provide information on "prices for household customers including prepayment systems, switching rates, disconnection rates, charges for and the execution of maintenance services"[18] including customer consumption data[19] and handling complaints. However, it only created the premises for access to the grid for users of microgeneration systems in order to "assist the active participation of consumers in the electricity supply market".[20] In addition, NRAs main objective is "helping to achieve, in the most cost-effective way, the development of secure, reliable and efficient non-discriminatory systems that are consumer oriented, energy efficiency as well as the integration of small-scale production of electricity from RES in distribution networks".[21]

---

[14] Article 1.

[15] Article 1.

[16] Preamble 1.

[17] Preamble 51.

[18] Article 37, (1), (j).

[19] Article 37, (1), (p).

[20] Annex I.

[21] Article 36, (d).

As a result, by creating the common framework for the internal electricity market, the IEM Directive addresses the issue of and helps boost competitiveness of the European energy sector, incentivizing at the same time the participation in the market of renewable energy suppliers.

4.2 RES Directive

The RES Directive created the "common framework for the promotion of energy from renewable sources"[22] in order to contribute to the European Union's objectives.[23] To do so, the Directive set national targets for the use of RES and encouraged all participants in the electricity market to engage in the production of energy through RES. In addition, the Directive acknowledged the positive impacts RES have on local communities[24] and end consumers[25] supporting the use of decentralised energy technologies.

Microgeneration systems are addressed directly on two levels. First of all, on the *energy level,* consumers/users of microgeneration technologies benefit from simplified procedures and support schemes. Secondly, on the *information level,* consumers/users of microgeneration technologies benefit from transparent and cost-related charges, information, training and guarantee of origin.

On the energy level, the Directive gives consumers/user of microgeneration technologies the right to benefit from simplified procedures (simple notifications instead of authorisations) when "installing small decentralised devices for producing energy from renewable sources"[26] as a way to encourage the active participation in the energy production market, having as legal basis Article 13, (f).[27] Furthermore, they can also benefit from support schemes[28] if offered by the MS according to Article 3, Para. 3, (a).

---

[22] Article 1 (Subject matter and scope).

[23] Preamble, Point (1): "Reduce greenhouse gas emissions", "promoting security of supply" and "promoting technological developments and innovation".

[24] Preamble, Point (4): "regional and local development, export prospects " and Preamble, Point (6): "utilisation of local energy sources, increased local security of supply, shorter transport distances and reduced energy transmission losses".

[25] Preamble, Point (4): "social cohesion and employment opportunities" and Preamble, Point (6): "income sources and creating jobs locally".

[26] Preamble, Point (43).

[27] Article 13 (Administrative procedures, regulations and codes), (f): "simplified and less burdensome authorisation Procedures, including through simple notification if allowed by the applicable regulatory framework, are established for smaller projects and for decentralised devices for producing energy from renewable sources, where appropriate".

[28] Defined by Article 2, (k) as "any instrument, scheme or mechanism applied by a Member State or a group of MS, that promotes the use of energy from renewable sources by reducing the cost of that energy, increasing the price at which it can be sold, or increasing, by means of a renewable energy obligation or otherwise, the volume of such energy purchased".

On the information level, consumers/users of microgeneration technologies have the right to transparent and cost-related administrative costs based on Article 13, (e)[29] and the right to be informed on support schemes[30] and benefits of using energy from renewable sources[31] as stated in Article 14, (1) and (6). Finally, the right to be guaranteed in an "objective, transparent and non-discriminatory" manner that a share of the received energy comes from RES is ensured by Article 15, (1).[32]

As a result, the RES Directive encourages the use of decentralised renewable energy technologies thus setting out the premises for involvement in *security of supply* while helping *sustainable development* through the use of RES as promoted through European energy policies. Nonetheless, lacking clear and concise details on how this can happen, it leaves MS a lot of room for manoeuvre to take necessary measures.

4.3 EPBD Directive

The Energy Performance of Buildings Directive focuses on energy efficiency in buildings as they "account for 40 % of total energy consumption in the Union"[33] and growing. Therefore they must have a share too in reducing both the energy imports and GHG emissions. In order to get concrete results in "achieving the great unrealised potential for energy savings in buildings",[34] the use of energy from RES together with reductions of consumption are needed.

The EPBD Directive is meant to promote and improve the energy performances of buildings, as defined by Article 2, (4),[35] "within the Union, taking into account outdoor climatic and local conditions, as well as indoor climate requirements and cost-effectiveness".[36] To this end, an energy performance certificate, as defined by

[29] Article 13, (e): "administrative charges paid by consumers … are transparent and cost related".

[30] Article 14, (1): "MS shall ensure that information on support measures is made available to all relevant actors, such as consumers…".

[31] Article 14, (6): "MS … shall develop suitable information, awareness raising, guidance or training programmes in order to inform citizens of the benefits and practicalities of developing and using energy from renewable sources".

[32] Article 15, (1): "For the purposes of proving to final customers the share or quantity of energy from renewable sources … MS shall ensure that the origin of electricity produced from renewable energy sources can be guaranteed as … in accordance with objective, transparent and non-discriminatory criteria".

[33] Preamble, Point (3).

[34] Preamble, Point (7).

[35] Article 2, (4) as "the calculated or measured amount of energy needed to meet the energy demand associated with a typical use of the building, which includes, inter alia, energy used for heating, cooling, ventilation, hot water and lighting".

[36] Article 1, (1).

Article 2, (12)[37] is issued, allowing the "owners or tenants of the building or building unit to compare and assess its energy performance"[38] when "offered for sale or for rent".[39]

Microgeneration systems are directly addressed also on both *energy* and *information* levels.

On the *energy level*, consumers/users of microgeneration systems are bound to reduce energy losses and must possess an energy performance certificate for "buildings or building units which are constructed, sold or rented out[40] to a new tenant".[41] Permission must also be given for "regular inspection of the accessible parts of systems used for heating buildings"[42] with an effective rated output of more than 20 kW and for "a regular inspection of the accessible parts of air-conditioning systems"[43] with an effective rated output of more than 12 kW. However, for certain categories of buildings presented in Article 4, Para. 2, the energy performance of the building is not compulsory.[44]

For new buildings, account must be taken to the "technical, environmental and economic feasibility of high-efficiency alternative systems",[45] inter alia technologies based on: (a) decentralised energy supply systems using energy from renewable sources; (b) cogeneration; (c) heat pumps, according to Article 6, (1), before construction starts and by the end of 2020 all new buildings must be "zero-energy buildings".[46] In what old/existing buildings are concerned, renovations are good occasions to adopt measures which would make the buildings "meet minimum energy performance requirements".[47] Moreover, the Directive supports the use of "intelligent metering systems whenever a building is constructed or undergoes

---

[37] Article 2, (12): 'energy performance certificate' means a certificate recognised by a Member State or by a legal person designated by it, which indicates the energy performance of a building or building unit, calculated according to a methodology adopted in accordance with Article 3.

[38] Article 11, (1).

[39] Article 12, (4).

[40] However, for single building units rented out, MS can defer the application until 31 December 2015 as stated by Article 28, Para. 4.

[41] Article 12, (1), (a).

[42] Article 14, (1).

[43] Article 15, (1).

[44] Article 4, Para. 2: " MS may decide not to set or apply the requirements referred to in paragraph 1 to the following categories of buildings: (a) buildings officially protected as part of a designated environment or because of their special architectural or historical merit, in so far as compliance with certain minimum energy performance requirements would unacceptably alter their character or appearance; (d) residential buildings which are used or intended to be used for either less than 4 months of the year or, alternatively, for a limited annual time of use and with an expected energy consumption of less than 25 % of what would be the result of all-year use; (e) stand-alone buildings with a total useful floor area of less than 50 m^2".

[45] Article 6, (1).

[46] Article 9, (1).

[47] Article 7, (1).

major renovation"[48] and in order to realise the energy performance standards, financial assistance could also be given.[49]

On the *information level*, the certificate for energy performance must provide all necessary information including references for additional sources such as energy audits, financial/other assistance.[50]

As a result, the EPBD Directive is certainly a step further in helping mitigate the challenges in the European energy sector as it imposes duties on consumers to obtain energy performance certificates for their households and use high-performance heating/cooling equipments. It tackles thus both *security of supply* and *sustainable development* through the use of energy efficiency.

4.4 EED Directive

Based on the Europe 2020 strategy for smart, sustainable and inclusive growth in which energy efficiency is one key priority in reducing the energy consumption with 20 % by 2020[51] and following the 2011 Energy Efficiency Plan (see Sect. 3.1), the give Energy Efficiency Directive came to "a more integrated approach to energy efficiency and savings" [73] with the purpose of establishing a "common framework for the promotion of energy efficiency within the Union"[52] and laying down the "rules designed to remove barriers in the energy market and overcome market failures".[53]

Since the European Union was "not on track to achieve its energy efficiency target"[54] and the existing legal framework was not fully covering the energy saving potential, the Energy Efficiency Directive came to repeal the CHP and ESD Directives taping the gaps and upgrading the energy efficiency legal framework.

Furthermore, the EED Directive is part of a comprehensive set of legislation comprising also Efficiency in Buildings (EPBD Directive), Efficiency in Products (appliances, lighting, ICT, motors) (Energy Labelling, EcoDesign, Energy Start) [74] all serving the purpose of implementing the Energy Efficiency Plan [56].

The Energy Efficiency Directive establishes four ways of action (General Measures promoting Energy Efficiency, Indicative national Energy Efficiency targets, Monitoring and Reporting and fully sectored measures (households, public sector, energy supply, industry, services) [74] successfully integrating and covering a non-exhaustive wide range of cogeneration technologies: combined cycle gas turbine with heat recovery, steam back pressure turbine, steam condensing

[48] Article 8, (2).

[49] Article 10, (6) and (7).

[50] Article 11, (4).

[51] Preamble 10.

[52] Article 1, (1).

[53] Article 1, (1).

[54] Preamble 8.

extraction turbine, gas turbine with heat recovery, internal combustion engine, microturbines, Stirling engines, fuel cells, steam engines, organic rankine cycles.[55]

The directive provides also extensive definitions to cogeneration (including high-efficiency cogeneration, small scale and microgeneration). In the sense of the directive, microgeneration systems use as input fossil fuels, RES or a combination of both and generate heat, electricity, heat and electricity as output, to microscale (below 50 kWe) or small scale (below 1 MWe).[56] Furthermore, the directive is rather flexible allowing MS to adopt other forms of definitions too.[57]

Microgeneration systems are deployed both in the energy use (Chap. Decentralized Poly-generation of Energy: Basic Concepts of EED) and in the energy supply (Chap. Combined Micro-Systems of EED). In the energy use, microgeneration systems are addressed to both the public sector at national, regional and local level and to final consumers, both residential and industry including SMEs. In the energy supply, the connection of microgeneration systems to the grid is facilitated by energy suppliers.

In the public sector, MS have to comply with a set of national energy efficiency targets[58] and assess the potential for microgeneration in energy use and energy supply, in order to establish a long-term strategy for building renovation in "residential and commercial buildings, both public and private".[59] In addition, public bodies must have an exemplary role[60] and implement microgeneration technologies when making comprehensive renovations to their building spaces and when making public purchasing.[61]

In the private sector, a significant untapped potential for energy efficiency and energy savings comes from final consumers. Therefore, improvements and benefits have been brought to both household consumers and SMEs. A bidirectional flow of energy is ensured as grid access is created by energy suppliers for "the connection and dispatch of generation sources at lower voltage levels".[62] Smart metres are provided to "accurately account for electricity put into the grid from the final customer's premises".[63] Moreover, MS have the possibility to take various measures to "promote and facilitate an efficient use of energy by small energy customers, including domestic customers".[64] Users of microgeneration systems, as small- and medium-sized producers of energy, will benefit from easier and faster administrative

55 Annex I Part II.

56 Art 2, Points (38), (39).

57 Preamble 38.

58 Article 3.

59 Article 4.

60 Article 5.

61 Article 6—"purchase only products, services and buildings with high energy-efficiency performance".

62 Annex XI, (2), (d).

63 Article 9, Para. 2 (c).

64 Article 12, Para. 1.

procedures, simple "install and inform" notifications.[65] Their connection to the grid is facilitated[66] as they can simply call for tender.[67]

In the energy supply sector, the national energy regulatory authorities (NRAs) have the task to provide "incentives for grid operators to make available system services to network users permitting them to implement energy efficiency improvement measures in the context of the continuing deployment of smart grids"[68] to be achieved through network tariffs, regulations and "efficiency in infrastructure design and operation"[69]

Financial incentives are also provided in order to increase the investments in energy efficiency covering also the development and large-scale deployment of microgeneration systems. The EED Directive proposes a series of measures to be taken by the MS varying from changing the behaviour of consumers, through fiscal incentives, access to finance, grants and subsidies, information provisions, exemplary projects, workplace activities, to engaging them in the roll-out of smart metres, through information on easy ways to change energy use and energy efficiency measures.[70] In addition, full use of support schemes is encouraged at both European and national level. At EU level, the general financing framework (cohesion, structural and rural development funds) and a dedicated instrument (European Energy Efficiency Fund). At national level, MS are encouraged to use Structural Funds and Cohesion Funds[71] and to create a dedicated Energy Efficiency National Fund.[72] The scope of these financing solutions is to provide support for energy efficiency initiatives, including therefore producers of microgeneration systems which can obtain resources for "research on and demonstration and acceleration of uptake of small- scale and micro- technologies to generate energy and the optimisation of the connections of those generators to the grid".[73]

Market access is strengthened to fully "remove barriers in the energy market and overcome market failures".[74] The last link in the energy chain is finally liberalised largely extending the energy market and providing final consumers' empowerment as they are ensured market integration and equal market entry opportunities.[75] This should lead to improved competition, economic boost and high-quality job creation

---

65 Article 19.

66 Article 15, Para. 5.

67 Article 15, Para. 7.

68 Article 15, Para. 1.

69 Article 15, Para. 4.

70 Article 12, Para. 2.

71 Preamble 49.

72 Article 20, Para. 4.

73 Preamble 53.

74 Article 1.

75 Preamble 45.

in several sectors,[76] development in demand and variety[77] of services market[78] including producers of microgeneration equipments.

As a result, the EED Directive uses an integrated approach to address security of supply by reducing the primary energy consumption through energy efficiency, consequently covering sustainable development by contributing to the reduction of GHG emissions and addressing the competitiveness of the internal energy market by removing remaining market barriers.

Moreover, by taking into account the "safeguards formulated by academia to avoid unintended negative consequences that would be adverse to the energy efficiency objectives" [75] and continuing to develop the legal framework implementing the European energy policies by paving the road to DEP, the EED Directive sets the basis for the widespread deployment of microgeneration systems in the European Union by reinforcing market access and creating grid access and financial incentives necessary to make the transition from traditional energy consumers to prospective energy prosumers. In making the transition, consumer empowerment is achieved allowing both a bidirectional flow of energy (consumption, production and energy export) and information (more accurate information available and greater interaction with the electricity market).

5 Trends

The successful resolution of the energy challenges, be it security of supply as the most pressing issue in the European energy sector due to resource scarcity and constant increase in the energy demands or mitigating climate changes to ensure a sustainable future, as viewed from political, legal and technological perspectives, is only possible through the use of RES. Moreover, competition issues are ancillary and interlinked.

In order to solve all the challenges present in the European energy sector, the European Union has the right policies in place [76]. The short-term targets are under implementation (Europe 2020). The final goal has been set (Roadmap 2050). In the meantime, intermediate milestone are being drafted (2030 framework). Furthermore, the legal analysis indicates that the policy targets are concretely being implemented through binding European legislation.

As a result of all the above, the trend in the European energy sector points out that the use of RES is increasingly encouraged as green electricity is considered a privately provided good with public benefits [77, 78]. Within RES, microgeneration systems in correlation with cogeneration have a great still untapped potential to help solve all the challenges, bringing benefits on multiple levels: first of all, they

[76] Preamble 1.

[77] Preamble 19.

[78] Preamble 33.

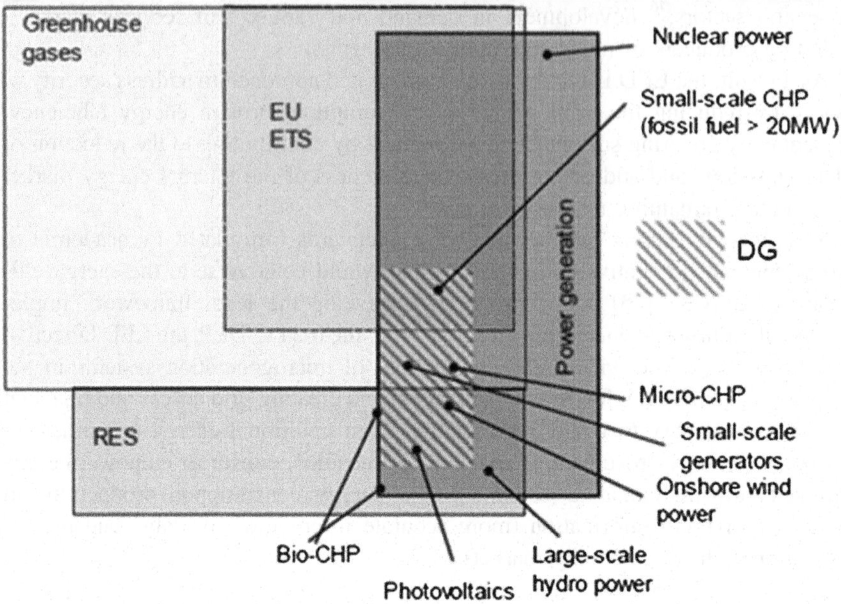

Fig. 15 The place for microgeneration in European context [79]

produce electricity as a by-product of heat implying that using the same resources they produce two products, being therefore much more energy efficient and have a potential to bring bill reductions and certain revenues, at the same time leading to a greater security of supply; secondly, using RES, these low-carbon technologies are more eco-friendly and help reduce GHG emissions and contribute towards a sustainable future; lastly, the widespread deployment of microgeneration systems would imply a vast amount of new market participants increasing the competition in the internal energy market. For this reasons, microgeneration systems are highly promoted through the European energy policies and recent binding legislation, having their place accordingly (Fig. 15):

Moreover, behavioural changes for energy consumers are encouraged by a plethora of incentives[79] in order to make energy consumers more active and engaged in the energy sector, this leading to a shift towards prosumers.

Connecting the dots between RES, microgeneration systems (micro-CHP) and prosumers leads us to DEP. Interconnecting the microgrids of DEP leads to the ultimate goal of Smart Grids. All of these have a huge potential to successfully solve both efficiently and effectively[80] [80] all the challenges in the European energy sector and are promoted through European energy policies and implemented through European legislation.

---

[79] See the incentives promoted by the Energy Efficiency Directive.

[80] "Efficiency is doing the things right; effectiveness is doing the right things".

References

1. European Commission (2012) Making the internal energy market work. COM 663 final, Brussels, p 2
2. Directive 2012/27/EU of the European Parliament and of the Council of 25 October 2012 on energy efficiency, amending Directives 2009/125/EC and 2010/30/EU and repealing Directives 2004/8/EC and 2006/32/EC (2012) OJ L315, pp. 1–56, Preamble 1
3. Pinder J (1998) The building of the European Union, 3rd edn. OUP, Oxford, p 3
4. Europea Union (2014), The history of the European Union, http://europa.eu/about-eu/euhistory/index_en.htm
5. Potocnik J (2006) European smart grids technology platform—vision and strategy for Europe's electricity networks of the future. European Commission, Brussels, foreword
6. European parliament (2010) Directorate general for internal policies, policy department a: economic and scientific policy—industry, research and energy. Decentralized Energy Systems, Brussels
7. DKE German Commission for Electrical, Electronic and Information Technologies (2010) The German national smart grid standardization strategy. CIM User Group, Milan, p 4
8. Potocnik J (2006) European Smart Grids Technology Platform—Vision and strategy for Europe's electricity networks of the future. European Commission, Brussels, p 6
9. Potocnik J (2006) European SmartGrids Technology Platform—Vision and strategy for Europe's electricity networks of the future. European Commission, Brussels, foreword, p 6
10. International Energy Agency (2011) Technology roadmap—smart grids. © OECD/IEA, Paris, p 6. (www.iea.org)
11. Badea GV et al (2013) The legal framework for microgeneration systems in the deployment of Smart Grids, MicrogenIII. In: Proceedings of the 3rd edn. of the international conference on microgeneration and related technologies, Naples, 15–17 April 2013, pp 834–841
12. Micropower Europe (2010) Mass market microgeneration in the European union—from vision to reality. Brussels, p 3
13. Wikipedia (2013) Microgeneration. Web. (en.wikipedia.org/wiki/Microgeneration)
14. European Commission (2007) An energy policy for Europe. COM 1 final, Brussels, p 3
15. University of Leipzig (2014) Energy fundamentals. Web (uni-leipzig.de/~energy/ef/01.htm)
16. Carr T (1974) Testimony to U.S. Senate Commerce Committee
17. European Commission (2013) Energy challenges and policy. Brussels, p 2
18. European Commission (2013) Energy challenges and policy. Brussels, pp 3–6
19. European Parliament, Directorate General for Internal Policies, Policy Department A: Economic And Scientific Policy—Industry, Research and Energy (2010) Decentralized energy systems. Brussels, pp 64–71
20. European Commission (2013) Energy challenges and policy. Brussels, pp 4–7
21. European Parliament, Directorate General for Internal Policies, Policy Department A: Economic And Scientific Policy—Industry, Research and Energy (2010) Decentralized energy systems. Brussels, pp 73–76
22. European Commission, European Smart Grids Technology Platform (2006) Vision and strategy for Europe's electricity networks of the future. Brussels, p 12
23. European Commission (2013) Energy challenges and policy. Brussels, p 1
24. See www.un-documents.net/wced-ocf.htm
25. Eszter T (2006) Different approached of sustainable development—the global level, the EU guideline and the success of local implementation. Budapesti Gazdasági Főiskola, pp 17–22
26. Taoiseach EK et al (2012) Our sustainable future—a framework for sustainable development for Ireland. p 10
27. European Commission (2013) Energy challenges and policy. Brussels, p 3
28. Map. Available at: http://unstats.un.org/unsd/environment/air_greenhouse_emissions.htm
29. European Commission. EU greenhouse gas emissions and targets. Available at: ec.europa.eu/clima/policies/g-gas/

30. European Environment Agency (2013) Trends and projections in Europe 2013—tracking progress towards Europe's climate and energy targets until 2020. Report No 10/2013. Publications Office of the European Union, Luxembourg. See also EU Emissions Trading System (ETS) data viewer, European Environment Agency. www.eea.europa.eu/data-and-maps/data/data-viewers/emissions-trading-viewer. Accessed 24 Sept 2013. And European Union Transaction Log (EUTL) (http://ec.europa.eu/environment/ets)
31. Available at: http://www.eea.europa.eu/data-and-maps/figures/absolute-change-of-ghg-emissions-2
32. Eurostat—EU 27 (2010) Production of primary energy. http://epp.eurostat.ec.europa.eu/statistics_explained/index.php/Energy_production_and_imports
33. European Commission (2013) Energy challenges and policy. Brussels, p 16
34. European Commission (2013) Energy challenges and policy. Brussels, p 17
35. European Commission (2011) Energy efficiency plan. COM/2011/0109 Final, Brussels, p 2
36. European Commission (2013) Energy challenges and policy. IEA, Brussels, p 19
37. European Commission (2007) An energy policy for Europe. COM (2007) 1 final, Brussels, p 4
38. Eurostat (2011) Energy, transport and environment indicators, 2011 edn. Eurostat Pocketbooks, ISSN 1725-4566, p 24
39. http://epp.eurostat.ec.europa.eu/statistics_explained/index.php/Energy_production_and_imports
40. European Commission (2007) An energy policy for Europe. COM (2007) 1 final, Brussels, p 3
41. http://www.iea.org/topics/energysecurity/
42. http://www.worldcoal.org/coal-society/coal-energy-security/
43. http://fsr-encyclopedia.eui.eu/energy-mix/
44. Ademe (2012) Energy efficiency trends in buildings in the EU—lessons from the Odyssee Mure project. Available at: http://www.odyssee-mure.eu/publications/br/energy-efficiency-in-buildings.html
45. Bohne E (2011) Conflicts between national regulatory cultures and EU energy regulations. Elsevier Util Policy J 19:1
46. European Commission (2009) Report on progress in creating the internal gas and electricity market. COM (2009) 115, p 13
47. European SmartGrids Technology Platform (2006) Vision and strategy for Europe's electricity networks of the future. European Commission, Brussels, p 4
48. European Commission (2011) Smart grid mandate—Standardization mandate to European standardisation organisations (ESOs) to support European smart grid deployment. Brussels, p 3
49. European Commission (2011) 2009–2010 Report on progress in creating the internal gas and electricity market. Brussels, Ibid, p 10
50. European Commission (2010) Europe 2020—A strategy for competitive, sustainable and secure energy. COM (2010) 639 final, Brussels
51. http://epp.eurostat.ec.europa.eu/statistics_explained/index.php/Energy_introduced
52. http://ec.europa.eu/clima/policies/package/index_en.htm
53. See European Commission (2007) An energy policy for Europe. COM (2007) 1 final, Brussels
54. See European Commission (2010) Europe 2020 a strategy for smart, sustainable and inclusive growth. COM (2010) 2020 final, Brussels
55. European Commission (2010) Energy 2020 a strategy for competitive, sustainable and secure energy. COM (2010) 639 final, Brussels
56. European Commission (2011) A resource-efficient Europe—flagship initiative under the Europe 2020 strategy. COM (2011) 21 final, Brussels
57. European Commission (2011) Energy efficiency plan. COM (2011) 109 final, Brussels
58. European Commission (2011) Smart grids: from innovation to deployment, COM (2011) 202 final, Brussels
59. Directive 2012/27/EU of the European Parliament and of the Council of 25 October 2012 on energy efficiency, amending Directives 2009/125/EC and 2010/30/EU and repealing Directives 2004/8/EC and 2006/32/EC
60. European Commission (2011) A roadmap for moving to a competitive low carbon economy in 2050. COM (2011) 112 final, Brussels

61. European Commission (2011) Energy roadmap 2050. COM (2011) 885 final, Brussels
62. http://ec.europa.eu/clima/policies/roadmap/index_en.htm
63. European Commission (2011) A roadmap for moving to a competitive low carbon economy in 2050. COM (2011) 112 final, Brussels, p 5
64. European Commission (2013) Green paper—a 2030 framework for climate and energy policies. COM (2013) 169 final, Brussels
65. European Commission (2014) A policy framework for climate and energy in the period from 2020 to 2030. COM (2014) 15 final, Brussels
66. http://ec.europa.eu/clima/policies/2030/index_en.htm
67. Directive 2004/8/EC of the European Parliament and of the Council of 11 February 2004 on the promotion of cogeneration based on a useful heat demand in the internal energy market and amending Directive 92/42/EEC (2004) OJ L 52:50–60
68. Directive 2006/32/EC of the European Parliament and of the Council of 5 April 2006 on energy end-use efficiency and energy services and repealing Council Directive 93/76/EEC (2006) OJ L 114:64–85. Amended by Regulation (EC) No 1137/2008 of the European Parliament and of the Council of 22 October 2008
69. Directive 2009/28/EC of the European Parliament and of the Council of 23 April 2009 on the promotion of the use of energy from renewable sources and amending and subsequently repealing Directives 2001/77/EC and 2003/30/EC (2009) OJ L 140:16–62
70. Directive 2009/72/EC of the European Parliament and of the Council of 13 July 2009 concerning common rules for the internal market in electricity and repealing Directive 2003/54/EC (2009) OJ L 211:55–92
71. Directive 2010/31/EU of the European Parliament and of the Council of 19 May 2010 on the energy performance of buildings (2010) OJ L 153:13–35
72. http://europa.eu/legislation_summaries/energy/internal_energy_market/index_en.htm
73. Proposal for a Directive of the European Parliament and of the Council on energy efficiency and repealing Directives 2004/08/EC and 2006/32/EC, COM (2011) 370 final, Brussels, 22.6.2011. Explanatory Memorandum, Point 5.1
74. Bertoldi P (2012) EU energy efficiency policies. Moscow
75. Directorate general for internal policies policy department a: economic and scientific policy. Effect of smart metering on electricity prices. IP/A/ITRE/NT/2011, 16 February 2012, p 11
76. European Commission (2013) Energy challenges and policy. Brussels, p 4
77. See Kotchen M (2006) Green markets and private provision of public goods. J Polit Econ 114 (4):816–834
78. Kotchen M, Moore M (2007) Private provision of environmental public goods: household participation in green-electricity programs. J Environ Econ Manage 53(1):1–16
79. Directorate General for Internal Policies, Policy Department A: Economic and Scientific Policy—Industry, Research and Energy (2010) Decentralized energy systems. IP/A/ITRE/ST/2009–16, Brussels, p 17
80. Drucker P (1967) The effective executive: the definitive guide to getting the right things done. HarperBusiness Essentials

Decentralized Poly-generation of Energy: Basic Concepts

Nicolae Badea

Abstract This chapter presents the basic concepts for primary energy forms, energy conversion, delivered energy, and energy needed by consumers to satisfy their needs (useful energy). The conversion of primary energy into useful energy is evaluated on the basis of the energy efficiency factor for separate energy generation, cogeneration, and trigeneration. The difference between the primary energy in the case of separate production and the primary energy in the case of combined production represents the primary energy corresponding to the saved fuel. The energy saving measure is achieved through the *primary energy savings (PES) or percent fuel savings*. Finally, a trigeneration energy conversion is exemplified and the performance indicators of the system are given.

1 Energy

Matter is characterized through two fundamental measures: mass and energy. Mass is the measure of inertia and gravity, and energy is the scalar measure of matter movement. The energy modification of a physical system is known as mechanical work. Mechanical work appears when the state of physical system is modified as result of a transformation, and the latter implies the modification of the system's energy.

Polygeneration describes an integrated process which has three or more outputs that include energy outputs, produced from one or more natural resources.

N. Badea (✉)
"Dunarea de Jos" University of Galati, Galati, Romania
e-mail: nicolae.badea@ugal.ro

N. Badea (ed.), *Design for Micro-Combined Cooling, Heating and Power Systems*,
Green Energy and Technology, DOI 10.1007/978-1-4471-6254-4_2

1.1 Forms of Energy

There are several forms of energy, the most frequent of which being:

- chemical;
- thermal;
- electrical;
- electromagnetic;
- nuclear.

Chemical energy is stored in the atomic links which form the molecules. When different chemical compounds react among themselves, these links are broken or modified, frequently generating or absorbing energy under the form of heat. At the micro level, the energy of fossil sources of energy (coal, petroleum, natural gases, wood etc.) may be considered as *potential energy* of atomic links which change in the burning of fuel, thus generating energy. When wood burns, the carbon contained by the wooden mass reacts with the oxygen in the air. Through burning, a new chemical product is obtained (the atomic links are modified)—carbon dioxide, CO_2—and energy, under the form of heat and light (radiation), is released simultaneously. Another example may be considered, the cell of an electric accumulator or the combustion cell, where various chemical products react among themselves, producing electric energy and other chemical products.

Thermal energy is the sum of the kinetic and potential energies of all the atoms and molecules which form a certain solid, liquid, or gaseous body. Thermal energy includes both kinetic energy (since atoms and molecules move) and potential energy (since, as a result of the movement of atoms and molecules—oscillatory movement—the linking forces modify, which results in the modification of the potential energy of each atom and molecule forming the respective body). The greater the movement speed of atoms and molecules, the greater the temperature of the body and vice versa. In a boiler, the chemical energy of fossil fuels is transmitted to the steam under the form of thermal energy which, in turn, transmits it to the turbine.

Electrical energy is the kinetic energy of a flux of particles with an electric charge (called electrons and ions), which move inside an electric field. The movement of particles is produced by the force of the electric field. In metals, the charge bearers are the electrons, and in gases and liquids the main charge bearers are the positive and negative ions. Once the electric charge bearers move, it means that they have kinetic energy.

Electromagnetic energy manifests itself under the form of electromagnetic waves, having different values for the wave length, starting with radio waves and ending with X-rays. A particular example of electromagnetic energy is solar energy, which takes the form of an electromagnetic wave spectrum, of different wave length. On the other hand, the electromagnetic wave has particle properties, moving at the speed of light. That is why, essentially, electromagnetic energy is kinetic energy—which indicates the movement of particles without substance transport.

Nuclear energy is the energy obtained following the fission reaction of the atom's nucleus, for instance uranium-235 or plutonium-239. The fission of a nucleus means its splitting into several fragments. The difference between the mass of the initial nucleus and the sum of the fragment masses is found in the kinetic energy that these fragments acquire. In the nuclear reactor, this energy is transformed into thermal energy. In the process of nuclear fission, only 0.1 % of the atom's total energy is emitted, the rest of 99.9 % remaining stored in the mass of the newly created fragments. But this quantity is millions of times larger than that obtained following the oxidation (burning) reaction of a fossil fuel (coal, for instance), where the chemical energy (that is the energy of the atomic and molecular links) is transformed into thermal energy.

1.2 Energy Units

The consumer is interested in being delivered energy to satisfy his needs, such as the heating and lighting of the residence. For him, essential are the quantity of *delivered energy*, the form in which it is delivered, and the amount due. The basic forms of delivered energy are heat and power. Afterward, the energy delivered is converted, transformed, or transmitted to other bodies, the result being called *useful energy*. For example, the heat is transmitted to the radiator in the room, for its heating, and the electricity is converted into light energy, for lighting the house area. These energies are derived from *primary energy* supplied fuel. Figure 1 exemplifies the three notions regarding energy: primary energy (obtained from natural gas, in a thermal station), delivered energy (by the electric energy distributor), and useful energy (necessary for the consumer).

The energy units have been defined either in keeping with the useful energy, or in keeping with the primary energy of the energy source.

(a) When the energy units are defined in keeping with *the useful energy*, the following cases occur:

- If the useful energy is electrical, its unit is the kilowatt-hour (kWh)
- If the useful energy is mechanical, its unit is the Joule (J).

Fig. 1 Primary, delivered, and useful energy

- If it is thermal, the calorie is used additionally. A calorie (cal) is the quantity of energy necessary to raise the temperature of a gram of pure water by a Celsius degree, or the quantity of energy lost by a gram of pure water, in which case temperature decreases by a Celsius degree. The conversion is 1 cal = 4,184 J.

On the other hand, in many areas of the Anglo-Saxon world, especially in the United States, the temperature scale used is that of Fahrenheit (°F). The melting temperature of pure frozen water at sea level is of 32 °F, and the boiling point of pure liquid water is of 212 °F. Thus, 32 °F corresponds to 0 °C and 212 °F corresponds to 100 °C. As a result, 1 °C = (5/9)(F − 32) °F or 1 °F = (1.8–32 °C) °C.

In the same areas, the BTU (British thermal unit) is used to measure thermal energy. A BTU is the energy necessary to raise the temperature of a lb of pure liquid water by a degree Fahrenheit. The conversion is 1 BTU = 1055 J. The energy transfer rate (the power) is defined as 1 BTU/h = 0.293 W. If the useful energy is electric, its unit is Wh. The conversion is 1 Wh = 3600 J.

(b) If the energy units are defined in keeping with *the primary energy* of the source, they depend on the calorific power of the fuel used as reference. Thus, if it is petroleum, the unit is the petroleum equivalent kg, whose symbol is kg oe. Changing the nature of the reference fuel changes the energy unit. For example, kg is used for coal and m^3 *gas equivalent* is used for pit gas. Table 1 presents the conversion factors for energy units in different systems of measurement.

1.3 Energy Conversion

One of the fundamental laws of physics is the law of energy conservation, according to which, in physical processes, energy cannot be created or destroyed, increased or

Table 1 Conversion factors for energy

Units	kJ	kcal	kWh	kg ce	kg oe	m^3 gas	BTU
1 kilojoule (kJ)	1	0.2388	0.000278	0.000034	0.000024	0.000032	0.94781
1 kilocalorie (kcal)	4.1868	1	0.001163	0.000143	0.0001	0.00013	3.96831
1 kilowatt-hour (kWh)	3,600	860	1	0.123	0.086	0.113	3412
1 kg coal equivalent (kg ce)	29,308	7,000	8.14	1	0.7	0.923	27,779
1 kg oil equivalent (kg oe)	41,868	10,000	11.63	1.428	1	1.319	39,683
1 m^3 natural gas	31,736	7,580	8.816	1.083	0.758	1	30,080
1 British thermal unit (BTU)	1.0551	0.252	0.000293	0.000036	0.000025	0.000033	1

decreased, but only transformed (converted) from one form of energy into another. The most common energy conversion processes are presented in Fig. 2.

The few possibilities of direct conversion of energy from one form to another are shown on Table 2 and Figure 2

According to energy conservation law, a system's total quantity of energy remains constant and is called the primary energy of that system. The elements that the primary energy of a system is stored in are called sources of energy.

Quantitatively, the primary energy of a system represents the sum of the quantities of energy contained in all its sources of energy.

- Today, the main sources of energy are:
- fossil fuels (coal, petroleum, and natural gases);
- bio fuels (firewood, wood residue, agricultural residue etc.);
- renewable as hydraulic, geothermal, solar, wind energy;
- nuclear energy.

In the case of fuels (whether fossil or bio), primary energy is obtained through burning and is evaluated by multiplying the quantity of burned fuel by its calorific power. Numerically, the calorific power of a fuel is the energy resulting from burning a unitary quantity of fuel (for example, a kg—if the fuel is solid or liquid, or an m^3—if it is gaseous).

The great majority of fuels contain water which, in the burning process, is released under the form of vapors. That is why, in the burning process, part of the energy resulting from the chemical reactions is used to evaporate the water. Consequently, the primary energy which is converted into heat during the burning of the fuel cannot be measured directly; it can only be evaluated, and the evaluation is done relative to the reference state of the water in the fuel.

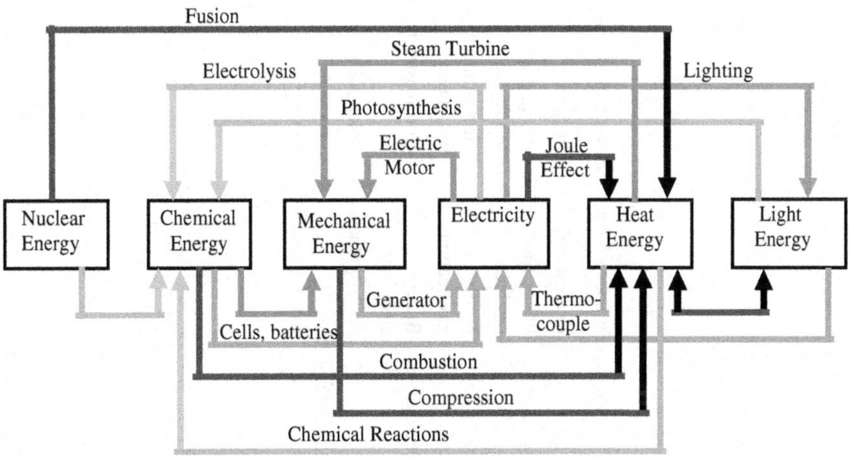

Fig. 2 Conversion processes

Table 2 A few possibilities of direct conversion of energy

Energy type	In chemical	In heat	In electricity	In electromagnetic	In mechanical
From chemical	Plants food products	Burning anaerobic fermentation	Battery fuel cell	Fluorescence lamp	Human muscle and animals
From heat	Pyrolytic gasification	Heat pump heat exchanger	Thermocouple	Fire	Gas turbine steam turbine
From electricity	Battery electrifier	Hob toaster flat iron	Transformer frequency converter	Fluorescent bulb light emitting diode	Electric motor electromagnet
From electromagnetic	Photosynthesis	Solar collector	Photovoltaic cell	Laser	Solar radiation pressure
From mechanical	Crystallization	Brake friction	Generator	Flint	Propeller wind Volant

Table 3 Heating values of fuels

	Units	LHV	HHV	LHV/HHV
Natural gas	BTU/CubicFoot	950	1,050	0.905
Fuel oil	BTU/Gallon	130,000	138,300	0.940
Propane	BTU/Gallon	84,650	92,000	0.920
Sewage/landfill	BTU/CubicFoot	350	380	0.921
Coal-bituminous	BTU/lbs	13,600	14,100	0.965

Standardized reference states may vary, and that is why several calorific powers of the same fuel have been defined. The most commonly used are the Lower Heating Value (LHV) and the Higher Heating Value (HHV) shown in Table 3.

For the inferior calorific power, the reference state of the water is the gaseous one, and for the superior calorific power, the reference state is the liquid one. For this reason, the inferior calorific power decreases with the increase in the water content (ignoring the fact that a large content of water implies a smaller content of combustive substances). (3) In practice, to be able to light the fuel and extract its energy, the maximum content of water is of 55 % (of the humid mass).

2 The Concept of Cogeneration

Today's energy supply system is still dominated by central generation connected to the high voltage level. Cogeneration is the combined production of two forms of energy—electric or mechanical power plus useful thermal energy—in one technological process. The electric power produced by a cogenerator can be used on-site or distributed through the utility grid, or both. The thermal energy usually is used on-site for space conditioning, and/or hot water. But, if the cogeneration system produces more useful thermal energy than is needed on-site, distribution of the excess to nearby facilities can substantially improve the cogeneration's economics and energy efficiency. Cogeneration is an old and proven practice. Because cogenerations produce two forms of energy in one process, they will provide substantial energy savings relative to conventional separate electric and thermal energy technologies. The principal technical advantage of cogeneration systems is their ability to improve the efficiency of fuel use. A cogeneration facility, in producing both electric and thermal energy, usually consumes more fuel than is required to produce either form of energy alone. However, the total fuel required to produce both electric and thermal energy in a cogeneration system is less than the total fuel required to produce the same amount of power and heat in separate systems. Hence, cogeneration is most likely to be competitive with conventional separate electric and thermal energy technologies when it can use relatively inexpensive, plentiful fuels.

Cogeneration unit in turn, can be large-scale centralized near big cities and small or medium scale nears the final consumer.

2.1 Centralized Versus Distributed Energy Generation

The importance of the cogeneration system is showing in 2012/27 EC Directive EED in which the MS for 2014 by National Energy Efficiency Action Plan must provide a description of measures, strategies, and policies, including programs and plans, at national, regional, and local levels to develop the economic potential of high-efficiency cogeneration and efficient district heating and cooling and other efficient heating and cooling systems as well as the use of heating and cooling from waste heat and renewable energy sources, including measures to develop the heat markets (EED Article 14(2), Article 14(4), Annex VIII 1(g)). In this, programs and plans include number of new micro-CHP and small-scale CHP installed and number of other new efficient heating systems and trends in their market uptake, (e.g., heat-pumps efficient boilers, solar equipment and), new or as replacement of old systems installed either as new installations or as replacement for old systems. In the Energy Efficiency Directive are given two types of energies, namely:

- "Primary energy consumption" means gross inland consumption, excluding nonenergy uses (Article 2.2)
- "Final energy consumption" means all energy supplied to industry, transport, households, services, and agriculture. It excludes deliveries to the energy transformation sector and the energy industries themselves (Article 2.3).

The conversion of primary energy into useful energy (or final energy consumption) is evaluated on the basis of the efficiency factor EFF, defined as a percentage ratio between the useful energy and the primary energy.

$$\mathrm{EFF} = (\mathrm{Eu}/\mathrm{Ep})100\,\% \qquad (1)$$

where

- Eu is the useful energy and
- Ep is the primary energy.

Energy efficiency has a major role to play in economically, environmentally, and socially sustainable energy policies. Energy efficiency can play a vital role in reducing the energy intensity of economic activity and avoiding the need for significant new supply. Energy savings are among the fastest, highest impacting and most cost-effective ways of reducing greenhouse gases (GHG) emissions. Over the years, the European Union has introduced a number of directives, regulations, and initiatives to encourage and support Member States, regional authorities, individuals, and so on to increase energy efficiency in the different sectors, including buildings, transport, and products. The span of policies, have yet to change our combined thinking, capacity, and ambition to capture significant savings. Although everyone agrees with the importance of saving energy, it has enjoyed little high level political attention and as such we are a long way from achieving the indicative 20 % energy savings target by 2020.

Energy saving means producing a larger quantity of material goods and offering a larger number of services, consuming a smaller quantity of primary energetic resources, to obtain useful energy, that is to increase the efficiency factor EFF.

Different terms are used, often with little precision or accuracy, to express targets in the area of energy efficiency policy. The definitions provided in the Energy Efficiency Directive establish a clear relation between "energy savings" and "energy efficiency".

The following definitions from Article 2 of the EED are worth recalling here as they are relevant:

- "Energy efficiency" means the ratio of output of performance, service, goods, or energy to input of energy (Article 2.4).
- "Energy savings" means an amount of saved energy determined by measuring and/or estimating consumption before and after implementation of an energy efficiency improvement measure, whilst ensuring normalization for external conditions that affect energy consumption (Article 2.5).

Specifically, energy savings are defined as the result of improvements of energy efficiency. Savings are measured as the difference in energy consumption before and after the efficiency improvement has taken place.

For years, the model of the electric energy industry development has been based on the idea that energy must be produced in a centralized way, in large electric plants, then delivered to the large consumption areas through electric transport lines and, finally, delivered to the consumers through the distribution infrastructure, at the lowest voltage levels. Thus, energy circulates in a unidirectional way, from high voltage to low voltage. This situation is indicated in Fig. 3a.

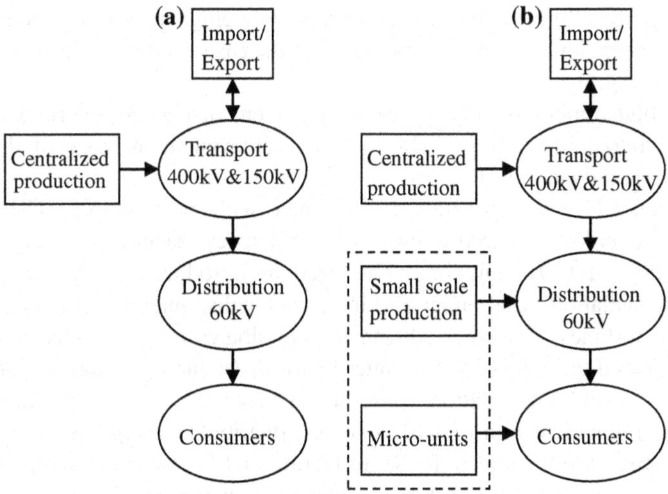

Fig. 3 Models of energy generation

In the first case, the cogeneration unit operates with high efficiency in winter when electricity production is injected into the grid and heat distribution is done to final consumers. Efficiency during the warm season, of the centralized cogeneration units is low, same as power separate systems, because the heat production is not used.

A very important role in the conversion of primary energy into useful energy is played by the capacity of each component of conversion, transformation, and the distribution. For example, the conversion of fuel primary energy into electric energy, in a thermal station, is achieved with an efficiency of maximum 35 %. The electric energy produced by the station is transported along the energy distribution line and reaches the consumer as delivered energy, with an efficiency of 85 %. Then, the consumer uses it to supply various devices, becoming useful energy. For example, the consumer converts electric energy into useful energy under the form of light, with an efficiency of 5 %—in the case of using an incandescent light bulb, of 20 %—in the case of using a fluorescent light bulb, or converts it into mechanical energy, with the aid of an electric engine, with an efficiency of 90 %.

In its turn, the efficiency factor is dependent on the efficiency of converting the energy into different component elements of the chain of energy production, distribution, and use.

Increasing the energy efficiency factor may be achieved in two different ways:

1. by *reducing the energy losses*, as a result of increasing the energy conversion efficiency in each component element of the chain of converting primary energy into useful energy;
2. by *recuperating the energy losses* in the component elements of the chain of converting primary energy into useful energy and transforming these losses into useful energy.

The first applies to systems of separate heat and power generation (SHP systems), where the effort is concentrated on achieving efficient systems (technologies) of energy conversion. This may be achieved in the case of centralized generation, as indicated in Fig. 3a.

The second implies modifying the management of primary energies and their conversion into useful energy. This may be achieved in the case of distributed generation, as indicated in Fig. 3b.

Decentralized power generation combined with heat supply (CHP) is an important technology for improving energy efficiency, security of energy supply, and reduction of CO_2 emissions. The need to introduce several "environment friendly" installations (like microturbines, fuel cells, photoelectric installations, small wind turbines, and other advanced technologies for distributed generation) have determined an increase in the interest for distributed generation, particularly for local ("on-site") generation. Introducing environment friendly installations implies the implementation of two concepts: distributed energy resources (DER) and renewable energy sources (RES). In fact, the DER concept encompasses three main aspects, whose focus is set on the electrical standpoint:

- *Distributed generation* (*DG*) that is local energy production from various types of sources [1–5]. Distributed generation has emerged as a key option for promoting energy efficiency and use of renewable sources as an alternative to the traditional generation.
- *Demand response* (*DR*) that is energy saving brought by the customer participation to specific programs for reducing the peak power or the energy consumption [6–9]. Demand response (DR implies not only satisfying the consumer's electric energy demand, but ensuring any form of energy demanded by the consumer (heat, air conditioning, etc.) also, at any moment, and in the specific quantities necessary to the consumer.
- *Distributed storage* (*DS*) local energy storage with different types of devices [10–13].

In the case of a residence, implementing the DER concept [14] consists in achieving the system for energy production on the basis of the convenient association of three key ideas:

- combined production of heat and power, *from the same fuel, in the same system*, resulting in a so-called Combined Heat and Power system (CHP system);
- the simultaneous use of *more sources of energy* (fossil fuel, sun, wind, geothermal sources, etc.) and their integration in a system;
- placing the cogeneration installations *as close to the final consumer* as possible and dimensioning them so that they may offer the amount of heat and electricity necessary for the consumer, resulting in a *local system* for producing electric and thermal energy, in the *specific quantities* necessary to satisfy the useful energy needs of each consumer *at any given moment in time* (Fig. 4);
- smart metering, with bidirectional way of energy in low voltage to create smart grid concept.

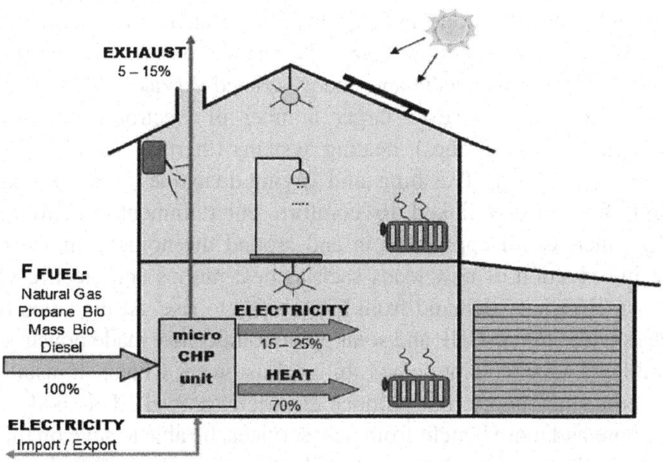

Fig. 4 Energy system for residences, based on the DER concept [15]

Usually, distributed generation uses low power generation units situated with the consumer or in his proximity. These units are installed so as to cover all of the consumer's needs, to ensure an economical functioning of the electric distribution grid, or to satisfy both conditions. Distributed generation complements the traditional centralized generation and distribution of electric energy. It ensures a favorable answer (as regards the cost of capital) to the increase in the energy demand, avoids the installation of supplementary transport and distribution capacities, localizes the generation of electric energy where it is necessary and has the flexibility of delivering it to the grid in the consumer's proximity.

Technological developments, the progressive rise of distributed generation, and the increasing need to manage the system close to consumption points all require a *shift towards* a more decentralized electricity system. Therefore, decentralized and centralized generation can coexist; smart grid development will enhance their complementarities. In this transition, a new figure appeared: the "prosumer", producing and consuming his/her own electricity. By covering on-site part of the final user's electricity needs, PV systems will generate new opportunities. Such decentralized electricity generation will have to be better incorporated in future strategies.

Small-scale power producers are also giving rise to new ownership structures and business models, some of which directly compete with conventional utilities. "More than 3 million households have started producing their own electricity with solar PV", says Eurelectric, while 133 "bio-villages" have emerged in Germany since 2000, generating more than 50 % of their electricity and heat from bioenergy resources.

In Italy, at the end of 2011, PV [16] already covered 5 % of the electricity demand, and more than 10 % of the peak demand. In Bavaria, a federal state in southern Germany, the PV installed capacity amounts to 600 W per inhabitant, or three panels per capita. In around 15 regions in the EU, PV covers on a yearly basis close to 10 % of the electricity demand; in Extremadura, a region of Spain, this amount to more than 18 %.

Market trends across Europe indicate that liberalization is bearing fruit, with wholesale markets becoming more *competitive and customers* increasingly benefiting from new types of products and energy-related services.

Customers use today, an ever larger number of electronic appliances (TV, computers, tablets, smart phones), heating systems (thermostats, air conditioning, heat pumps), green goods (washing and drying machines, dishwashers, ovens, refrigerators), for reasons linked to comfort, entertainment, environment, and security. This increase of appliances in and around the home combined with the progressive introduction of new loads such as heat pumps and electric vehicles is likely to cause electricity demand from households to rise. At the same time, new technologies such as micro-CHP and solar photovoltaic have made power generation at household level a real economic possibility. Customers with such installations no longer only consume energy, but produce electricity as well. This is the way customers will more and more benefit from new services, be able to save on their energy bill and potentially become electricity producers themselves. But these developments also pose new challenges to the electricity industry, which has to cope with a higher share of variable and decentralized generation, in which customers will progressively move to the center of the electricity system.

Competitiveness of the new technologies can be analyzed from three points of view:

- "Wholesale competitiveness", which compares PV's generation cost (the Levelised Cost of Electricity, or LCOE) with wholesale electricity prices, requires examining much more complex boundary conditions (such as market design);
- "Dynamic grid parity", when comparing PV's LCOE with PV revenues (earnings and savings);
- "generation value competitiveness", when comparing PV's generation cost to that of other electricity sources.

A meaning of these terms is as follows:

"Dynamic grid parity" is defined as the moment at which, in a particular market segment in a specific country, the present value of the long-term net earnings (considering revenues, savings, cost, and depreciation) of the electricity supply from a PV installation is equal to the long-term cost of receiving traditionally produced and supplied power over the grid.

"Generation value competitiveness" is defined as the moment at which, in a specific country, adding PV to the generation portfolio becomes equally attractive from an investor's point of view to investing in a traditional and normally fossil fuel based technology.

"Wholesale competitiveness" is defined as the moment at which—in a particular segment in a country—the present value of the long-term cost of installing, financing, operating, and maintaining a PV system becomes lower than the price of electricity on the wholesale market.

Competitiveness of the PV is analyzed in EPIA study [15] for the residential, commercial, and industrial segments—involving the local consumption of PV electricity, when a user goes from being a consumer to a "prosumer" as "dynamic grid parity", when comparing PV's generation cost (the Levelised Cost of Electricity, or LCOE) with PV revenues (earnings and savings).

In the future, decentralized electrical generation units (DG units) of small size will be connected to the low voltage grid with an increasing number and generation capacity. With a rapidly increasing number of DG units, also the rated power installed increases. In the future, the consumer (operator of DG unit) will decide at which time and at which level electric power is fed into the grid.

Cogeneration (CHP) solutions can exhibit excellent overall energy efficiency and allow for significant primary energy saving with respect to the separate production of heat and electricity. As a consequence of the primary energy saving, CHP systems can also be an effective means to pursue the objectives of the Kyoto's Protocol in terms of greenhouse gas emission reduction

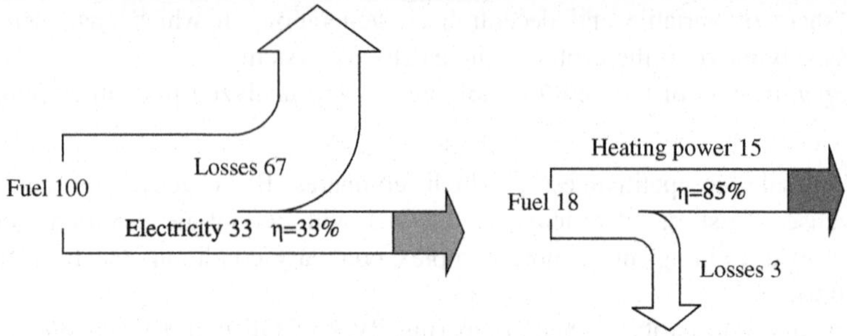

Fig. 5 Energetic balance of separate energy production

2.2 Performance Indicators of Cogeneration Systems

To define the performance indicators of cogeneration systems, let us firstly analyze the energetic efficiency in the case of the separate production of heat and power (Fig. 5).

Heat may be produced with the aid of a boiler, through the transformation of the primary energy of a fuel (chemical energy) into thermal energy useful to the consumer. Boilers (both condensing and noncondensing) are by far the most widely adopted technology for separate production (i.e., excluding district heating) of space heating and domestic hot water heating.

The efficiency of transforming fuel energy into heat (EFF_H) can be defined by means of the First Law of Thermodynamics as the net energy output (Q_H) divided by the fuel consumed (Q_F) in terms of kWh thermal energy content.

$$EFF_H = \frac{Q_H}{Q_F} \tag{2}$$

The efficiency of transforming fuel energy into heat is dependent on the type of fuel used and the performance of the burning chamber. If the boiler uses pit gas, the standard efficiency is of 85 %, and if it uses biomass, the efficiency is considered to be of 65 %, assuming that the heat of condensation cannot be recovered (Fig. 5).

Power is generated through the multiple transformation of energy: the primary energy of a fuel (chemical energy) is transformed into thermal energy (Q_H) which, in turn, is transformed into mechanical energy (W_M), to be then transformed into electric energy (E). The efficiency of electric energy generation (EFF_P) is defined by the relation:

$$EFF_P = \frac{E}{Q_F} = \frac{Q_H}{Q_F}\frac{W_M}{Q_H}\frac{E}{W_M} \tag{3}$$

Its value is much lower (of approximately 33 %) than that of the efficiency of thermal energy generation.

On the other hand, according to the Directive 2012/27/EC, the overall efficiency shall mean the annual sum of electricity and mechanical energy production and useful heat output divided by the fuel input used for heat and electricity and mechanical energy, produced in a cogeneration process (for heat produced in a cogeneration process and gross electricity and mechanical energy production, produced in a cogeneration process).

In the case of energy generation with a SHP system (Fig. 5), the overall efficiency (EFF_{SHP}) is defined as the sum of net power (E) and useful thermal energy output (Q_H) divided by the sum of fuel consumed to produce each and may be calculated with the relation:

$$EFF_{SHP} = \frac{E + Q_H}{\frac{E}{EFF_P} + \frac{Q_H}{EFF_H}} \tag{4}$$

where: E = Net power output from the SHP system; Q_H = Net useful thermal energy from the SHP system; EFF_P = Efficiency of electric generation; EFF_H = Efficiency of thermal generation.

2.2.1 Energy Efficiency

In the case of cogeneration in a CHP system, the overall efficiency (EFF_{CHP}) is defined as the sum of the net power (E) and net useful thermal output (Q_H) divided by the total fuel (Q_F) consumed, in terms of kWh thermal energy content, and may be calculated with the relation:

$$EFF_{CHP} = \frac{E + Q_H}{Q_F} \tag{5}$$

What may be observed is that, as compared to a SHP system, the energetic efficiency of a CHP system is greater, as a result of recuperating the lost energy and transforming it into useful energy. For example, let us consider the situation presented in Fig. 6, where the consumer needs 33 units of electric energy and 15 units of thermal energy. In the case of a SHP system, 118 units of primary energy are needed, while only 100 units are needed if the cogeneration is implemented in a CHP system.

It results that the *effect of recuperating energy losses using the concept of cogeneration is the increase in overall efficiency*. To evaluate this effect, two indicators are used: fuel utilization efficiency and percent fuel savings.

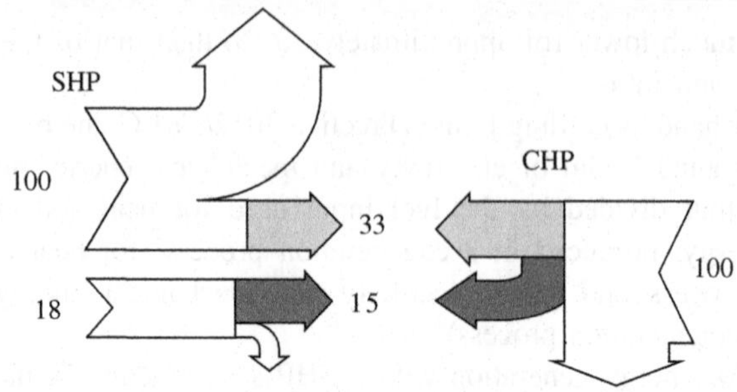

Fig. 6 Energy balance of separate vs. combined production

2.2.2 Fuel Utilization Efficiency

Fuel utilization efficiency (*FUE*) is defined as the ratio of net power output to net fuel consumption, where net fuel consumption excludes the portion of fuel used for producing useful heat output. Fuel used to produce useful heat is calculated assuming that the typical boiler efficiency, EFF_H, is 85 %. FUE can be calculated with the relationship:

$$FUE = \frac{E}{Q_F - \frac{Q_H}{EFF_H}} \tag{6}$$

The difference between the primary energy in the case of separate production and the primary energy in the case of combined production represents the primary energy corresponding to the saved fuel.

2.2.3 Primary Energy Saving

Primary energy saving or Percent fuel savings (*PES*) with combined production is obtained relating this energy to the primary energy in the case of separate production and is calculated with the relation:

$$PES = \left(1 - \frac{Q_F}{\frac{E}{\eta_{Eref}} + \frac{Q_H}{\eta_{Href}}}\right) \times 100\,\% \tag{7}$$

Similarly, fuel saving compares the fuel used by the CHP system to a separate heat and power system. Positive values represent fuel savings, while negative values indicate that the CHP system is using more fuel than SHP.

Recoverable thermal energy from the various prime movers is available in one or both of the following two forms, namely hot exhaust gases and hot water.

Two options for recovering heat from the hot exhaust gases from the prime movers could be considered:

1. Direct use of the exhaust for providing process heat;
2. Indirect use via heat exchangers for producing hot water. Hot water produced can be used to meet the needs for space heating. In applications that require more thermal energy or higher temperatures than that available from power generation equipment, supplementary heat is supplied using a duct burn .The possibilities and the level of energy loss recuperation are dependent on the energy conversion technology for CHP systems.

3 The Trigeneration Concept

Trigeneration is a basic and most popular form of polygeneration. The term describes an energy conversion process with combined heat, cooling, and power generation. Today, availability of CHP technologies with good electrical and excellent overall efficiency has been adopted on a small-scale [17] and even on a microscale [18] basis, with suitable applications ranging from residential houses to schools, restaurants, hotels, and so forth. The trend toward distributed micro-cogeneration could be significant in terms of increasing the local energy source availability, reducing both the energy dependency [19] and the vulnerability of the electrical system to the effects of grid congestions, reducing service interruptions, blackouts, vandalism or external attacks [19, 20] through the formation of self-healing energy areas [21, 22]. The advantage of combined production of heating and power in a cogeneration (or CHP) system is obvious: the waste heat which is always produced when electricity is generated using thermodynamic cycles is not released into the environment—as in large-scale centralized power plants—but can be used. Typical use of this heat is to heat buildings or to produce domestic hot water. Depending on the building site and building standard, the heating season often lasts for only 6 months or less. But for the economic viability of CHP systems, it is important that they are used as much as possible. Therefore, other uses of the waste heat are awakening more interest. One of the possible uses of waste heat during the nonheating season is cooling. The concept of trigeneration is an extension of the CHP concept through adding cold producing equipment for the summer. Classical trigeneration solutions are represented by coupling a CHP prime mover to an absorption chiller fired by cogenerated heat. In this scheme, the produced thermal power is exploited also in the summertime to produce cooling. In this way, one of the biggest shortcomings that often make cogeneration unprofitable, that is the lack of adequate thermal request throughout the whole year, is made up for by transforming the cooling demand into thermal demand.

One difference between various CCHP systems resides in the connection mode of the cold producing device (Fig. 7), where: PM-primary mover; EG-electrical generator; MCP-monitoring control and protection system; PC-personal computer.

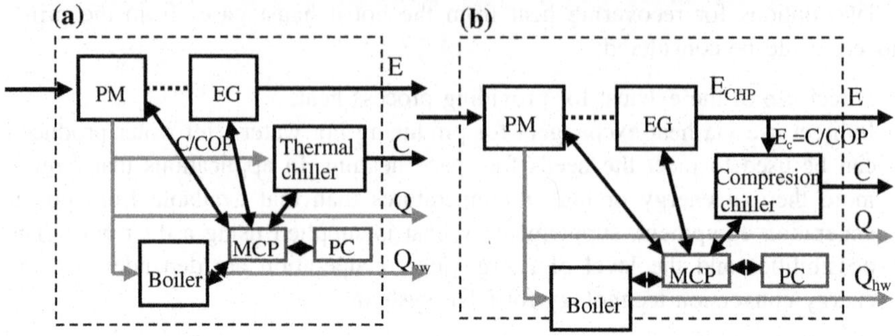

Fig. 7 Trigeneration processes

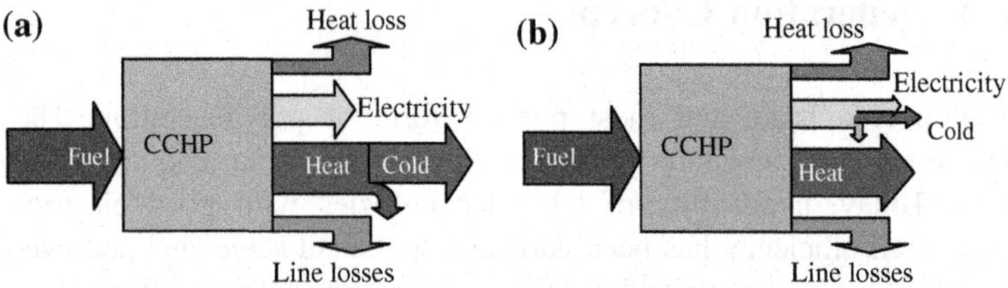

Fig. 8 Energy balance of trigeneration processes

The first version presupposes transforming heat into cold (thermally activated chiller) and the second presupposes obtaining cold by transforming electric energy (mechanical compression chiller). The energy balance related to these two trigeneration processes is given in Fig. 8.

A standard mode of achieving CCHP system does not exist but, generically, such a system consists of a cogeneration unit CHP (Fig. 9), a refrigerating device (a Thermal Driven Chiller—TDC or a Mechanical Compression Chiller—MCC), heat storage unit (hot water storage—HW), and electricity storage unit.

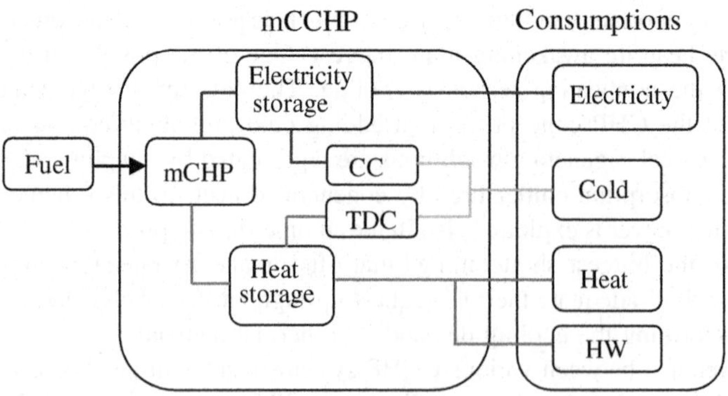

Fig. 9 Trigeneration system

The energies produced by the CCHP system are dependent on the type of refrigerating device, but also on whether the system is connected to the national electricity grid or not.

3.1 Energy Conversion in the Trigeneration

Note electric energy as E, the thermal energy for heating the residence as Q, the energy necessary for the air conditioning of the residence as C, and the thermal energy for preparing domestic hot water as Q_{hw}. To satisfy the demand for electric energy of the residence, the following systems may be used:

- *Centralized energy producing system.* In this case, the residence is connected to the electricity grid. The thermal energy demand for heating the residence or for cooling the air in the residence must be ensured with a system which contains equipment installed in the residence. This system includes a conventional condensing boiler (with 90 % thermal efficiency) providing heat for space heating and sanitary uses (hot water), and a conventional compressing refrigerator which supplies cold for air conditioning (Fig. 10).

Regarding the cooling equipment, performance is usually described by means of the specific coefficient of performance (COP). The COP_C can be generally defined as the ratio of the desired cooling energy output to the relevant input (electrical energy for electric chillers).

The energetic balance of this system is:

- for the electric subsystem:

$$E_{\text{grid}} = E + \frac{C}{\text{COP}_C} \tag{8}$$

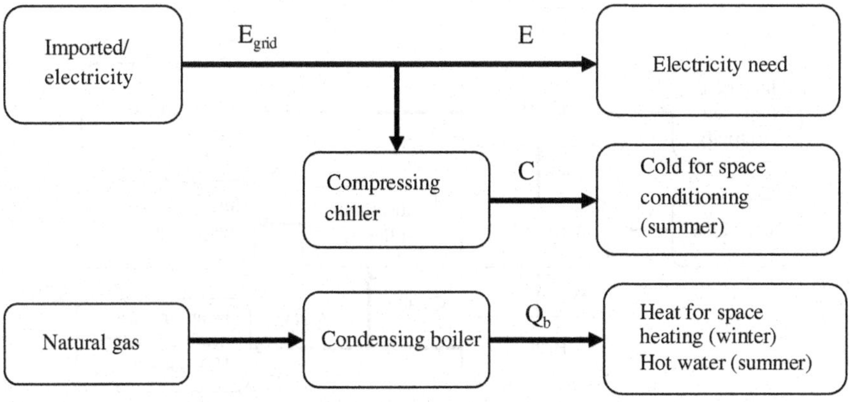

Fig. 10 Centralized energy producing system

- for the thermal subsystem:

$$Q_b = Q + Q_{\text{hw}} \tag{9}$$

where:

E_{grid} the quantity of electricity consumed from the grid;
COP_C the coefficient of performance of the compression chiller;
Q_b the heat produced by the boiler

- *Decentralized energy producing system.* In the case of the decentralized system, two solutions are applied for the supply with electricity:

(a) On-grid (or open) system

This solution produces combined heat and power by using a CHP technology [23, 24], and cold for air conditioning is generated by means of an absorption refrigerator making use of the "cogenerated" heat (Fig. 11). The CHP are sized in order to satisfy the maximum heat demand, so that they generate power in excess of customer needs. This excess power is exported to the utility grid.

The characteristics of the CHP prime movers can be effectively and synthetically described by means of the electrical efficiency and the thermal efficiency. For an absorption refrigerator, the COP_a can be defined as the ratio of the desired cooling energy output to the relevant input (thermal energy for steam-fed, hot water-fed, or exhaust-fed absorption chillers).

The energetic balance of this system is:

- for the electric subsystem:

$$E_{\text{CHP}} = E - E_{\text{grid}} \tag{10}$$

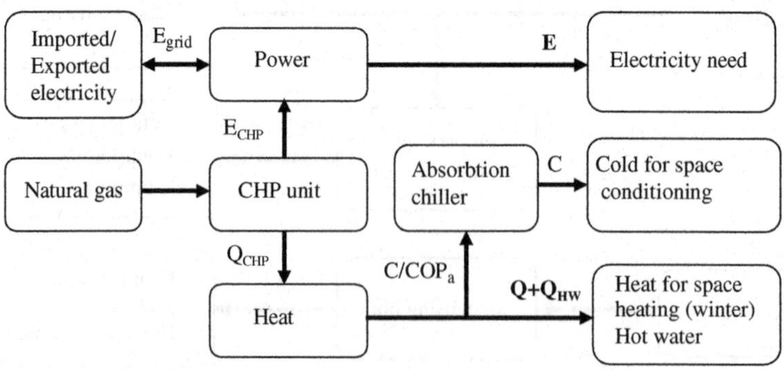

Fig. 11 Decentralized on-grid energy producing system

- for the thermal subsystem:

$$Q_{\text{CHP}} = Q + Q_{\text{hw}} + \frac{C}{\text{COP}_a} \tag{11}$$

where:
COP_a the coefficient of performance of the ad/absorption chiller;

(b) Off-grid (or "isolated") system

Since they do not involve importing/exporting electricity from/to the utility grid, small power plants are sized in order to satisfy the maximum customer needs of electricity. This implies that the amount of "cogenerated" heat is yes or not sufficient to satisfy energy needs for domestic heating and air conditioning. Heat and cold demands of the residence can be covered by adding a boiler and an absorption chiller (Fig. 12). Energy balance of the system is the following:

- for the electric subsystem:

$$E_{\text{CHP}} = E + \frac{C'}{\text{COP}_C} \tag{12}$$

- for the thermal subsystem:

$$Q_b + Q_{\text{CHP}} = Q + Q_{\text{hw}} + \frac{C - C'}{\text{COP}_a} \tag{13}$$

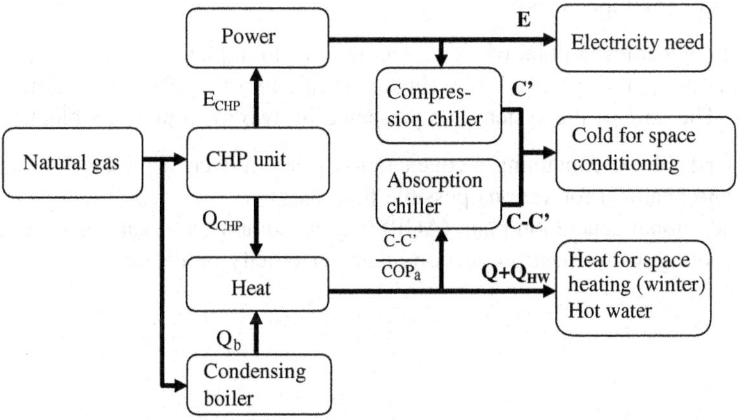

Fig. 12 Decentralized system with off-grid energy production

where:

C' is the quantity of cold produced by the mechanical compression chiller.

The term "cogeneration" is traditionally adopted with reference to the combined production of heat and electricity from fossil fuels. However, other types of cogeneration sources can be adopted [24, 25], for example from solar power, that allow for clean high-performance solutions. In this respect, increasing interest is being lately gained by applications of solar technologies for multigeneration. More specifically, photovoltaic (PV) modules, thermal collectors, and hybrid photovoltaic/thermal (PV/T) systems can be effectively coupled to bottoming cooling/heating equipment. In particular, although in theory electricity can be produced in a PV system and then utilized to feed an electric chiller, it results more energetically and economically effective to adopt heat-fired cooling technologies (namely adsorption, absorption, or desiccant systems) to be fed by cogenerated heat in a PV/T solar system (solar trigeneration), or by heat produced in a solar collector. In particular, PV/T solar units for cogenerative and trigenerative applications bear the additional intrinsic energy benefit that the optimized cooling of the solar modules brought by the heat recovery system brings along an increase in the PV electrical generation efficiency owing to decrease of the module temperature. The rationale of the utilization of solar systems for both heating and cooling generation is the same as for adopting conventional CHP unit in trigeneration applications. In fact, the solar thermal power is optimally exploited throughout the year, namely, in the summertime for cooling generation, and in the wintertime for heat generation.

What results from the presentation of these systems is that the applications developed in the decentralized energy production area can be categorized in structural and functional terms and in the increasing order of complexity, as:

(a) classical cogeneration (single input fuel, double output, single site);
(b) trigeneration (single/multiple input fuel, manifold output energy vectors, single site);
(c) distributed multigeneration (single/multiple input fuel, manifold output energy vectors, multiple sites).

From a generalized point of view, with trigeneration planning [25] it is possible to look at the plant as a black box (Fig. 13) with an array of inputs and manifold outputs. The core of the system is represented by two main physical blocks:

the CHP block, containing a cogeneration prime mover. It produces electricity (E) and heat (Q) for various possible final uses,

the additional generation plant (AGP) may be composed of various equipments for cooling and/or heat production and/or electricity production.

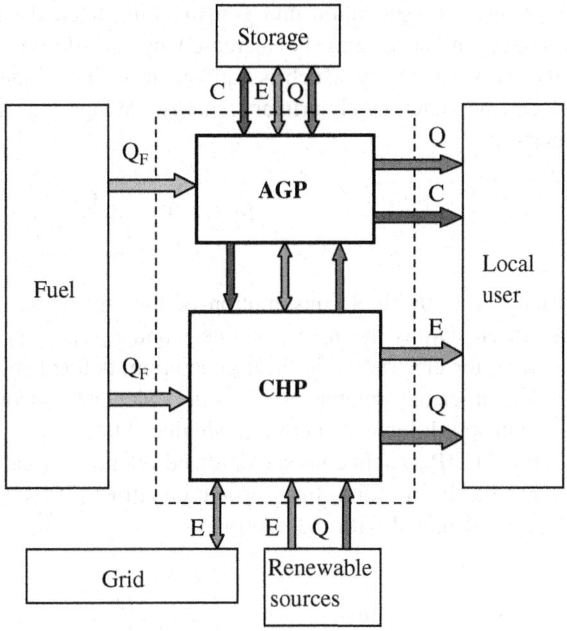

Fig. 13 Trigeneration black box system

3.2 Performance Indicators of the Trigeneration Systems

3.2.1 Primary Energy Saving to Trigeneration Systems

The CCHP system is compared to separate heat and power (SHP) systems in order to define its efficiency. This comparison is similar to the one between CHP and SHP systems. For SHP systems, Directive 2012/27/EC defines primary energy saving (PES), which may be calculated with the relation:

$$\mathrm{PES} = 1 - \frac{1}{\frac{\eta_H}{\eta_{\mathrm{Href}}} + \frac{\eta_E}{\eta_{\mathrm{Eref}}}} \tag{14}$$

where:

η_H is the heat efficiency of the cogeneration production, defined as annual useful heat output divided by the fuel input used to produce the sum of useful heat output and electricity from cogeneration;

η_{Href} is the efficiency reference value for separate heat production;

η_E is the electrical efficiency of the cogeneration production defined as annual electricity from cogeneration divided by the fuel input used to produce the sum of useful heat output and electricity from cogeneration;

η_{Eref} is the efficiency reference value for separate electricity production.

Where a cogeneration unit generates mechanical energy, the annual electricity from cogeneration may be increased by an additional element representing the amount of electricity which is equivalent to that of mechanical energy.

On the other hand, primary energy saving may also be calculated with the relation:

$$\mathrm{PES} = 1 - \frac{Q_F}{\frac{E}{\eta_{\mathrm{Eref}}} + \frac{Q_H}{\eta_{\mathrm{Href}}}} \tag{15}$$

where Q_F is the fuel consumption of the cogeneration plant, E is the electricity generated, Q_H is the heat generated, and η_{Href}, η_{Eref} are the two reference efficiencies for electricity and heat generation, defined separately.

The efficiency reference values are calculated according to fuel categories and the climatic differences between Member States.

For CCHP systems, no standardized relation for calculating the primary energy saving has been found, but in some countries [26] where this technology is used, PES is calculated with the relation:

$$\mathrm{PES} = \left(1 - \frac{Q_F}{\frac{E}{\eta_{\mathrm{Eref}}} + \frac{Q_H}{\eta_{\mathrm{Href}}} + \frac{C}{\mathrm{COP}_{\mathrm{ref}}\eta_{\mathrm{Eref}}}}\right) \times 100 \tag{16}$$

where C is the cooling energy generated and COP_{ref} is the performance reference chiller. Reference thermal efficiency is set to 0.8 for civil cogeneration and 0.9 for other cases. The reference chiller performance COP_{ref} is set to 3.0. Reference values for electrical efficiency used to calculate the energy saving index [26] is presented in Table 4.

The relation for calculating PES may be rearranged as:

$$\mathrm{PES} = \left(1 - \frac{Q_F}{\frac{E + \frac{C}{\mathrm{COP}_{\mathrm{ref}}}}{\eta_{\mathrm{Eref}}} + \frac{Q_H}{\eta_{\mathrm{Href}}}}\right) \times 100 \tag{17}$$

Table 4 Reference values for electrical efficiency [26]

Electrical efficiency for Italian regulation energy calculation				
Nominal power (MW)	Natural gas, liquid gas	Oil, naphtha, diesel fuel	Solid fossil fuels	Solid refuse fuels (organic and inorganic)
<1 MWe	0.38	0.35	0.33	0.23
>1–10 MWe	0.4	0.36	0.34	0.25

Or

$$\mathrm{PES} = \left(1 - \frac{Q_F}{\frac{E_{\mathrm{sys}}}{\eta_{\mathrm{Eref}}} + \frac{Q_{\mathrm{sys}}}{\eta_{\mathrm{Href}}}}\right) \times 100$$

where:

$$E_{\mathrm{sys}} = E + \frac{C}{\mathrm{COP_{ref}}}$$

Under this form, the relation is similar to the relation for calculating PES defined in cogeneration, when the cooling energy is generated through electric energy consumption. Generalizing to thermal systems (activated thermally), where cooling energy is generated through thermal energy consumption, we may define PES with the similar cogeneration relation:

$$\mathrm{PES} = \left(1 - \frac{Q_F}{\frac{E}{\eta_{\mathrm{Eref}}} + \frac{Q_H + \frac{C}{\mathrm{COP_{ref}}}}{\eta_{\mathrm{Href}}}}\right) \times 100\,\% \qquad (18)$$

or

$$\mathrm{PES} = \left(1 - \frac{Q_F}{\frac{E_{\mathrm{sys}}}{\eta_{\mathrm{Eref}}} + \frac{Q_{\mathrm{sys}}}{\eta_{\mathrm{Href}}}}\right) \times 100\,\%$$

where:

$$Q_{\mathrm{sys}} = Q_H + \frac{C}{\mathrm{COP_{ref}}}$$

3.2.2 Energy Efficiency of the Trigeneration Systems

PES refers basically to the percentage of fuel saved from the energy production of the CCHP system compared to the same energy produced separately. PES does not point if this production of the CCHP system is useful or not. Energy can be exceedingly produced and dissipated in the environment (especially for thermal energy). The second indicator sets the correlation between the produced and useful energy. This is also called CCHP efficiency. The second performance indicator is CCHP system efficiency given by relation:

$$EFF_{CCHP} = \frac{E_{sys} + Q_{sys}}{Q_F + E_{PV} + Q_{TP}} \times 100\,\%$$

where:

E_{PV} is the annual specific electricity production from photovoltaic panels (PV);
Q_{TP} is the annual specific heat production from thermal panels (TP);
E_{sys} is the annual specific electricity production from CCHP system;
Q_{sys} is the annual specific heat production from CCHP system;
Q_F is the specific fuel consumption of CCHP system (sum of useful fuel input for CHP unit and additional boiler)

If we use the renewable energy, the primary energy is considered the amount of energy produced. In other words, if the electricity production is from renewable sources, is not taken into account the efficiency of the conversion process.

References

1. Jenkins N, Allan R, Crossley P, Kirschen D, Strbac G (2000) Embedded generation, IEE power and energy series 31. The IEE, London, UK
2. Willis HL, Scott WG (2000) Distributed power generation: planning and evaluation. Dekker, New York
3. Ackermann T, Andersson G (2001) Distributed generation: a definition. Electr Power Syst Res 57(3):195–204
4. Pepermans G, Driesen J, Haeseldonckx D, Belmans R, D'haeseleer W (2007) Distributed generation: definition, benefits and issues. Energy Policy 33(6):787–798
5. Alanne K, Saari A (2006) Distributed energy generation and sustainable development. Renew Sustain Energy Rev 10:539–558
6. Kirschen DS (2003) Demand-side view of electricity markets. IEEE Trans Power Syst 18 (2):520–527
7. Levy R (2006) Avision of demand response—2016. Electr J 19(8):12–23
8. Valero S, Ortiz M, Senabre C, Alvarez C, Franco FJG, Gabaldon A (2007) Methods for customer and demand response policies selection in new electricity markets. IET Gener Transm Distrib 1(1):104–110
9. Sezgen O, Goldman CA, Krishnarao P (2007) Option value of electricity demand response. Energy 32(2):108–119
10. Ribeiro PF, Johnson BK, Crow ML, Arsoy A, Liu Y (2001) Energy storage systems for advanced power applications. Proc IEEE 89(12):1744–1756
11. Clark W, Isherwood W (2004) Distributed generation: remote power systems with advanced storage technologies. Energy Policy 32(14):1573–1589
12. Ise T, Kita M, Taguchi A (2005) A hybrid energy storage with a SMES and secondary battery. IEEE Trans Appl Superconduct 15(2):1915–1918
13. Bolund B, Bernhoff H, Leijon M (2007) Flywheel energy and power storage systems. Renew Sustain Energy Rev 11(2):235–258
14. Micro-Map (2002) Mini And Micro CHP—Market Assessment And Development Plan, European Commission SAVE programme, Faber Maunsell Ltd
15. EPIA (2012) Solar photovoltaics on the road to large-scale grid integration report, Sept 2012

16. Small-scale cogeneration, why? In which case? A guide for decision makers. In: European commission directorate general for energy DGXVII, July 1999
17. Onovwiona HI, Ugursal VI (2006) Residential cogeneration systems: review of the current technology. Renew Sustain Energy Rev 10(5):389–431
18. Pehnt M, Cames M, Fischer C, Praetorius B, Schneider L, Schumacher K et al. (eds) (2006) Micro cogeneration: towards decentralized energy systems. Springer, Berlin, Heidelberg
19. Asif M, Muneer T (2007) Energy supply, its demand and security issues for developed and emerging economies. Renew Sustain Energy Rev 11(7):1388–1413
20. Costantini V, Gracceva F, Markandya A, Vicini G (2007) Security of energy supply: comparing scenarios from a European perspective. Energy Policy 35(1):210–226
21. Haibo Y, Vittal V, Zhong Y (2003) Self-healing in power systems: an approach using islanding and rate of frequency decline-based load shedding. IEEE Trans Power Syst 18 (1):174–181
22. Amin SM, Wollenberg BF (2005) Toward a smart grid: power delivery for the 21st century. IEEE Power Energ Mag 3(5):34–41
23. Badea N, Paraschiv I, Oanca M (2013) Micro CHP with fuel cell for boiler supply. In: 4th international symposium on electrical and electronics engineering (ISEEE), Galaţi, Romania. ISBN 978-1-4799-2442-4/13/©2013 IEEE doi:10.1109/ISEEE.2013.6674335. Accessed 11–13 Oct 2013
24. Badea N, Cazacu N, Voncila I, Uzuneanu K (2010) Optimal architectures of domestic mCCHP systems based on renewable sources. In: 4th WSEAS international conference on renewable energy sources (RES'10), pp 95–100, Kantaoui, Sousse, Tunisia. ISBN 978-960-474-187-8, ISSN 1790-5095, Accessed 3–6 May 2010
25. Chicco G, Mancarella P (2009) Distributed multi-generation: a comprehensive view. Renew Sustain Energy Rev 13:535–551
26. PolySmart-Poly-generation www.polysmart.org
27. Soderman J, Pettersson F (2006) Structural and operational optimisation of distributed energy systems. Appl Therm Eng 26:1400–1408

Combined Micro-Systems

Nicolae Badea

Abstract This chapter presents the technologies for achieving Combined Heat and Power (CHP) and micro-Combined Cooling Heat and Power (mCCHP) micro-generation systems. Section 1 presents based on the relevant data from literature the main primary thermal motors used in CHP micro systems. A comparative analysis of cogeneration technologies based on performance indicators is presented in Sect. 2. Section 3 describes the mCCHP systems from the points of view of architectural achievement and operation modes in order to satisfy the residential consumers' energy needs.

1 The Micro-CHP Technologies

CHP technologies refer to the energy conversion, recuperation, and management in view of obtaining heat and power from burning a fuel. The term CHP (Cooling, Heating, and Power) describes all electrical power generation systems that utilize recoverable waste heat for space heating, cooling, and domestic hot water purposes (Fig. 1). The term CHP describes all electrical power generation systems that utilize recoverable waste heat for space heating, cooling, and domestic hot water purposes. Micro-CHP encompasses all systems in the range of 15–35 kW of electrical production or less. These systems range from single family homes to small apartment complexes to small office buildings. In a typical micro-CHP system, electricity is generated on-site from the combustion of a fuel source in an electrical generation set (prime mover and generator). This combustion produces recoverable heat in the form of heated engine coolant and high temperature exhaust. The use of the recoverable thermal energy for space heating and cooling purposes is the driving factor behind the increased overall energy usage from conventional power generation systems. Cogeneration system [1] consists of four basic elements: a prime mover (engine), an electricity generator, a heat recovery system and a control system. In the systems based on these technologies, primary engines play an extremely important role; they

N. Badea (✉)
"Dunarea de Jos" University of Galati, Galati, Romania
e-mail: nicolae.badea@ugal.ro

N. Badea (ed.), *Design for Micro-Combined Cooling, Heating and Power Systems*,
Green Energy and Technology, DOI 10.1007/978-1-4471-6254-4_3

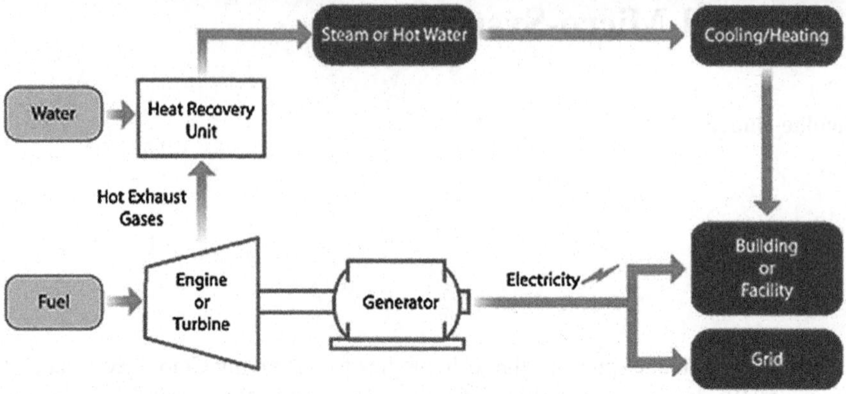

Fig. 1 CHP technologies requirements [1]

represent the basic components and, to a certain extent, determine the architecture of these systems.

The performance characteristics of a CHP system are overall efficiency, electric efficiency, power, power-to-heat ratio, and start-up time. The overall efficiency is dependent on many factors, such as technology used, fuel types, operation point, size of the unit, and also on the heat potential. All these characteristics are closely linked to the primary engine of the CHP system. That is why cogeneration technologies for residential, commercial, and institutional applications can be classified according to their prime mover and to where their energy source is derived from.

Present day technologies of cogeneration in mCHP systems [2] are based on the recuperation of the thermal energy of the following prime movers:

- Steam turbines, which are capable of operating over a broad range of steam pressures. They are custom designed to deliver the thermal requirements of CHP applications through use of backpressure or extraction steam at the appropriately needed pressure and temperature.
- Gas turbines, which produce a high quality (high temperature) thermal output.
- Reciprocating engines, which are well suited for applications that require hot water or low-pressure steam.
- Stirling engines.
- Fuel cells, where the waste heat can be used primarily for domestic hot water and space heating applications.

1.1 Steam Turbines

Steam turbines represent the widest used technology in industrial electric plants [3]. The thermodynamic cycle which lies at the basis of conventional steam plant functioning may be the one with overheated steam (the Hirn cycle) or the one with

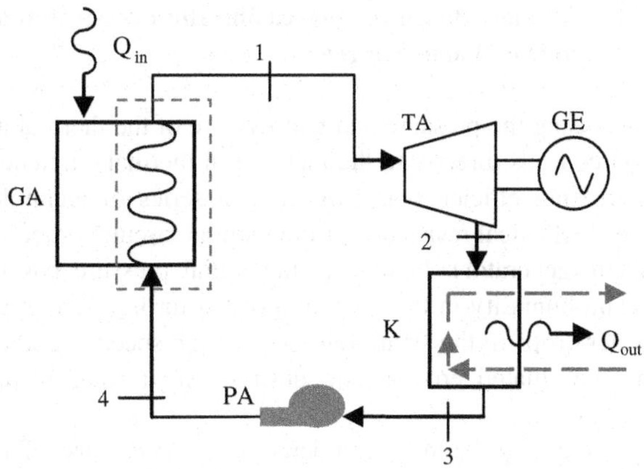

Fig. 2 CHP with a condensation steam turbine [3]

saturated steam (the Rankine cycle). In Fig. 2 is present a CHP system with a steam turbine [3],

where

GA—steam generator, TA—steam turbine, GE—electric generator, K—condenser, PA—charging pump.

The steam generator has the role of vaporizing water and transforming it into saturated or overheated steam. This process is achieved using the heat from burning a fuel. The steam is released in the turbine, producing mechanical energy, which the electric generator transforms into electric energy.

The condenser ensures the condensation of the water vapors released from the turbine and represents the cold source of the thermodynamic cycle. To evacuate the heat toward the exterior, water or, rarely, atmospheric air may be used as cooling agents.

The thermal efficiency can also be expressed in terms of the heat transfer terms or in terms of temperatures in the following manner:

$$\eta = \frac{\dot{Q}_{\text{in}} - \dot{Q}_{\text{out}}}{\dot{Q}_{\text{in}}} \tag{1}$$

or

$$\eta = 1 - \frac{\vartheta_{\text{out}}}{\vartheta_{\text{in}}} \tag{2}$$

where:

ϑ_{out}—is the temperature on the steam side of the condenser

ϑ_{in}—is the average temperature of heat addition at the boiler.

To increase the turbine thermal efficiency, the methods presented in what follows are used.

1.1.1 Method Based on Increasing the Pressure and Temperature in the Warm Source

Increasing the pressure and temperature of the thermal agent delivered by the warm source leads directly to the increase in thermal efficiency. However, this method of increasing efficiency is subject to a series of technological restrictions, such as the mechanical resistance of the thermal circuit components (especially those of the steam generator). Increasing the initial pressure has the effect of increasing the steam humidity in the final area of the turbine. The presence of a large number of water drops in the steam released at high speed (>200 m/s) leads to a phenomenon of accentuated erosion and destruction of rotor blades in the final area of the turbine.

Increasing the initial temperature has a contrary effect on the release humidity in the steam turbine. Consequently, increasing the initial pressure must be accompanied by increasing the initial temperature.

This method of increasing efficiency is achieved through the intermediary overheating of the thermal agent, which comes from the warm source of the thermodynamic cycle. The method presupposes the interruption of the steam release in the turbine, for it to be then sent back to the steam generator. Here, it is overheated once again, up to a temperature comparable with the initial one and then it continues to be released in the steam turbine. Figure 3 presents the simplified scheme of an intermediary overheating system, where: GA—steam generator, SÎI—intermediary over-heater, CIP—high pressure body, CMJP—medium and low pressure body, GE—electric generator, K—condenser, PA—circulation pump, PR—regenerative pre-heater.

Intermediary overheating presupposes a complication of the thermal circuit and of the steam generator, with direct effects on the initial investment. Consequently, the scheme is generally justified only for high power groups (>100 MW) with a sufficiently high annual duration of installed power use.

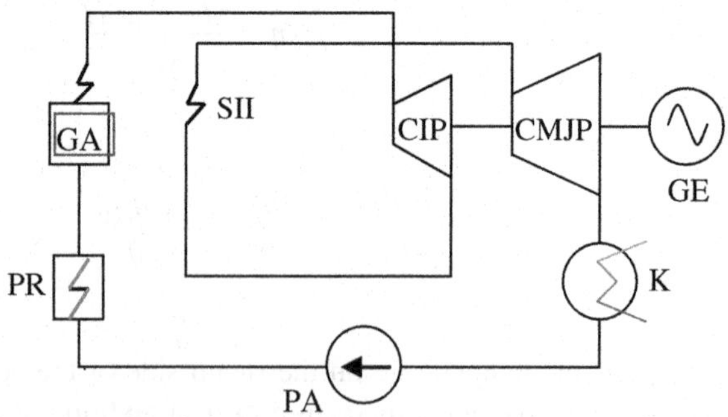

Fig. 3 Intermediary overheating system scheme

1.1.2 Method Based on Decreasing the Temperature and Pressure in the Cold Source

Decreasing the condensation temperature and/or pressure represents a method of increasing efficiency, since the lower the steam temperature in the condenser, the higher the thermal efficiency is. We are positing that the effect produced by a decrease (by 1 °C) of the condensation temperature may be equivalent with the effect produced by the increase (by 10–15 °C) of the initial temperature of the thermal agent. Therefore, this method of increasing thermal efficiency is very productive. A low condensation temperature is conditioned by the existence of several cooling fluids, with an appropriate flow and thermal level.

Figure 4 presents the simplified thermal scheme corresponding to such a system, which uses a backpressure steam turbine. In such a turbine, the release pressure is much higher than in the case of condensation turbines and depends on the type of consumer, as 0,7…2,5 bar for urban consumers (heating, sanitary hot water preparation, etc.);

Steam exits the turbine at a pressure higher or at least equal to the atmospheric pressure, which depends on the needs of the thermal load. This is why the term back—pressure is used. After the exit from the turbine, the steam is fed to the load, where it releases heat and is condensed. The condensate returns to the system with a flow rate which can be lower than the steam flow rate, if steam mass is used in the process or if there are losses along the piping. Make–up water retains the mass balance.

Being a mature technology, long studied and improved, steam turbines have a large life cycle and, if properly maintained, are very reliable. However, there are aspects which limit the use of these turbines: the thermo-electric conversion efficiency, the relatively long barring process, and the weak charge performances.

The characteristics and performance indicators of CHP systems with industrial turbines and of mCHP systems with steam micro-turbines are synthetically given in Table 1.

Based on an Rankine cycle engine, the new existing technology [5] produces clean and cost-effective heating and electricity from a low temperature solar thermal system. The product provides households with 1 kW of electricity and 8.8 kW of heating from free solar energy (Fig. 5).

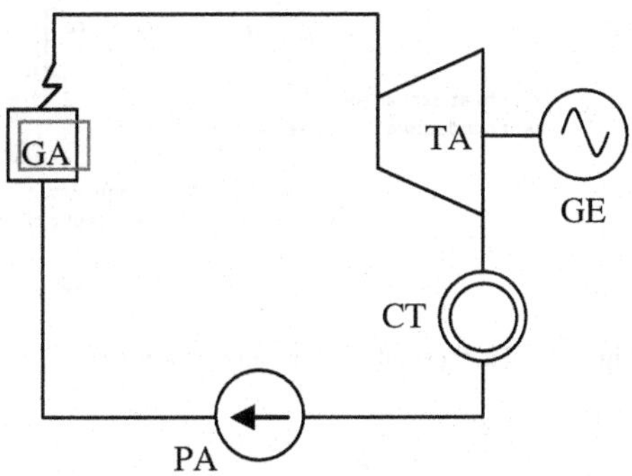

Fig. 4 Backpressure steam turbines

Table 1 Turbines—indicators of performance [4]

Features	Industrial steam turbines	Micro steam turbines
Electrical efficiency	15–38 %	18–27 %
Thermal efficiency	80 %	65–75 %
Global efficiency	75 %	50–75 %
Power to Heat ratio	0.1–0.3	0.4–0.7
Start-up time	1 h–1 day	60 s
Power range	50 kW–250 MW	5–250 kW
Using thermal energy	Low and high pressure steam	Heat, hot water, low pressure steam
Advantages	Report electricity /heat depending flexible operating system;	High reliability (few moving parts); easy installation;
		Compact; Light weight;
		Acceptable noise level
	Ability to adapt to the requirements of several types of heat consumers; variety of types and sizes; long life	High temperature heat recovery
Disadvantages	High report heat/electricity;	High costs
	High investment cost;	
	Slow start	

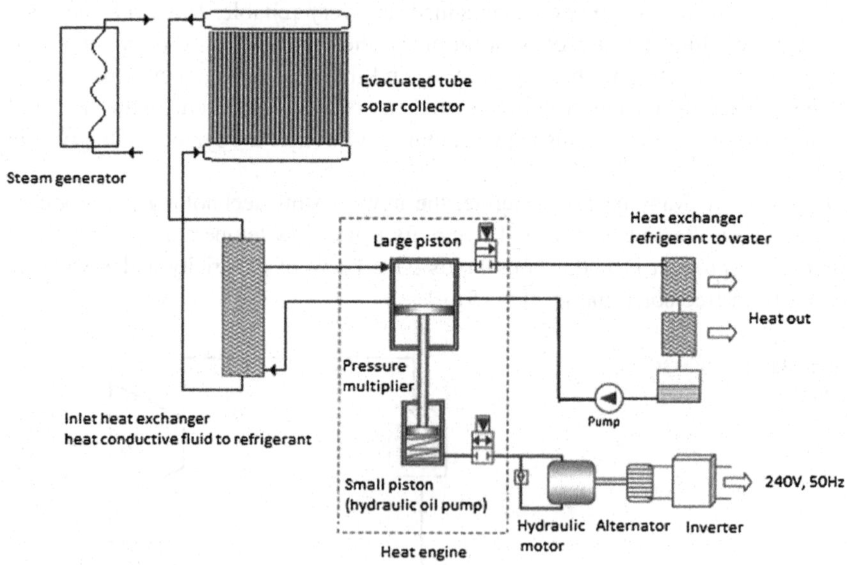

Fig. 5 Solar thermal heating and power [5]

1.2 Gas Micro Turbines

Internal combustion turbines are primary engines, frequently used in cogeneration due to the high level of safety and reliability and to the wide power range.

A CHP system with a gas turbine contains a compressor, a combustion chamber and a turbine on the same axis with the compressor. On this axis, the turbine provides the useful mechanical work.

Microturbines are small gas turbines in which hot and pressurized gases (obtained from the combustion of fuel added to compressed air) are expanded through a rotating turbine, thereby acting on the motor shaft. These are used to produce electricity. The majorities of microturbines have an exit power between 25 and 300 kW and may be fuelled with natural gas, diesel, petrol or alcohol.

The process of a microturbine is similar to that of an industrial turbine, which has a recovery to recuperate part of the exhaust heat and to use it to preheat the combustion air. The scheme in Fig. 6 shows the main components of a microturbine.

Most turbines have a rotation speed of 90,000–120,000 rpm [7] Most producers use one shaft (with the compressor, the turbine and the electric generator on it), for which bearing greasing is easy. The electric generator has permanent magnets, and a group redresser-converter is needed to adapt the electric charges to the grid. The air in the compressor is preheated and, in the combustion chamber, the air combines with the fuel and the resulting combustion mix burns and expands in the turbine. The rotation of the turbine engages the compressor and generator shaft. The gases exhausted from the turbine are sent back, through the recovery, to preheat the combustion air.

Nonrecuperative turbines produce electricity from natural gas, with an efficiency of approximately 15 %. Microturbines equipped with a recovery have an electric efficiency between 20 and 30 %. The difference in electric efficiency is generated by the air preheating achieved in the recovery, which reduces the quantity of necessary fuel. The global efficiency of the system may be of 85 % and may be reached when

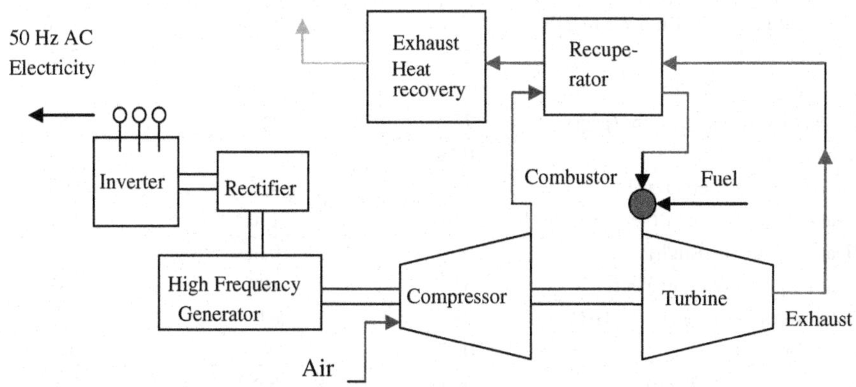

Fig. 6 Scheme of a gas turbine [6]

the microturbines are coupled to thermal components to recuperate the heat. The heat lost in a microturbine is mainly contained in the hot exhaust gases. This heat should be used to fuel a steam generator, to heat a residence or to fuel various cooling absorption systems. The way in which the heat lost may be used depends on the configuration of the turbine. In a nonrecuperative turbine, the exhaust gas comes out at a temperature of 538–594 °C. A recuperative turbine capitalizes the heat lost by using it to heat spaces or for the thermal agent in a cooling absorption system where the exhaust temperature is around 271 °C.

The exploitation of a microturbine has several advantages. The first is that it has less moving parts than a combustion engine. The limited number of moving parts and the reasonable demands regarding the greasing make microturbines have a long life cycle. Consequently, microturbines have low exploitation costs (seen as cost per kW of power produced). The second advantage of microturbines is their relatively small size, as compared to the power produced. Microturbines are light and have a low level of noxe. The third, and maybe the most important advantage of microturbines, is their capacity to use more types of fuel, including recuperated fuel or bio fuel. The main disadvantages of microturbines are due to the fact that they have low levels of electric efficiency. Moreover, in conditions of increased altitude and environment temperature, microturbines observe a decrease in exhaust power and efficiency. Environment temperature directly affects the intake air temperature. A gas turbine will work more efficiently when colder air is available at intake. The performance characteristics of microturbines are given in Table 2.

Micro Turbine Technology (MTT) is developing recuperated micro turbines up to 30 kW electrical powers for CHP and other applications [9]. Automotive turbocharger performance and efficiency have increased significantly during recent

Table 2 Micro-turbine cogeneration system performance characteristics [8]

	Capstone model 330 micro-turbine	IR energy systems 70LM (two shafts)	Turbec T 100
Nominal electricity capacity (kW)	30	70	100
Electrical heat rate (Btu/kWh) HHV	15,075	13,540	12,639
Electrical efficiency (%) HHV	22.6	25.2	27.0
Fuel input (MMBtu/h)	0.422	0.948	1.264
Required fuel gas pressure (psig)	75	55	75
Exhaust flow (lbs/s)	0.69	1.40	1.74
GT exhaust temperature (F)	530	435	500
Heat exchanger exhaust temperature (F)	150	130	131
Heat output (MMBtu/h)	0.17	0.369	0.555
Heat output (kW equivalent)	51	108	163
Total overall efficiency (%) HHV	73	64	71
Power/heat ratio	0.47	0.65	0.62
Net heat rate (Btu/kWh)	5509	6952	5703
Effective electrical efficiency (%) HHV	46.7	49	60

years, even for very small sizes. This has created an opportunity to develop low-cost micro turbines that can be produced in large volumes at low prices. Coupled with a generator electric power can be produced. By adding a recuperator, a large portion of the exhaust gas heat is recovered and electrical efficiency can be substantially increased. With low-cost configurations, 16 % electrical efficiency can be realized offering an excellent solution for a micro CHP application. With more advanced versions, levels up to 25 % can be realized opening a wide variety of additional applications. Based on the MTT, a 3 kW electrical/15 kW thermal micro CHP system is being developed to replace heating boilers for small business and households. Major focus is given to low-cost price, reliability, noise reduction, and low maintenance.

1.3 Thermal Engines with Internal Combustion

Internal combustion engines based on fossil fuel are widely spread and have diverse uses. An internal combustion engine is activated by the explosion of the fuel mix which burns in direct contact with the engine. To function, internal combustion engines need fuel, air and a mechanism which can achieve the compression of the air-fuel mix.

The basic elements of a reciprocating internal combustion engine based on a cogeneration system are the engine, generator, heat recovery system, exhaust system, and controls and acoustic enclosure. The generator is driven by the engine, and the useful heat is recovered from the engine exhaust and cooling systems. The main scheme of a CHP device with thermal combustion engines is shown in Fig. 7.

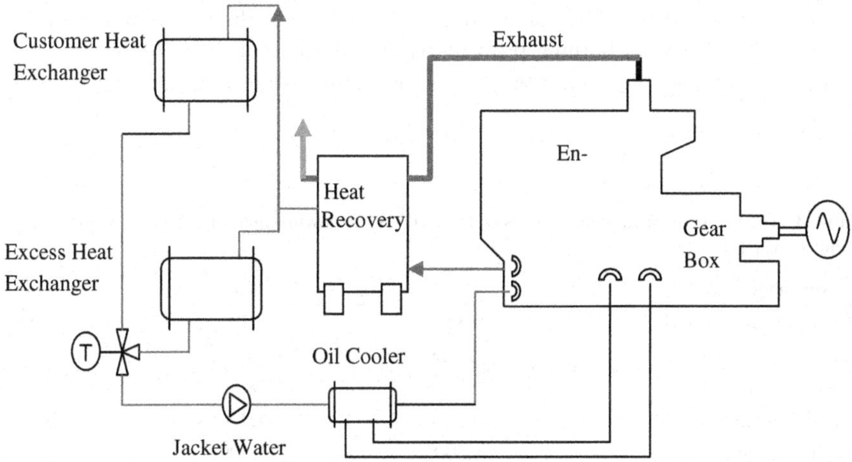

Fig. 7 CHP system with internal combustion engines

Two types of internal combustion engines are used to generate energy:

- spark ignition engines (with plugs, Otto);
- internal combustion engines ignited by compression, where the mix is brought to a high pressure before entrance into cylinders (Diesel).

Piston engines are available, from those coupled to small generators (0.5 kW) to those coupled to large generators (3 MW). These use common fuels like petrol, natural gas, diesel, and are convenient in a multitude of applications, due to their small sizes and their low costs. In energy production, the functioning of piston engines includes both continuous functioning and functioning at peak charges, as spare source. Piston engines are ideal for applications where there is a substantial need for hot water or for low pressure steam.

In the case of a CHP system used for producing electricity on a wide scale, efficiency is around 30 % while, during a combined exploitation cycle, efficiency is of approximately 48 %. The heat in the burning gases is lost, together with their exhaust into the atmosphere. Unlike this case, with a mCHP system based on internal combustion thermal engines, the heat from the cooling water, from the machine oil and from the exhaust circuit of burning gases is recuperated. From the recuperated heat, low pressure steam or hot water may be obtained, which may then be used for heating, obtaining domestic hot water and refrigerating.

The heat from the cooling circuit is able to produce hot water at 90 °C and represents approximately 30 % of the fuel input energy. Engines which work at high pressure are equipped with a cooling system for very high temperatures and can work to up to a temperature of 455–649 °C. Since the temperature of the exhaust gases must be kept above the condensation limit, only part of the heat contained by these gases can be recuperated. Heat recuperation units are generally designed for a temperature of 150–175 °C, to avoid corrosion and condensation in the exhaust pipes. Low pressure steam and hot water 110 °C is produced using the machine's heat exhaust. The heat recuperated from the cooling fluid and from the gases resulted from the burning process may reach approximately 70–80 %.

The performance characteristics of internal combustion engines are given in Table 3.

Table 3 The performance characteristics of the CHP systems with internal combustion engines as prime mover [10]

Power range (kW_e)	10	100	3,000
Power to heat ratio	0.50	0.79	0.97
Electrical efficiency (%)	25–28	34	36
Total efficiency (%)	79	78	73
Fuel Input (MMBtu/hr)	0,5	4,9	28
Engine Speed (rpm)	1,500	1,500	750
Fuel type	A variety of gaseous and fluid fuels		

1.4 Stirling Engines

Stirling engine is the most promising technology on the short medium term. It can be installed in urban environment every type of fuel can be utilized (methane, hydrocarbons, hydrogen, biomass, or heat from renewable sources). The Stirling technology is mainly mechanical, is well established, does not require special infrastructure.

A Stirling engine is an internal combustion engine which uses a difference in temperature to create movement in the shaft. The functioning of a Stirling engine is based on the behavior of a fixed quantity of air or gas (like helium or hydrogen) inside the engine cylinders. Two properties of gases lie at the basis of a Stirling engine functioning.

- in the case of a constant quantity of gas, with a fixed volume, the more the temperature increases, the more the pressure increases also;
- when a fixed quantity of gas is compressed, the temperature of that gas will increase.

The applications for the Stirling engine (an example being given in Fig. 8) include units of energy generation for space ships, small planes, refrigeration, mCHP systems and, on a smaller scale, residential or portable energy generators.

Main characteristics of the CHP systems with Stirling engine is given in Table 4.

Stirling engines have an electric efficiency of 10–25 %. If, however, the heat lost is recuperated in CHP type systems, the overall efficiency of these systems may increase significantly. Typical normal temperatures for operating vary between 650 and 800 °C. The heat may be recuperated by using the heat exchanger in the cold source of the engine, as well as by using the heat exchanger through which the burnt gases are exhausted into the atmosphere.

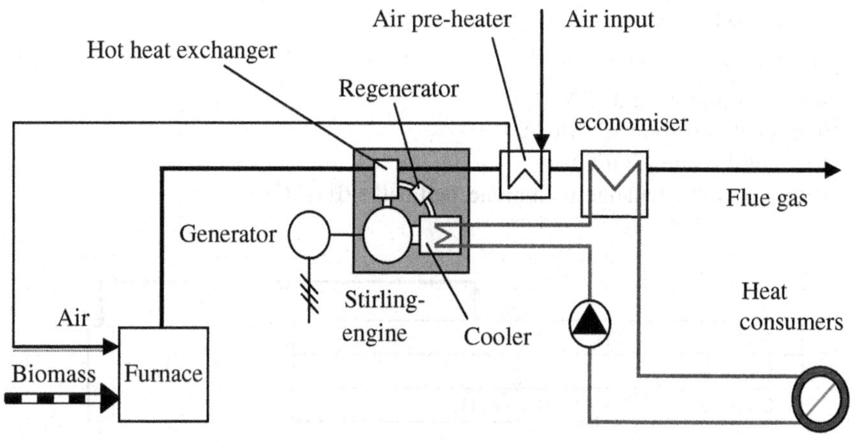

Fig. 8 Scheme of a biomass CHP system based on a Stirling engine [11]

Table 4 Main characteristics of the CHP systems with Stirling engine

Power range (kW_e)	0.003–100
Power to heat ratio	0.33
Electrical efficiency (%)	10–25
Thermal efficiency (%)	40–80
Total efficiency (%)	70–90
Fuel type	All fuels

1.5 Fuel Cell

Fuel cells use an electrochemical process that release the energy stored in natural gas or hydrogen fuel to create electricity. Heat is a by-product. Fuel cells that include a fuel reformer use hydrogen from any hydrocarbon fuel. Fuel cells transform the electrochemical energy into electricity and heat, through combining hydrogen with oxygen in the presence of a catalyst (Fig. 9).

Through a catalytic reaction, the hydrogen at the anode produces ions and electrons. Ions may pass through the electrolyte and reach the cathode through the external electric circuit. The reaction at the cathode leads to heat and water generation.

Besides hydrogen, other gases (like pit gas) may be used, especially for high power cells. The scheme for this case is shown in Fig. 10. The fuel–oxygen mix may be achieved outside or inside the cell, depending on the work temperature of the fuel cell.

Fuel cells are similar to electric batteries and have the capacity to produce continuous current through an electro-chemical process. While an electric battery produces energy during a time span which is limited by the stored chemical energy, fuel cells may operate indefinitely.

Five main types of cells are known today, being classified according to the type of the electrode used:

- alkaline (AFC)
- with phosphoric acid (PAFC)
- with melted carbonate (MCFC)
- with solid oxide (SOFC)
- with a proton exchange membrane fuel cell (PEMFC).

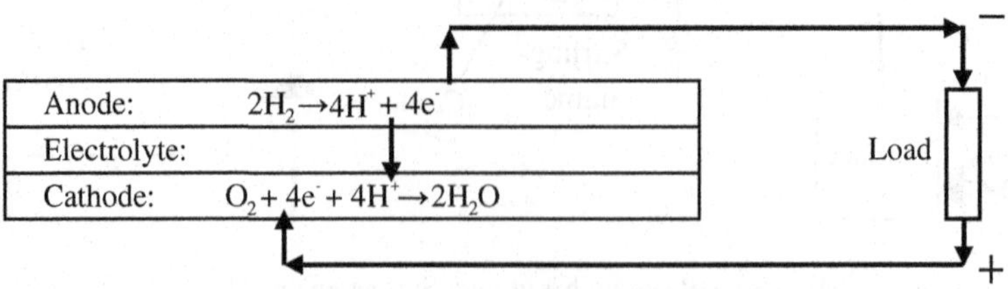

Fig. 9 Electro-chemical conversion in a fuel cell [12]

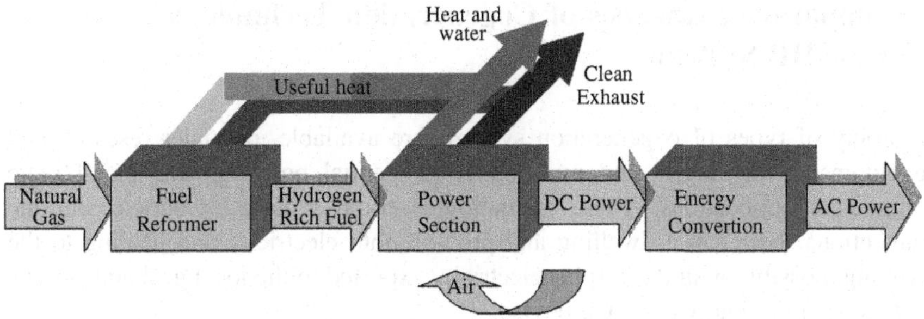

Fig. 10 Conversion system with a fuel cell [13]

Fuel cells offer the advantage of nearly 1-to-1 power—to-heat ratio, suited for modern low-energy buildings. Fuel cells provide a higher proportion of electricity than other CHP technologies. The fuel cell systems most widely commercially deployed are PAFC and PEMFC, with 5,000 PEMFC fuel cells having been installed in Japan [14] in 2009 alone. Most operational experience is therefore limited to these fuel cells, although there are increasing numbers of SOFC and MCFC systems installed. Combined heat and power systems of building scale is 1–10 kWe. Traditional CHP systems are mature and a useful transitional technology, while micro-CHP, biomass CHP and even fuel cell systems (using CO_2-free hydrogen) may emerge as abatement option. The advantages of fuel cell CHP over traditional mechanical CHP are the greater degree of modulation and the heat-to power ratio is typically much lower. In general, fuel cells have high electric efficiency, under variable charge, and reduced emissions. However, energy generation seems to represent another promising market that fuel cells could conquer. Actually, with the exception of PAFC cells, fuel cells are not completely viable for sale yet. World wide, there is a total capacity of over 40 MW, PAFC fuel cells. A detailed comparison of the characteristics of these fuel cells is given in Table 5.

Table 5 Performance characteristics for representative commercially available and developmental natural gas fuel cell based cogeneration systems [15]

Fuel cell type	PEM	PEM	PAFC	SOFC	MCFC
Nominal electricity capacity (kW)	10	200	200	100	250
Electrical efficiency (%) HHV	30	35	36	45	43
Fuel input (MMBtu/hr)	0.1	2.1	1.9	1.0	2.4
Operating temperature [°C]	70	70	200	950	650
Cogeneration characteristics					
Heat output (MMBtu /hr)	0,42	0,76	0,85	0,34	1,9
Heat output (kW equivalent)	11,7	211	217	100	556
Total overall efficiency (%) HHV	65	72	81	77	65
Power/heat ratio	0.85	0.95	0.81	1.25	1.95
Effective electrical efficiency (%)	55	65.0	81	65.6	56.5

2 Comparative Analysis of Cogeneration Technologies in mCHP Systems

A variety of types of cogeneration systems are available, or under research and development, for single- and multifamily residential buildings and small scale commercial applications. These technologies could replace or supplement the conventional boiler in a dwelling and provide both electricity and heating to the dwelling, possibly with the surplus electricity exported to the local grid and surplus heat stored in a thermal storage device.

A review of existing residential cogeneration systems performance assessments and evaluation can be found in [16] and the main advantages and disadvantages of each of the prime mover options for cogeneration can be found in [17]. A summary of typical cost and performance characteristics by CHP technology type [18] is presented in Table 6. Data represent illustrative values for typically available systems.

Table 6 Typical cost and performance characteristics by CHP technology type [18]

Technology	Steam turbine	Diesel engine	Nat. gas engine	Gas turbine	Microturbine	Fuel cell
Power efficiency (HHV)	15–38 %	27–45 %	22–40 %	22–36 %	18–27 %	30–63 %
Overall efficiency	80 %	70–80 %	70–80 %	70–75 %	65–75 %	65–80 %
Typical capacity (MW_e)	0.2–800	0.03–5	0.03–5	1–500	0.03–0.35	0.01–2
Typical power to heat ratio	0.1–0.3	0.5–1	0.5–1	0.5–2	0.4–0.7	1–2
CHP Installed costs ($/kW$_e$)	300–900	900–1,500	900–1,500	800–1,800	1,300–2,500	2,700–5,300
Availability	near 100 %	90–95 %	92–97 %	90–98 %	90–98 %	>95 %
Hours to overhauls	>50,000	25,000–30,000	24,000–60,000	30,000–50,000	5,000–40,000	10,000–40,000
Start-up time	1 h–1 day	10 s	10 s	10 min–1 h	60 s	3 h–2 day
Fuels	all	Diesel, residual oil	Gas, biogas, propane, landfill gas	Natural gas, biogas, propane, oil	Natural gas, biogas, propane, oil	Hydrogen, natural gas, propane, methanol
Uses for thermal output	LP-HP steam	Hot water, LP steam	Hot water, LP steam	Hot water, LP–HP steam	Heat, hot water, LP steam	Hot water, LP–HP steam

From among the CHP technologies presented, the most appropriate for residential use are the technologies presented in Table 7. The electric efficiency of reciprocating internal combustion engines is higher compared to micro-turbines and Stirling engines. On the other hand, fuel cells promise to offer the highest electric efficiency for residential and small-scale cogeneration applications in comparison with the other technologies, but are challenged by lack of demonstrated performance.

To better understand the potential benefits of different micro-CHP technologies, a comparative study [19, 20] involved a major field trial of micro-CHP units in domestic applications, with a corresponding trial of A-rated condensing boilers to provide a baseline for comparison.

As expected, the measured thermal efficiencies for the micro-CHP units are around 10–15 % lower than for the condensing boilers. This is primarily a consequence of some of the heat generated by the engine being used to generate electricity.

Table 7 Technical features of small-scale CHP Devices [8, 10, 15, 19]

	Reciprocating engines	Microturbines	Stirling engines	PEM fuel cells
Electric power (kW)	10–200	25–250	2–50	2–200
Electric efficiency, full load (%)	24–45	25–30	15–25	40
Electric efficiency, half load (%)	23–40	20–25	25	40
Total efficiency (%)	75–85	75–85	75–85	75–85
Heat/electrical power ratio	0.9–2	1.6–2	3–3.3	0.9–1.1
Output temperature level (°C)	85–100	85–100	60–80	60–80
Fuel	Natural or biogas, diesel fuel oil	Natural or biogas, diesel, gasoline, alcohols	Natural or biogas, LPG, several liquid or solid fuels	Hydrogen, gases, including hydrogen, methanol
Interval between maintenance (h)	5,000–20,000	20,000–30,000	5,000	N/A
Investment cost ($/kW)	800–1,500	900–1,500	1,300–2,000	2,500–3,500
Maintenance costs ($/kW)	1.2–2.0	0.5–1.5	1.5–2.5	1.0–3.0

The net carbon benefit of the electricity generated by the domestic micro-CHP systems is significantly higher than for the condensing boilers at 88 %. Conclusion on market potential [20] is micro-CHP can be competitive with condensing boiler if their investment costs and maintenance are comparable.

An important characteristic of mCHP cogeneration systems is the ratio between the electrical energy and the thermal energy of the primary engine. Between these powers depends on the primary thermal engine and is called cogeneration index:

$$\gamma = \frac{E}{Q} \tag{3}$$

where

- E—is the amount of electricity from cogeneration, γ is the power to heat ratio, and
- Q—is the amount of useful heat from cogeneration (calculated for this purpose as total heat production minus any heat produced in separate boilers or, in case of turbines, by live steam extraction from the steam generator before the turbine).

The power to heat ratio is of special importance, since the calculation of electricity from cogeneration is based on the actual power to heat ratio.

An over-unitary value of the cogeneration index shows the predominance of the electricity produced by the mCHP system as compared to the thermal one. The majority of cogeneration technologies have a subunitary cogeneration index.

The dependence of the mCHP system electric and thermal power on this index may be determined by imposing that the cogeneration index should vary between 0.1 and 4, and assuming that the sum of electric power and thermal power is the power unit.

$$E + Q = 1 \tag{4}$$

Measured with this unit, electric power may be calculated in keeping with this ratio, with the relation:

$$E = \frac{\gamma}{\gamma + 1} \tag{5}$$

and the thermal power with the relation:

$$Q = \frac{1}{\gamma + 1} \tag{6}$$

The cogeneration index has a great influence on the mCHP performance indices (PES and overall efficiency). If we assume that the cogeneration system has a fuel saving of 5 % then, to produce the same output power and admitting an electric efficiency of 40 % and a thermal one of 80 %, the input energy has the following dependency on the cogeneration index:

- in the case of separate production:

$$Q_{\mathrm{SHP}} = \frac{E}{\eta_e} + \frac{Q}{\eta_{th}} = \left(\frac{10}{4}\frac{\gamma}{\gamma+1} + \frac{10}{8}\frac{1}{\gamma+1}\right) \tag{7}$$

- in the case of cogeneration using relation:

$$\mathrm{PES} = 1 - \frac{Q_{\mathrm{CHP}}}{Q_{\mathrm{SHP}}} \tag{8}$$

Resulting is the following dependency:

$$Q_{\mathrm{CHP}} = 0,95 \cdot Q_{\mathrm{SHP}} = 0,95\left(\frac{10}{4}\frac{\gamma}{\gamma+1} + \frac{10}{8}\frac{1}{\gamma+1}\right) \tag{9}$$

Figure 11 presents the dependence of powers on the cogeneration index. What may be noticed is that, practically, for high indices, the cogeneration system produces only electric energy, and for very low indices, it produces only thermal energy.

The efficiency of the separate energy production may be compared to the efficiency of the combined system only on the basis of the cogeneration index value.

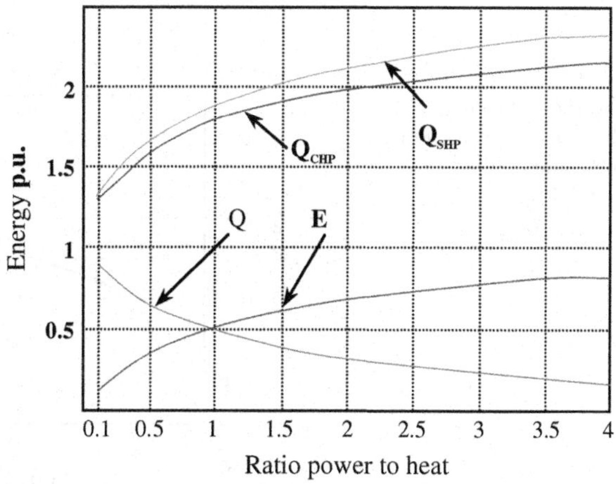

Fig. 11 Power dependence of the cogeneration index

Thus, the dependency of the overall efficiency of separate energy production on the cogeneration index may be mathematically expressed with the relation:

$$EFF_{\text{SHP}} = \frac{1+\gamma}{\frac{\gamma}{\eta_e} + \frac{1}{\eta_{th}}} \tag{10}$$

The overall cogeneration efficiency, defined by the relation:

$$EFF_{\text{CHP}} = \frac{E+Q}{Q_{\text{CHP}}} = \frac{E+Q}{Q_{\text{SHP}}}\frac{Q_{\text{SHP}}}{Q_{\text{CHP}}} = \eta_{\text{SHP}}\frac{Q_{\text{SHP}}}{Q_{\text{CHP}}} \tag{11}$$

leads to the following dependency on the cogeneration index:

$$EFF_{\text{CCHP}} = \frac{1}{0,95}\frac{1+\gamma}{\frac{\gamma}{\eta_e} + \frac{1}{\eta_{th}}} \tag{12}$$

Figure 12 represents the dependency of efficiency on the power ratio.

According to the Directive 2012/27/EC, the promotion of high-efficiency cogeneration based on a useful heat demand is a Community priority given the potential benefits of cogeneration with regard to saving primary energy, avoiding network losses and reducing emissions, in particular of greenhouse gases. In addition, efficient use of energy by cogeneration can also contribute positively to the security of energy supply and to the competitiveness of the European Union.

Directive 2012/27/EC on the promotion of cogeneration based on a useful heat demand in the internal energy market shows that the power to heat ratio is a technical characteristic that needs to be defined in order to calculate the amount of electricity from cogeneration. Electricity from cogeneration shall mean electricity generated in a process linked to the production of useful heat and is calculated according to the following formula: $E = \gamma \cdot Q$

Fig. 12 Efficiency dependency of the power to heat ratio

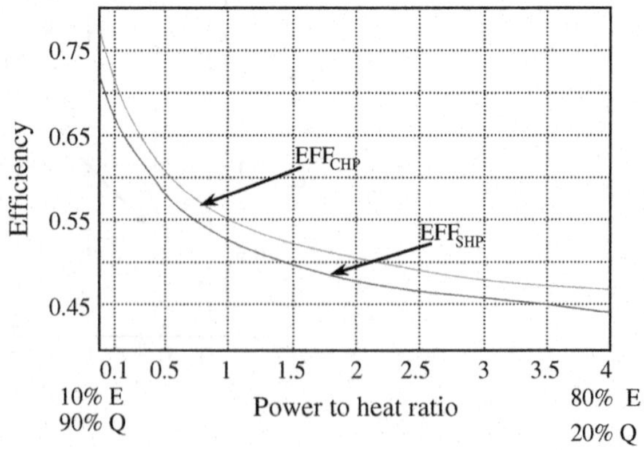

Table 8 The default values of the power to heat ratio [21]

Type of the unit	Default power to heat ratio γ
Combined cycle gas turbine with heat recovery	0.95
Steam backpressure turbine	0.45
Steam condensing extraction turbine	0.45
Gas turbine with heat recovery	0.55
Internal combustion engine	0.75

If the actual power to heat ratio of a cogeneration unit is not known, the default values given in Table 8, may be used.

Cogeneration applications in buildings can be designed to:

- satisfy both the electrical and thermal demands,
- satisfy the thermal demand and part of the electrical demand,
- or satisfy the electrical demand and part of the thermal demand
- or, most commonly, satisfy part of the electrical demand and part of the thermal demand.

In addition, cogeneration in buildings can be designed for peak shaving applications, i.e., the cogeneration plant is used to reduce either the peak electrical demand or thermal demand.

In the case of single-family applications, the design of systems poses a significant technical challenge due to the potential noncoincidence of thermal and electrical loads, necessitating the need for electrical/thermal storage or connection in parallel to the electrical grid.

3 The mCCHP Systems

3.1 Architecture of the mCCHP Systems

Trigeneration applications in buildings have to satisfy either both the electrical and the thermal demands, or to satisfy the thermal demand and part of the electrical demand, or to satisfy the electrical demand and part of the thermal demand. Architecture of the mCCHP systems depends on the magnitude of the electrical and thermal loads, whether they match or not, and on the operating strategy.

The trigeneration system can run at part-load conditions, the surplus energy (electricity or heat) may have to be stored or sold, and deficiencies may have to be made up by purchasing electricity from other sources such as the electrical grid [6]. The surplus heat produced may be stored in a thermal storage device, such as a water tank, or in phase change materials, while surplus electricity may be stored in electrical storage devices such as batteries or capacitors.

All these must ensure the necessary heat for the residence. It results that, to achieve a conceptual scheme of a mCCHP system, the energy demand of the residential consumer must be known in advance. Depending on this demand (load curves), the optimal architecture of the mCCHP system may be determined. In establishing the architecture of the mCCHP system, the satisfaction of the residence energy demand and of the system's being connected or not to the electricity grid, must be had in view.

The general architecture of a micro-CCHP system based on renewable energy (Fig. 13) might include the following sources:

- a mCHP cogeneration unit
- auxiliary heating unit/units of the solar collector type and back-up heater
- photovoltaic panels
- thermal and/or electric energy storage unit
- an air-conditioning system thermally/mechanically activated.

In the architecture structure diagram, two subsystems may be identified:

- *Electrical energy supplying subsystem*. The mathematical model of the electrical subsystem is the one that describes the dynamic storage in the battery. It can be written as follows:

$$\frac{dW_B}{dt} = P_{CHP}(t) + P_{PV}(t) - P_{al}(t) - P_{air}(t) \tag{13}$$

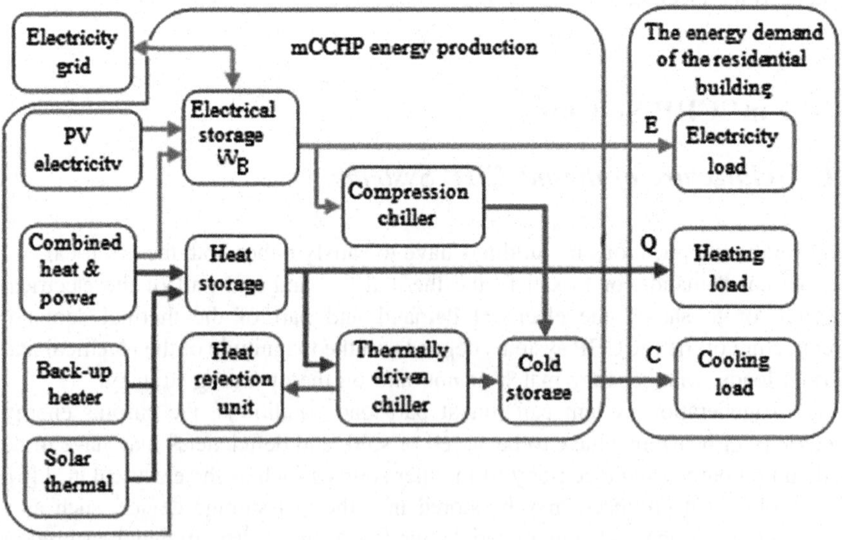

Fig. 13 General architecture of the mCCHP systems

where W_B is the energy accumulated in the battery [J]; P_{CHP} and P_{PV}, are the powers produced by the CHP and photovoltaic source, respectively [W]; P_{al} is the powers corresponding to the useful load and internal intake of the entire system, respectively [W] (the pump engines etc.), and P_{air} is the power consumed by the air-conditioning [W].

- *Thermal energy supplying subsystem.* The mathematical model of the thermal subsystem has as core the thermal balances at the level of the thermal accumulator and of the building. The thermal balance equation for the thermal accumulator is the following:

$$m_w c_w \frac{d\vartheta}{dt} = P_{CHP,t}(t) + P_{PT}(t) + P_{ph}(t) - P_{hl}(t) - P_{acl}(t) - P_{dhw}(t) \tag{14}$$

where: m_w, c_w represent the mass and the specific heat of the water accumulated in the heat storage, ϑ—the water temperature in the heat storage,

$P_{CHP,tt}$—the CHP thermal power, P_{PT}—the thermal solar collectors power, P_{ph}—the power provided by the back-up boiler, P_{hl}—the power consumed in the building to cover losses through transmission and ventilation, P_{acl}—the power consumed by the air conditioning, P_{dhw}—the power consumed by the domestic hot water circuit.

In steady state (equilibrium electric and thermal) by integration of the system of Eqs. (13) and (14) conduct to energy balance equation.

3.1.1 The mCCHP System with a Mechanical Compression Chiller

The electricity demand of the residence is composed of the electricity intake of the residence appliances, to which the electricity necessary to produce cold is added. This demand must be satisfied by the mCHP system, according to Fig. 14 and the relation (15):

$$E_{CCHP} = E + \frac{C}{COP_c} - E_{grid} \tag{15}$$

The thermal energy demand of the residence is composed of the energy necessary for heating the residence and preparing domestic hot water, and must be supplied by the CCHP system according to the relation:

$$Q_{CCHP} = Q + Q_{hw} \tag{16}$$

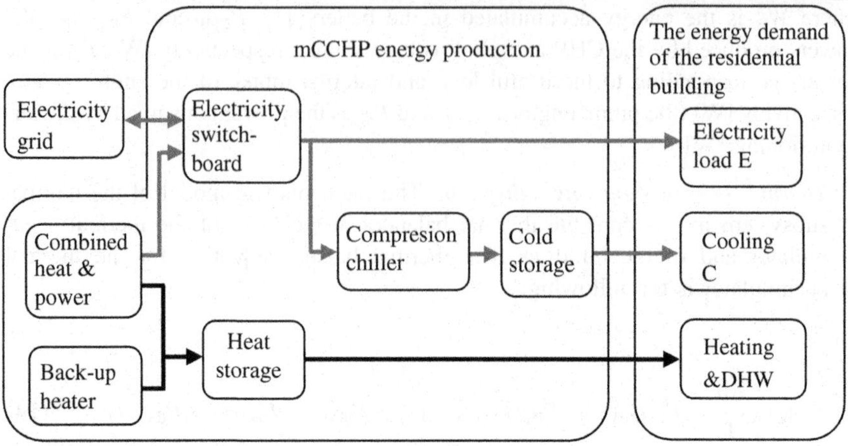

Fig. 14 The mCCHP system with a mechanical compression chiller

3.1.2 The mCCHP System with a Thermal Compression Chiller

The thermal energy demand of the residence (Fig. 15) is composed of the energy intake necessary for heating (heating the residence and preparing domestic hot water), to which the thermal energy necessary to produce cold is added, and must be satisfied by the CCHP system according to the relation:

$$Q_{\text{CCHP}} = Q + Q_{\text{hw}} + \frac{C}{COP_a} \tag{17}$$

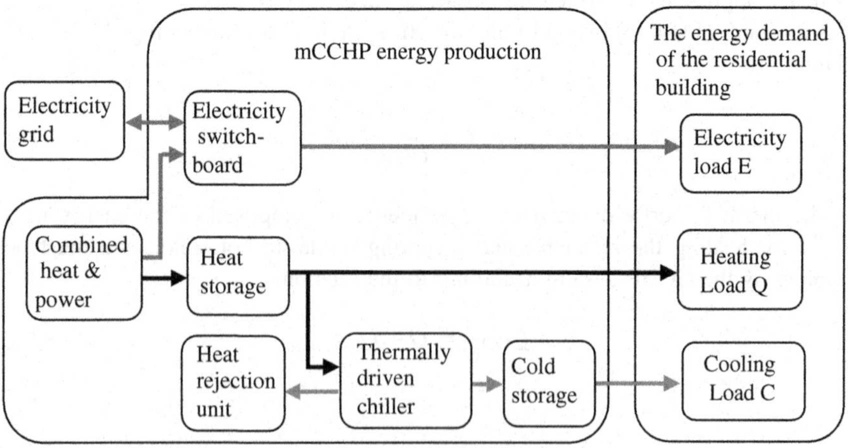

Fig. 15 The mCCHP system with a thermal compression chiller

The electricity demand of the residence is composed of the electricity intake of the residence appliances. The electricity production of the CCHP system is the difference between the electricity demand of the residence and the energy received from the grid:

$$E_{\text{CCHP}} = E - E_{\text{grid}} \tag{18}$$

3.2 Operation Modes of the mCHP Unit

In satisfying these demands, the mCHP unit may have various modes of operation. The mode of operation is characterized by the criterion on which the adjustment of the electrical and useful thermal output of a trigeneration system is based. The following operation modes can be applied:

(a) *Heat-match mode*, where the total thermal energy for the residence is ensured by the cogeneration unit (Fig. 16). As a consequence, the useful thermal output of a cogeneration system at any instant of time is equal to the thermal load (without exceeding the capacity of the cogeneration system), that is:

$$Q_{\text{CHP}} = Q_{\text{CCHP}} \tag{19}$$

If the generated electricity is higher than the load, excess electricity is sold to the grid; if it is lower, supplementary electricity is purchased from the grid.

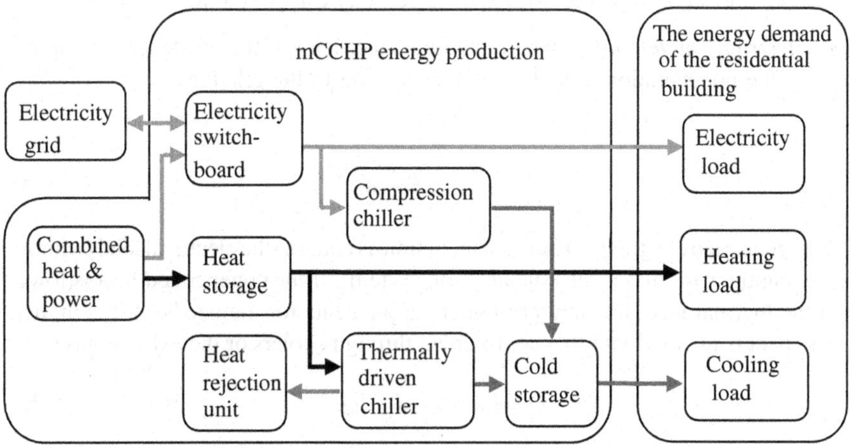

Fig. 16 Heat-match mode

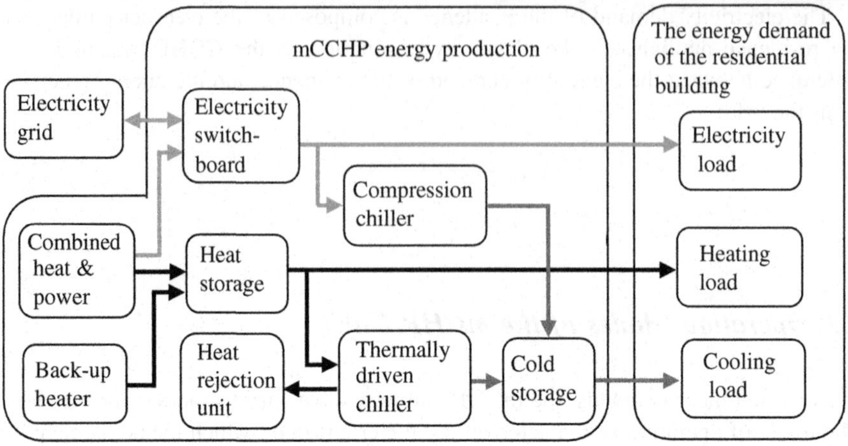

Fig. 17 Base thermal load matching mode

(b) *Base thermal load matching mode*, (Fig. 17) where the total thermal energy of the residence is provided by the cogeneration unit and by a supplementary boiler which ensures the peak thermal charges, according to the relation:

$$Q_{CHP} = Q_{CCHP} - Q_{ad} \tag{20}$$

Here, the cogeneration system is sized to supply the minimum thermal energy requirement of the site. Stand-by boilers or burners (Q_{ad}) are operated during periods when the demand for heat is higher. The prime mover installed operates at full load at all times. If the electricity demand of the site exceeds that which can be provided by the prime mover, then the remaining amount can be purchased from the grid. Likewise, if local laws permit, the excess electricity can be sold to the power utility.

(c) *Electricity-match mode*, where the total electricity of the residence is supplied by the cogeneration unit (Fig. 18), according to the relation:

$$E_{CHP} = E_{CCHP} \tag{21}$$

The generated electricity at any instant of time is equal to the electrical load (without exceeding the capacity of the cogeneration system). If the cogenerated heat is lower than the thermal load, an auxiliary boiler supplements for the needs; if it is higher, excess heat is released into the environment through coolers or the exhaust gases.

$$Q_{CHP} = Q_{CCHP} - Q_{ad} \tag{22}$$

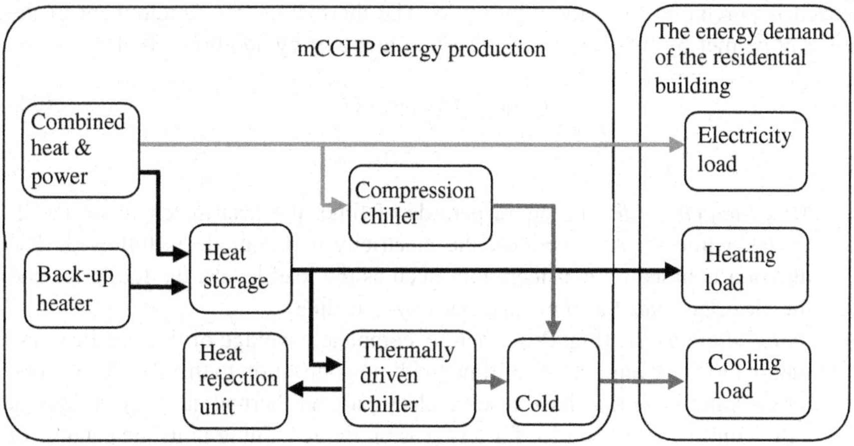

Fig. 18 Electricity-match mode

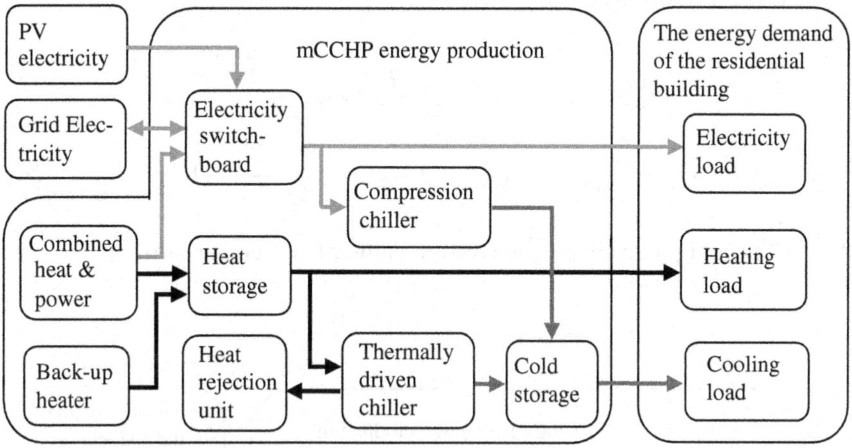

Fig. 19 Base electrical load matching mode

(d) *Base electrical load matching mode*, where the total electricity of the residence is provided by the cogeneration unit and a supplementary system of producing electric energy, and/or from the grid, which ensures the peak loads (Fig. 19), according to the relation:

$$E_{\mathrm{CHP}} = E_{\mathrm{CCHP}} - E_{\mathrm{ad}} - E_{\mathrm{grid}} \tag{23}$$

In this configuration, the CHP plant is sized to meet the minimum electricity demand of the site, based on the historical demand curve. The rest of the power needed is purchased from the utility grid. The thermal energy requirement of the site may be met by the cogeneration system alone or by additional boiler.

$$Q_{\mathrm{CHP}} = Q_{\mathrm{CCHP}} - Q_{\mathrm{ad}} \tag{24}$$

(e) *Mixed-match mode*, In certain periods of time, the heat-match mode is followed, while in other periods, the electricity-match mode is followed. The decision is based on considerations such as the load levels, the fuel price and the electricity tariff at the particular day and time.
(f) *Stand-alone mode*, (Fig. 20). There is complete coverage of the electrical and thermal loads at any instant of time, with no connection to the grid. This mode requires the system to have reserve electrical and thermal capacity, so that in case a unit is out of service for any reason, the remaining units are capable of covering the electrical and thermal load. This is the most expensive strategy, at least from the point of view of the initial cost of the system.

The energetic balance equations are:

$$E_{\mathrm{CHP}} = E_{\mathrm{CCHP}} - E_{\mathrm{ad}} = E + \frac{C'}{COP_c} - E_{\mathrm{ad}} \tag{25}$$

$$Q_{\mathrm{CHP}} = Q_{\mathrm{CCHP}} - Q_{\mathrm{ad}} = Q + Q_{\mathrm{hw}} + \frac{C - C'}{COP_a} - Q_{\mathrm{ad}} \tag{26}$$

where C' is cold produced by compression chiller, C is cooling load.

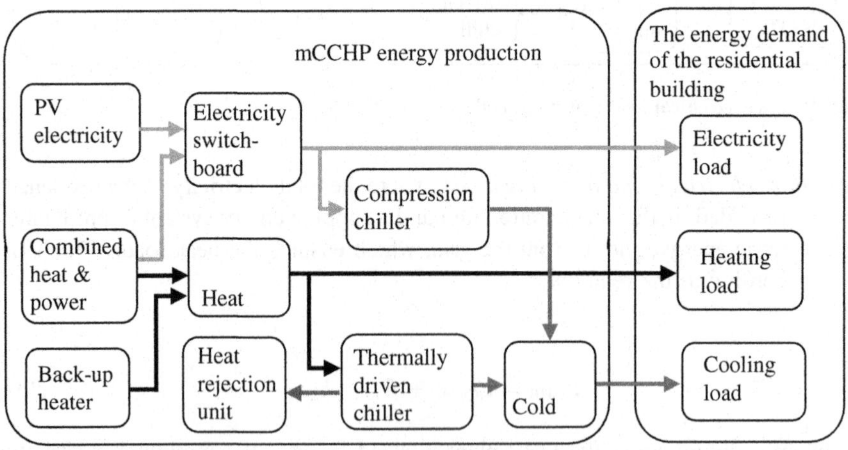

Fig. 20 Stand-alone mode

In general, the heat-match mode results in the highest fuel utilization rate (fuel energy savings ratio—PES) and perhaps in the best economic performance for trigeneration in the industrial and building sectors.

In the utility sector, the mode of operation depends on the total network load, the availability of power plants and the commitments of the utility with its customers, regarding supply of electricity and heat. However, applying general rules is not the most prudent approach in trigeneration.

Every application has its own distinct characteristics. There is a variety of trigeneration systems (according to the type of the technology, size, and configuration). The design of a trigeneration system can be tailored to the needs of the user. The design of a trigeneration system affects the possible modes of operation, and vice versa. Moreover, the technical and economic parameters may change with the day and time during the operation of the system. For example (Fig. 21), the mCCHP system type off-grid, for satisfy energy need the control can be divided between mCHP unit and boiler. The mCHP is back-up for electricity balance and the boiler is back-up for thermal energy balance. The equilibrium between the power provided by the sources and the power consumed (power consumed for building and the powers corresponding to the useful load and internal intake,

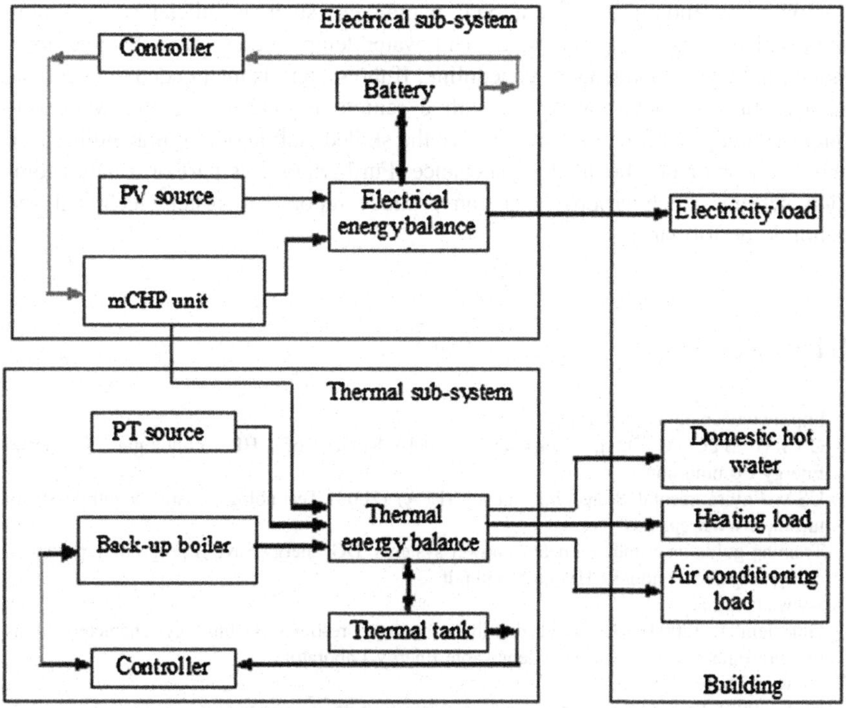

Fig. 21 Control of the thermal and electrical sub-systems

inclusive for air conditioning) is achieved through the voltage battery control and storage tank temperature, respectively.

All these aspects make it necessary to reach decisions not by rules of thump only, but by systematic optimization procedures, based on mathematical programming, for both the design and the operation of the system. For the operation of cogeneration systems, in particular, microprocessor-based control systems are available.

They may provide the capability to operate in a base load mode, to track either electrical or thermal loads, or to operate in an economic dispatch mode (mixed-match mode).

In the latter mode, the microprocessor can be used to monitor trigeneration system performance, including:

- the system efficiency and the amount of useful heat available;
- the electrical and thermal requirements of the user, the amount of excess electricity which has to be exported to the grid, and the amount of heat that must be released into the environment;
- the cost of purchased electricity and the value of electricity sales, as they may vary with the time of the day, the day of the week, or season.

Using the aforementioned data, the microprocessor can determine which operating mode is the most economical, even whether the unit should be shut down. Moreover, by monitoring operational parameters such as efficiency, operating hours, exhaust gas temperature, coolant water temperatures, etc., the microprocessor can help in maintenance scheduling. If the system is unattended, a telephone line can link the microprocessor with a remote monitoring centre, where the computer analysis of the data may notify the skilled staff about an impending need for scheduled or unscheduled maintenance. Furthermore, as part of a data acquisition system, the microprocessor can produce reports of system technical and economic performance.

References

1. http://www.epa.gov/chp/basic/
2. ONSITE SYCOM Energy Corporation (1999) Review of CHP technologies. California Energy Commission
3. USA: Environmental Protection Agency (EPA) (2008) Technology characterization: steam turbines. www.epa.gov
4. Training guide on combined heat & power systems. DG Energy Save II program http://www.cres.gr/kape/education/3.CHP_en_small.pdf
5. www.ata.org.au
6. Goldstein L (2003) Gas fired distributed energy resource technology characterizations microturbine systems. National Renewable Energy Laboratory
7. www.capstone.com
8. EPA (2008) Technology characterization: Microturbine. www.epa.gov
9. http://www.mtt-eu.com/en/home

10. EPA (2008) Technology characterization: reciprocating engines. www.epa.gov
11. BIOS Bioenergy Systems, Austria (2003). http://www.bios-bioenergy.at/en/electricity-from-biomass/stirling-engine.html
12. http://www.ballard.com/files/images/aboutballard/How-a-fuel-cell-works.jpg
13. U.S. Environmental Protection Agency (2002) Combined heat and power partnership
14. http://www.prototech.no/index.cfm?id=168592
15. EPA (2008) Technology characterization: fuel cells. www.epa.gov
16. Dorer V Annex 42 of the international energy agency energy conservation in buildings and community systems programme. http://www.ecbcs.org. ISBN No. 978-0-662-46950-6
17. EDUCOGEN (2001) A guide to cogeneration The European Association for the Promotion of Cogeneration SAVE II Programme of the EC-2001
18. Wu DW, Wang RZ (2006) Combined cooling, heating and power. Prog Energy Combust Sci 32:459–495
19. Guy R (2011) Micro-CHP accelerator. Final report—CTC726-carbon trust
20. Roads2HyCom Hydrogen and Fuel Cell Wiki (2014) Case study: micro-CHP (1–5 kW). http://www.ika.rwth-aachen.de/r2h
21. Directive 2012/27/EU of the European Parliament and of the Council of 25 October 2012 on energy efficiency, amending Directives 2009/125/EC and 2010/30/EU and repealing Directives 2004/8/EC and 2006/32/EC (OJ L 315, 14.11.2012, p 1)

Renewable Energy Sources for the mCCHP-SE-RES Systems

Nicolae Badea, Ion V. Ion, Nelu Cazacu, Lizica Paraschiv, Spiru Paraschiv and Sergiu Caraman

Abstract The mCCHP-SE-RES system is defined as a particular combined cold, heat, and power system, which is distinguished in that it is a microgeneration system (mCCHP) dedicated to residential building, the CHP unit is a Stirling engine (SE), and the primary energy is obtained from renewable energy sources (RES). In this chapter, the last feature is presented in detail, aiming to recall the basic data and information needed to design such a system. First, it shows the physical fundamentals of the solar energy conversion into electricity or thermal energy, and then the construction and operation of the photovoltaic and thermal solar panels, as well as of the electrical and thermal energy storages assigned to these panels. Also it shows the technical processes of obtaining and burning the biomass, as well as the construction and operation of the Stirling engine and the boiler that can be fueled by biomass.

N. Badea (✉) · I.V. Ion · N. Cazacu · L. Paraschiv · S. Paraschiv · S. Caraman
"Dunarea de Jos" University of Galati, Galati, Romania
e-mail: nicolae.badea@ugal.ro

I.V. Ion
e-mail: ion.ion@ugal.ro

N. Cazacu
e-mail: nelu.cazacu@ugal.ro

L. Paraschiv
e-mail: lizica.paraschiv@ugal.ro

S. Paraschiv
e-mail: spiru.paraschiv@ugal.ro

S. Caraman
e-mail: sergiu.caraman@ugal.ro

N. Badea (ed.), *Design for Micro-Combined Cooling, Heating and Power Systems*, Green Energy and Technology, DOI 10.1007/978-1-4471-6254-4_4

1 Primary Energy for Building's Energy Systems

Nicolae Badea

Private households are one of the world's largest energy consumers. The practice of the separate generation of electrical energy in electric supply stations and heat in the home, which is still the norm today, leads to primary energy losses, and contributes considerably to global warming.

To reduce this negative effect lately has been developed microgeneration systems and technologies, which use renewable energy sources (RES) and are conceived according to a new paradigm.

1.1 Microgeneration Systems

They are divided in three categories:

- Microheat generation systems, based on heat pumps (air, water, and ground source), biomass, and solar thermal panels;
- Microelectricity generation systems, based on solar PV panels, microwind turbines, and microhydro stations;
- Microcombined heat and power generation systems (micro-CHP systems or cogeneration systems) based on internal combustion engines, Stirling engines, and fuel cells.

Microgeneration systems use as intake either fossil fuels, RES or a combination of both in order to generate heat, electricity or a cumulus of both heat and electricity.

The most common type of micro-CHP system is the internal combustion engine, which has been adapted to run on mains gas and is connected to an electricity generator. Due to their large size and the fact they are quite noisy, they are not normally used to domestic properties.

Fuel cell based m-CHP systems are the new technology kind on the market. As it is such a new technology, it is not yet available to homeowners although it is hoped it will come onto the market within the next few years.

The Stirling engine (SE) micro-CHP is the most appropriate type model that can be installed into building because are smaller and quieter than the internal combustion models. Stirling engines produce power not by explosive internal combustion, but by transferring heat from an external source, which produces heat. This source may be an external combustion (when the SE can be fuelled with a wide variety of fuels, including all fossil fuels, e.g., natural gas), or even solar energy. In the context of moving toward new low carbon solutions such microgeneration systems are essential.

1.2 The New Paradigm

Energy produced by the microgeneration systems can be used not only for consumers needs but can also be delivered to the grid in exchange for certain revenues.

In present the energetic classes of the building by Energy Performance Certificates (EPCs) give information only on building properties, resulting in an energy need. Not possible to compare only envelope performance of building without assessing systems energy supply and their performances. In present the EPCs provide information on how to make a building, so that to decrease the loss energy. The energy efficiency of the property is calculated using ratings which range from "A" (the most energy efficient) to "G" (the least efficient). EPCs compare the current energy efficiency of the property, and carbon dioxide emissions with potential house if the home could achieve energy saving measures by thermal insulation.

The recast of the Energy Performance of Buildings Directive (EPBD) [1] lays down the requirements regarding the general framework of methodology for calculating the integrated energy performance of buildings. Two quotes from the EPBD on this are given below:

- Art 3: "Primary energy factors used for the determination of the primary energy use may be based on national or regional yearly average values and may take into account relevant European standards".
- Annex I: "The energy performance of a building shall be expressed in a transparent manner and shall include an energy performance indicator and a numeric indicator of primary energy use, based on primary energy factors per energy carrier, which may be based on national or regional annual weighted averages or a specific value for on-site production. The methodology for calculating the energy performance of buildings should take into account European standards and shall be consistent with relevant Union legislation, including Directive 2009/28/EC".

The relevant standard for primary energy factors [2] provides the calculation procedure to determine the annual overall energy used for heating, cooling, hot water, ventilation, and lighting (Fig. 1).

There are two definitions of the primary energy factor:

- *Total primary energy factor.* All the energy overheads of delivery to the point of use are taken into account in this version of the conversion factor, including the

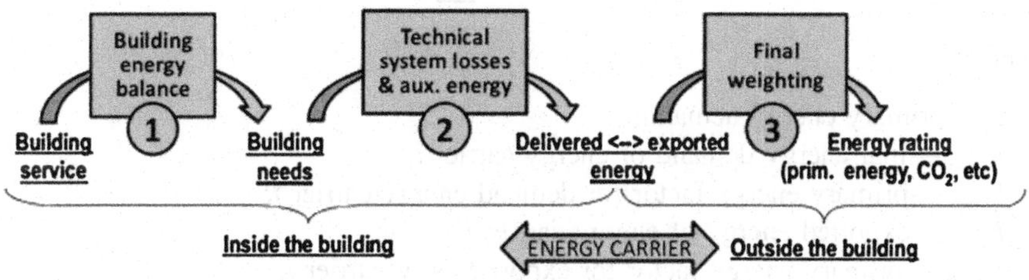

Fig. 1 Main steps in energy performance of building [3]

energy from renewable energy sources. Consequently, this primary energy conversion factor always exceeds unity.

- *Nonrenewable primary energy factor.* As above, but excluding the renewable energy component of primary energy. The renewable portion of delivered energy is considered as zero contribution to the primary energy use. Consequently, for a renewable energy carrier, this normally leads to a factor less than unity (ideally, zero).

Primary energy factor is a new concept and has three components:

- Total (fp,tot);
- Nonrenewable (fp,nren);
- Renewable (fp,ren).

Integrated energy performance of building ($E_{p,tot}$) is the sum of the nonrenewable ($E_{p,nren}$) and renewable energy performance ($E_{p,ren}$) namely:

$$E_{p,tot} = E_{p,nren} + E_{p,ren} \quad (1)$$

The integrated energy performance of building depends on:

- building properties, resulting in a value of the energy need;
- technical systems losses, resulting in a required delivered/exported energy;
- weighting of the delivered energy.

By renewable energy use for buildings, the new standard defines the ratio between renewable and integrated energy performance, namely share of renewable or Renewable Energy Ratio (RER) by equation:

$$RER = \frac{E_{p,ren}}{E_{p,tot}} \cdot 100\ \% \quad (2)$$

A new generation of EN standards [2] for implementing the Energy Performance of Building Directive introduced a clear distinction between renewable and nonrenewable primary energy and their connection with total primary energy factor, to support the renewable energy share evaluation. This is possible by introducing the weighting factor, namely the primary energy conversion factor to define the integrated energy performance of building, so:

$$E_p = \sum E_{del,i}\, f_{p,del,i} - \sum E_{\exp,i}\, f_{p,\exp,i} \quad (3)$$

where:

- E_p—primary energy demand;
- $E_{del,i}$—final energy demand of energy carrier *i;*
- $f_{p,del,i}$—primary energy factor for demand energy carrier *i;*
- $E_{\exp,i}$—exported energy of energy carrier *i;*
- $f_{p,\exp,i}$—primary energy factor for export energy carrier *i;*
- *i*—the current number of the carrier.

However, energy is not only consumed in (or near) buildings but can also be produced. The new proposal says that, per energy carrier, the exported energy can be subtracted from the imported energy. EN standards will have a significant impact on energy balance of building. These standards indicate the following:

- The assessment of buildings is not limited only to the building alone, but includes the energy supply systems. Also, the standard defines a new concept namely *integrated energy performances* by integrating into a building the energy sources to satisfy demands (heating, cooling, and power).
- The assessment of the technical systems and energy supply chain is fair and transparent.
- The positive impact of the CHP system and renewable energy integrated into building can be shown (common European certification scheme).
- Facilitates the microgeneration and can change the behaviors of the building's owners.

1.3 The Renewable Sources Used in Building's Energy Systems

The fossil fuels are exhaustible and, in essence, represent stored energy sources, formed along millions of years. Unlike these, RES are "energies obtained from the fluxes which exist in the environment and have a continuous and repetitive nature" [4]. In short, the renewable energy refers to forms of energy resulting from natural renewable processes.

The Earth is a vast flow-through system for the input and output of energy. The overwhelming majority of the input to the Earth's energy comes from the Sun in the form of solar irradiation, from the Earth—under the form of geothermal energy, or from the movement of the planets—under the form of tides. The flows of energy passing continuously as renewable energy through the Earth are shown in Fig. 2.

The Sun radiates electromagnetic energy. Although it covers the entire electromagnetic spectrum, the Earth receives energy from the Sun as short-wavelength irradiation. This is because the solar energy that enters the Earth system is shorter in wavelength (and thus higher in energy level) than the energy returned to space by the Earth.

Geothermal energy or the planet's internal heat energy is a source of energy that comes from heat deep below the surface of the Earth. This heat produces hot water and steam from rocks that are heated by magma. To use this type of energy, get to the heat by drilling wells into these rocks. The hot water or steam comes up through these wells. This source of energy is clean and safe and is an excellent resource in some parts of the world. For example, Iceland is a country that gets about one-fourth of its electricity from geothermal sources. Many parts of the world do not have underground sources of heat that are close enough to the surface for building geothermal power plants.

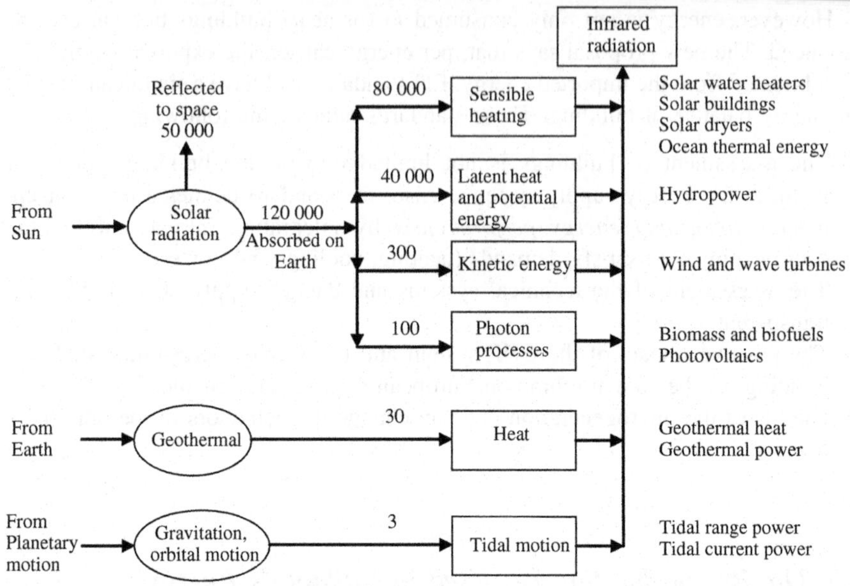

Fig. 2 Natural energy currents on the Earth, showing a renewable energy system [5]. *Note* Units are terawatts (10^{12})

Whereas the principal form of the energy comes from the Sun and the secondary source from the Earth itself, the third type of energy input to the Earth system comes chiefly from the Moon. The Moon has more influence over the movements of our planet's ocean waters. The gravitational pull of the Moon (and, to a lesser extent, that of the Sun) on the Earth causes the oceans to bulge outward on the side of the Earth closest to the Moon. At the same time, the oceans on the opposite side of the planet bulge in response. This gravitational pull creates a torque that acts as a brake on the Earth's rotation, producing a relatively small amount of energy that is dissipated primarily within the waters of the ocean.

Thus, renewable energy is obtained by capturing the energy of solar irradiation, of winds, of running waters, of biological processes, or of geothermal heat.

One of the most interesting components of the entire energy picture is the relationship of energy input, energy output, and the human beings. Human beings are entitled to a healthy and productive life in harmony with nature and are at the center of concerns for sustainable development. In essence, sustainable development [6] is a process of change in which the exploitation of resources, the direction of investments, the orientation of technological development and institutional change are all in harmony and enhance both current and future potential to meet human needs and aspirations.

The right to development must be fulfilled so as to equitably meet developmental and environmental needs of present and future generations. Human beings shall cooperate in a spirit of global partnership to conserve, protect, and restore the health and integrity of the Earth's ecosystem.

2 Solar Energy

Nicolae Badea

Solar energy is available all over the world in different intensities. According to the widely accepted terminology [7], the two concepts of solar (i.e., short wave) irradiation are used. The term irradiance is used to consider the solar power (power density) falling on unit area per unit time [W/m^2] or irradiance is a measure of the rate of energy received per unit area, and has units of watts per square meter (W/m^2). The term irradiation (or radiation) is used to consider the amount of solar energy falling on unit area over a stated time interval [Wh/m^2]. This interval can be one minute, day, month, or year. Irradiation is radiant exposure and is a time integral (or sum) of irradiance. Thus, one minute radiant exposure is a measure of the energy received per square meter over a period of one minute. Therefore, a 1 min radiant exposure = mean irradiance (W/m^2) × 60(s), and has units of joule(s) per square meter (J/m^2). For example: a mean irradiance of 500 W/m^2 over 1 min yields a radiant exposure of 3,000 J/m^2 or 3 kJ/m^2. Both irradiance and irradiation have the same symbols, namely H. The two concepts can be differentiated by context or by the attached units. $H_{h,i}$ is the monthly or yearly average of daily global irradiation on the horizontal or inclined surface. (Note: Depending on the country and manufacturer of solar equipment is also used notation G [W/m^2] for irradiance and for irradiation is used notation E [Wh/m^2].)

Irradiation is essential when looking at how much power can be derived from a certain area of real estate of an energy source. Low irradiation indicates that too much real estate is required to provide the power we demand at reasonable prices. Irradiation for most people is from 150 to 300 W/m^2 or 3.5 to 7.0 kWh/m^2/day. The global mean irradiance is 170 W/m^2.

Theoretically, the solar energy available on the surface of the Earth is enough to support the energy requirements of the entire planet. In reality, progress and development of solar science and technology depends to a large extent on human desires and needs.

The potential for using solar energy in EU is relatively important (Fig. 3). There are areas where the annual solar energetic flux is up to 1,800–2,200 kWh/m^2/year, like the Black Sea coast and Mediterranean countries.

The solar energy potential depends primarily on the latitude and local climatic conditions like cloudiness and, secondly, on air temperature. Simulations with the Photovoltaic Geographical Information System [8] show that the average sum of global irradiation per square meter received by optimally orientated modules lies between 1,000 kWh/m^2/year for Oslo region and 2,200 kWh/m^2/year in Andalucía region of Spain. In the majority of the countries, the annual solar energetic flux is of 1,000…1,400 kWh/m^2/year.

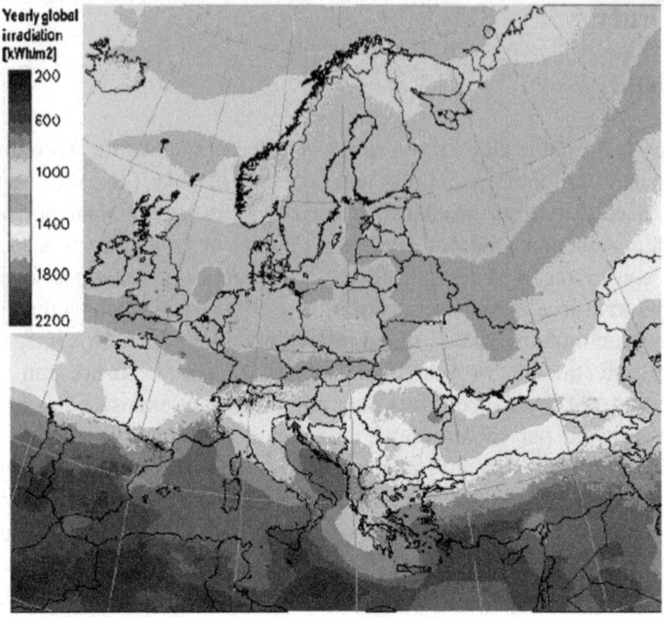

Fig. 3 EU Solar energy potential [8]

2.1 The Principle of Solar Energy Conversion

Solar irradiation is made up of photons with different energies. When absorbed by a material, they transfer their energy. The individual energy of a photon is directly linked to a certain characteristic, namely the wave length. To a smaller wave length corresponds a greater energy. Any photon transfers its energy, completely or not. The energy transfer from the photon to the material is called absorption (Fig. 4). If the photon is not completely absorbed, two types of interaction may be distinguished:

- the photon passes though the material (transmission);
- the photon is reflected by the material's surface (reflection).

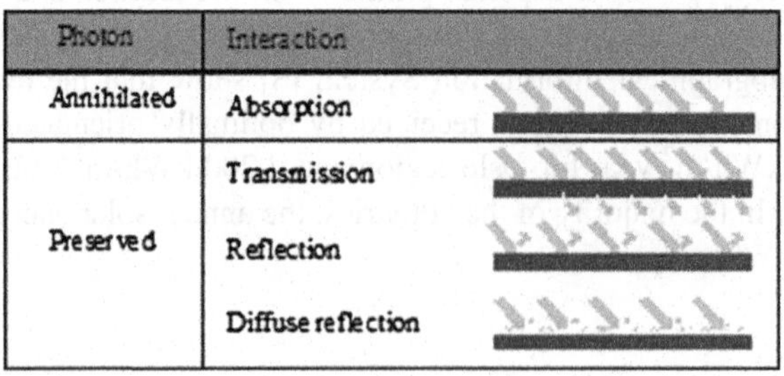

Photon
Interaction
Annihilated
Absorption
Transmission
Preserved
Reflection
Diffuse reflection

Fig. 4 The interaction of light with the illuminated body

If the photon is transmitted or is reflected in an arbitrary direction which is not correlated with the incidental direction, the process is called dispersion (as in the case of reflection, where a ray of light is reflected in a single direction by a mirror and is dispersed by a rough surface).

Taking into account the energy transfer, the following case may be underlined:

- *The energy transferred to the material leads to the increase in the kinetic energy of atoms*. This is underlined by an increase in the temperature of the respective body. During the absorption process, the radiating energy is converted into heat. In this case, although the energy of an atom in the visible and ultraviolet spectral domain is greater than the energy of a photon in infrared, the infrared energy is more important since a large part of the total solar energy (35–40 %) belongs to this domain. The thermal conversion is the oldest and the most widely spread form of using solar energy. Any black surface, called absorbing surface, exposed to the rays, transforms solar energy into heat. This surface represents the simplest example of a direct converter of solar energy into thermal energy, called flat thermal solar collector. Thermal solar collectors are usually used to prepare domestic hot water in individual residences. In the households, there is a demand for space heating and water heating as well as for cooling, (air conditioning) with low-temperature heat (<100 °C).
- *The transfer of energy causes a change in the internal state of atoms*. This process lies at the basis of solar irradiation conversion into electric energy. The occurrence of an electric tension, under the influence of solar energy is called *photovoltaic effect*. The photovoltaic effect is due to the discharge of negative electric loads (electrons) and of positive electric loads (gaps) in a solid material, called *photovoltaic cell*, when its surface is lighted. This electric polarization of the cell's material produces an electric tension, which may generate current in a closed circuit. To allow supplying a reasonable electric power, the photovoltaic cells do not operate individually, but linked in series and in parallel, in larger numbers, making up *photovoltaic panels* (*PV panels*) *or solar electric panels* not to be mistaken for solar panels, also called solar collectors or solar thermal panels (ST panels). The direct conversion of solar irradiation into electrical energy occurs by means of the photovoltaic effect, in which photons induce the emergence of an electrical potential as a result of a separation of charge carriers in semiconducting materials. The single solar PV cells are made of amorphous, poly or monocrystalline silicon, cadmium telluride, or copper indium selenide/sulfide. They are interconnected to solar panels producing a few hundred Watts. PV panels are very well suited for use in buildings because they are maintenance-free and emit neither noise nor pollution. At present, the use of crystalline silicon-based materials dominates the photovoltaic technology. Thus, the market share of global crystalline silicon-based module production capacity is estimated to amount to 82 % in 2009, but the share of thin film based technologies is expected to increase in the future [9].

2.2 Performances of Solar Energy Conversion

To design a device for converting solar energy into thermal or electric energy, it is necessary to know the incident solar irradiation on the collecting/active surface of the PV or ST panel. On the other hand, solar irradiation depends on several factors, such as the latitude and the altitude of the area, the season, the day, the hour, the degree of nebulosity and the dust content, the water vapors, and the greenhouse gases in the atmosphere.

That is why, to evaluate the performance level, the notion of multiannual average value of the horizontal daily solar irradiation is used; this is a unit of the primary energy existing in this source and is defined as the solar irradiation energy, which reaches a horizontal surface of 1 m^2 daily. The value of H is determined from statistical data, depending on the location of the PV or ST panel.

To exemplify, consider the case in which the location is in the southeastern part of Romania, where $H = 3.95$ kWh/m^2/d.

If, in an ST panel, this energy is converted into heat, the efficiency of the conversion might be of approximately 0.7. The thermal energy delivered daily by an ST panel whose active surface is of 1 m^2, noted q_{PT}, has the following value:

$$q_{PT} = H \cdot \eta = 3.95 \cdot 0.7 = 2.772 \qquad \left[\mathrm{kWh/m^2/d}\right] \tag{4}$$

If we consider that this specific energy is delivered to a boiler, where it heats the water, it results then that the mass m of the heated water by 1 m^2 of the ST panel, with a difference in temperature of $\Delta\theta$, is:

$$m = \frac{q_{PT}}{c_P \cdot \Delta\theta} \quad \mathrm{Kg/d}, \tag{5}$$

where $c_p = 4.187$ kJ/(kg °C) is the heat specific to water.

Considering the concrete case in which the water is heated from 10 to 90 °C, it results that an ST panel, with a surface of 1 m^2 may heat of approximately 25 kg water daily.

Similarly, the specific electric energy delivered by a PV panel depends on the multiannual average value of the horizontal daily solar irradiation and on the conversion efficiency of the PV panel.

To exemplify, consider once again the case in which the location of the panel is in the southeastern part of Romania, where $H = 3.95$ kWh/m^2/d. As the efficiency is of about 0.15, it results that the electric energy delivered daily by a panel with a surface of 1 m^2, noted e_{PV}, has the following value:

$$e_{PV} = H \cdot \eta = 3.95 \cdot 0.15 = 0.5925 \quad \left[\mathrm{kWh/m^2/d}\right] \tag{6}$$

The decision to use equipment for converting solar energy into thermal or electric energy is determined by the level of solar irradiation energy flux and by the

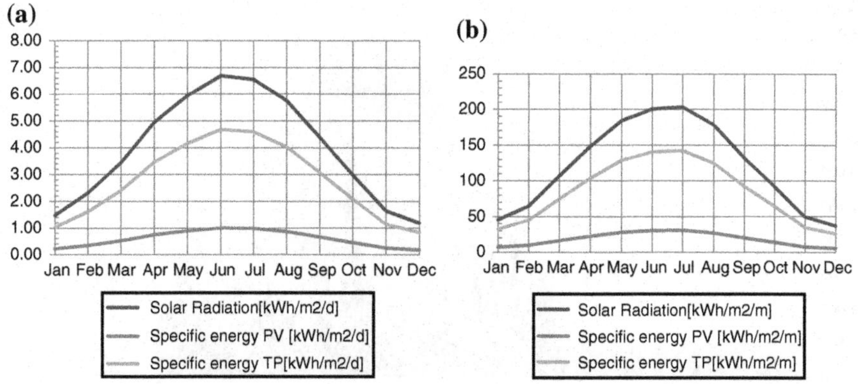

Fig. 5 Specific energy production of PV and ST panels, **a** Specific energy production, in kWh/m^2/day, **b** Specific energy production in, kWh/m^2/month

efficiency of the conversion technologies. The trend in PV panel is building-integrated photovoltaic (BIPV) systems incorporate photovoltaic properties into building materials such as roofing, siding, and glass and thus offer advantages in cost and appearance as they are substituted for conventional materials in new construction. Moreover, the BIPV installations are architecturally more appealing than classical PV structures. The new enclosure technology has the ability to regulate the panel temperature (cooling or heating).

Depending on the location of the building, the specific daily and monthly production of solar thermal energy and of photovoltaic energy can be determined. Solar energy in kWh/m^2/day on a south-facing horizontal surface for Galati city, Romania, for each month is shown in Fig. 5a.

If the yearly irradiation is of 1,440 kWh/m^2/year, then the specific annual energy produced by a PV panel (with 15 % efficiency) calculated as sum of the monthly specific power production in all months of the year, is of 216 kWh/m^2/year, while the heat produced by an ST panel (with 70 % efficiency), calculated as sum of the monthly specific heat production in all months of the year, is of 1,000 kWh/m^2/year. The monthly distribution of these specific productions is shown in Fig. 5b.

2.3 *Solar Energy Storage*

Collecting and storing the energy of the Sun for rainy days seems obvious. The photovoltaic technology and solar thermal energy systems offer reliable and cost-efficient methods of generating electricity and heat. Today, these systems are applied with success in commercial as well as residential buildings.

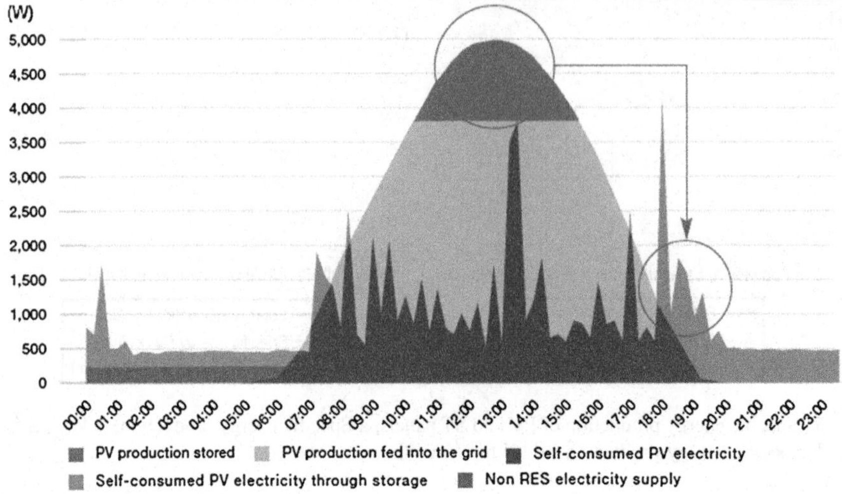

Fig. 6 Peak shaving strategy using storage at household level [10]

2.3.1 Electrical Energy Storage

PV systems are not producing during the evening peak consumption (Fig. 6). This is especially the case at the household level, where the daily peak consumption is at around 20:00.

To achieve a cost-effective PV deployment in household systems, the optimized strategies to solve the problem of peak production and/or consumption will have to be implemented using smart metering measures and/or storage at the household and/or local level.

Electrical energy storage (EES) can play many roles in the system, including:

- Storage can act as a Distributed Energy Resource (DER) by providing electricity and competing with other technologies based on cost and depending on application.
- Energy shifting of Variable Renewable Energies (VRE), increases the capacity factor of the system by displacing the excess production when there is high demand and low production, especially during peak consumption (e.g., evening hour-peak shaving).
- Storage can smooth load curves and enable the customer to shift its peak loads. This is also known as arbitrage or load leveling and generally implies that the storage unit is being charged at off-peak times and discharged at peak times when the load is higher.

In this case the energy management system can be the "ideal supplier" who attenuates smoother load curves with no peaks and ramps, making changing demand constant and transforming it in "ideal consumer".

Storage solutions with relatively high discharge duration and high capacity are best suited to these applications, are shown in Fig. 7.

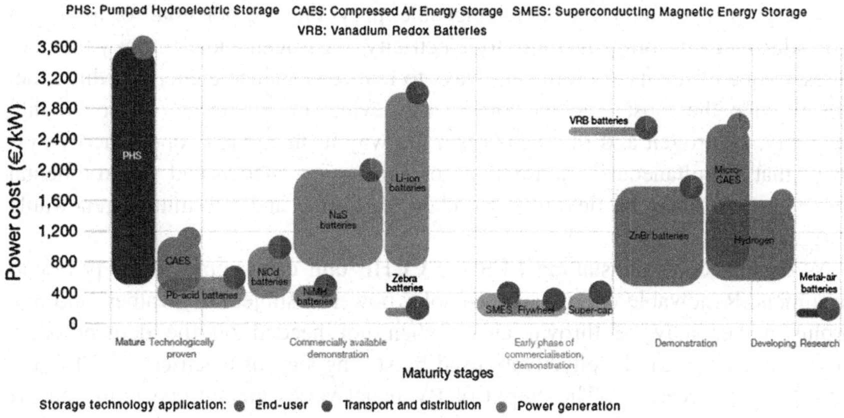

Fig. 7 Available energy storage technologies (€/kW) [11]

There is no single energy storage option to cover all requirements. Applications, technological development, and market evolution will define the share of decentralized storage solutions of the electrical energy.

The PV and daily storage is well suited for PV peak generation. This is because it is guaranteed that PV generation is available every day with a perfectly predictable peak that always occurs around midday. As the peak demand in house occurs in the evening hours the time shift between peak PV generation and the peak demand never exceeds 8 h.

Batteries are integrated into the classical power conversion systems with the purpose of providing additional energy as required by the load. The battery (secondary generation) is rechargeable by basic chemical reaction reversibility. Autonomous power systems supplied by RES need energy storage for their operation. Traditional energy storage system is lead-acid batteries. In addition to the storage device itself, a variety of energy management components are required, for example the charge controller. Storing energy produced by RES is dependent on the state of battery charge. If the battery is charged, RES energy production is not used. In addition, storage energy in batteries can be made only in the short term. There is no technology to long-term energy storage in batteries.

Weekly and seasonal storage capacities can be achieved by hydrogen technology. Hydrogen is not a primary energy source like coal and gas. It is a clean energy carrier with the unequaled advantage of being storable in various forms and transportable in various modes. For instance, compressed at 700 bar, H_2 is well over 100 times more energy dense than a Li-ion battery (33 kWh/kg for H_2 vs. 0.2 kWh/kg for battery). Hydrogen can be produced from carbon-free or carbon-neutral RES by electrolysis. The hydrogen production can be made using electrolysers powered by PV panels. The hydrogen produced by the electrolyser during summer provides the necessary stock for the fuel cell to function the longest period of time possible, in winter. Using fuel cell—hydrogen technology, the hydrogen and electricity are two energy forms effectively interchangeable, providing the potential for hydrogen to be a large-scale storage medium for electricity. Hydrogen and electricity together represent one of the most promising ways to realize sustainable energy, whilst fuel cells provide the most efficient conversion device for converting hydrogen into electricity. Hydrogen and fuel cells open the way to integrated "open energy systems" that simultaneously address all of the major energy and environmental challenges and have the flexibility to adapt to diverse and intermittent renewable energy sources.

As conclusion, by installing EES the CCHP unit can supply stable power to consumers. Renewable energy such as solar power is subject to weather, and any surplus power may be thrown away when not needed on the demand side. Therefore, energy can be effectively used by storing surplus electricity in EES and using it when necessary. The output of PV panel varies greatly depending on the weather, which can make connecting them to the grid difficult. EES used for time shift can absorb this fluctuation more cost-effectively than other single-purpose mitigation measures (e.g., a phase shifter). It is clear that energy storage can enable a more efficient and flexible operation of the energy supply system of building. For off-grid photovoltaic systems in the power range 1–500 kW lead-acid batteries for EES are commonly used.

2.3.2 Thermal Energy Storage

The analysis of the daily DHW sequence of the thermal profile in a building indicates the demand for necessary heat. By integration over 1 day and comparison with solar heat generation of the ST panel, we can obtain the minimum storage heat capacity from solar energy (Fig. 8).

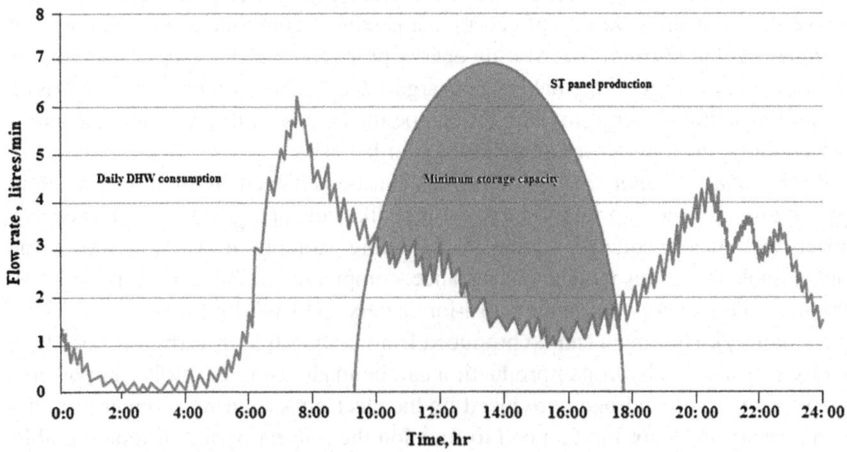

Fig. 8 DHW production versus consumption

Because the time interval of the ST panel production does not coincide with the use of the DHW, it follows as a direct consequence, need of thermal energy storage.

Thermal energy storage systems store available heat by different means in an insulated repository for later use in residential applications, such as space heating or cooling, and hot water production. Thermal storage systems are deployed to overcome the mismatch between demand and supply of thermal energy and thus they are important for the integration of renewable energy sources. Thermal storage can be subdivided into different technologies: storage of sensible heat, storage of latent heat, and thermochemical ad and absorption storage [12]. The storage of sensible heat is one of the best known and most widespread technologies, with the domestic hot water tank as an example. The storage medium may be a liquid such as water or thermo-oil, or a solid such as concrete or the ground. Thermal energy is stored solely through a change of temperature of the storage medium. The capacity of a storage system is defined by the specific heat capacity and the mass of the medium used. Latent heat storage is accomplished by using phase change materials (PCMs) as storage media. There are organic (paraffin) and inorganic PCMs (salt hydrates) available for such storage systems. Latent heat is the energy exchanged during a phase change such as the melting of ice. It is also called "hidden" heat, because there is no change of temperature during energy transfer. The advantage of latent heat storage is its capacity to store large amounts of energy in a small volume and with a minimal temperature change, which allows efficient heat transfer.

Sorption (adsorption, absorption) storage systems work as thermochemical heat pumps under vacuum conditions and have a more complex design. Heat from a high-temperature source heats up an adsorbent (e.g., silica gel or zeolite), and vapor (working fluid, e.g., water) is desorbed from this adsorbent and condensed in a condenser at low temperatures. The heat of condensation is withdrawn from the system.

The dried adsorbent and the separated working fluid can be stored as long as desired. During the discharging process the working fluid takes up low-temperature heat in an evaporator. Subsequently, the vapor of the working fluid adsorbs on the adsorbent and heat of adsorption is released at high temperatures. Depending on the adsorbent and working fluids, the temperature level of the released heat can be up to 200 °C and the energy density is up to three times higher than that of sensible heat storage with water. However, sorption storage systems are more expensive due to their complexity. For residential use the thermal energy storage with the domestic hot/cold water tank is one of the best known and most widespread technologies.

3 Biomass

Ion V. Ion

Biomass, as solar energy chemically accumulated in vegetable or animal matter is one of the most precious and diversified resources on the Earth. It provides not only food but also energy, building materials, paper, medicines, and chemicals. The term biomass covers a wide range of products, by-products, and wastes from forestry, agriculture, livestock, and municipal and industrial wastes. According to the Renewable Energy Directive "biomass is the biodegradable fraction of products, waste and residues from biological origin from agriculture (including vegetal and animal substances), forestry and related industries including fisheries and aquaculture, as well as the biodegradable fraction of industrial and municipal waste". Biomass energy or "bioenergy" includes any solid, liquid, or gaseous fuel, or electricity derived from biomass.

Under different forms, solid, liquid, and gaseous, the biomass can be converted through various technologies in heat, electricity, and biofuels for transportation. Biomass as raw material is presented in various forms, which are abundant in all parts of the world including Europe. In recent years have been developed advanced technologies for the conversion of biomass into fuels and for efficient combustion. Of course, not all resources can be used to produce energy. Biomass is also a source of food, timber, paper, and some valuable chemicals. For this reason, the use for energy production should be integrated with other priority applications.

All life on the Earth is based on green plant that converts atmospheric carbon dioxide and water into organic matter and oxygen using the energy provided by the Sun. This process is called photosynthesis. When the biomass is burned, the carbon contained in plant is released into atmosphere in form of carbon dioxide. Therefore, biomass is considered a neutral energy source in terms of the greenhouse gas emissions. The chemical composition of biomass varies greatly depending on the species, but it can be said that plants contain, on dry basis, (15–30) % lignin and carbohydrates. The representative categories of carbohydrates are (40–45) % cellulose and (20–35) % hemicellulose.

3.1 Biomass Sources and Technologies

The actual EU biomass potential is estimated to be 314 Mtoe and it will increase to 429 Mtoe in 2020, and will fall to 411 Mtoe by 2030 [13]. The agricultural residues class (manure, straw, and cuttings/pruning from permanent crops) seems to have the largest potential and the most substantial increase is envisaged for dedicated perennial crops.

At present, the biomass can be used to produce three different types of energy: electricity, heating, and transport fuel (Fig. 9).

Electricity can be generated from following biomass sources:

- *solid biomass* which was mechanically pretreated (wood chips, pellets, briquettes, straw, dry manure, fuel derived from municipal solid wastes) can be fired or cofired in conventional coal fired power plants. This is a low-cost option due to comparatively low investment. The conversion efficiency is practically the same as for the fossil fuel. Small-scale boilers using biomass are often integrated in cogeneration systems (having higher conversion efficiency) to compensate for lower electric efficiency and higher costs;

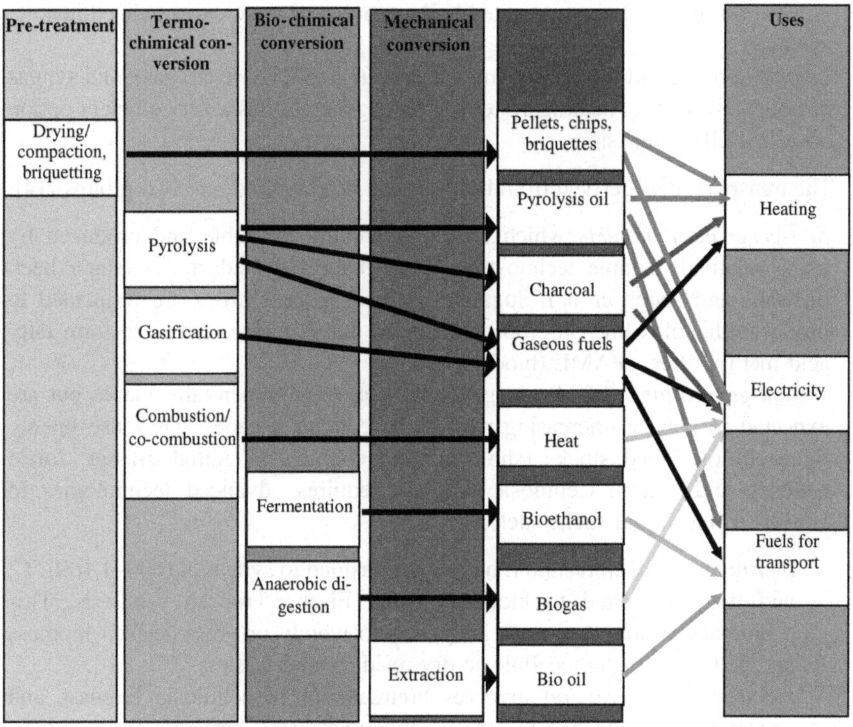

Fig. 9 Routes for converting biomass to energy

- *biogas and biomethane* generated by anaerobic digestion (and biogas upgrading) of biodegradable fraction of municipal solid waste, sewage sludge, manure, wet wastes farm and food wastes, macroalgae can be used for electricity generation or for cogeneration and as substitute for natural gas in gas grid. Electricity generation from these sources is quite efficient and low polluting;
- *syngas* produced from lignocellulosic biomass (wood, straw, energy crop, municipal solid wastes) by gasification and gas upgrading can be used both for electricity generation and as substitute for natural gas in gas grid;
- *pyrolysis products* (gas, charcoal and oil) generated by pyrolysis of lignocellulosic biomass can be used for electricity generation. The pyrolysis gas can also be used as substitute for natural gas in gas grid.

The best options for heat generation from biomass are:

- *burning of wood pellets/briquettes or logs* in small-scale specialized heating systems. This requires high capital investment compared with fossil fuel heating due to the necessity of adequate emission control;
- *burning of woodchips* in boilers for larger heating systems such as multifamily houses;

- *combustion of pellets* in automated boilers for small-scale decentralized heating systems;
- *combustion* of solid biomass, pyrolysis products (oil, charcoal, gas), and syngas in boilers for district heating systems with cogeneration is a very efficient option with low GHG emissions.

The transport fuels derived from biomass can be classified into two groups [14]:

- *first-generation biofuels* which are commercially available and produced by using relatively simple technologies from dedicated feed stocks (sugar beet, oilseeds, and starch crops). Sugars contained in these crops are fermented to produce ethanol, while oil crops provide oil that is transesterified to form fatty acid methyl ester—FAME (biodiesel);
- *second-generation biofuels* are generally not yet commercially viable but are expected to play an increasing role in the coming decades. They use mainly lignocellulosic feed stocks (short rotation coppice, perennial grasses, forest residues and straw). Cellulosic biomass requires advanced technologies to convert it into liquid fuels, such as:
  - *thermochemical conversion*: biomass is gasified to syngas at (600–1,100) °C, and then converted to biodiesel using Fischer-Tropsch synthesis. This '"biomass-to-liquid' process is applied to woody or grass-derived biomass and cellulosic or lignocellulosic dry residues and wastes;
  - *biochemical conversion* involves pretreatment of cellulosic biomass and enzymatically enhanced hydrolysis and subsequent fermentation to convert hemicellulose and sugar into ethanol.

Combustion properties of biomass can be classified as physical, chemical, thermal, and mineral. The physical properties vary greatly and properties such as density, porosity, inner surface are connected to the species of wood, and properties such as particle size and bulk density are related to the methods of biomass treatment. Chemical properties of combustion are elemental analysis, proximate analysis, and calorific value. The values of thermal properties such as specific heat, thermal conductivity, and emissivity vary with moisture content, temperature, and degree of thermal decomposition. The products of thermal decomposition of wood-based fuels are: moisture content, volatiles, fixed carbon, and ashes. Some properties vary with species and growing conditions of biomass. Other properties are dependent on the conditions under which combustion takes place. Characteristics influencing the biomass combustion are: grain size, ash content, moisture content, and the content of the structural constituents (cellulose, hemicellulose, and lignin).

Elemental composition of biomass is a very important property that determines the *energy content* and influences the clean and efficient use of biomass. The biochemical composition of biomass (cellulose, hemicellulose, and lignin) influences the calorific value. It increases with increase of lignin content. Cellulose and hemicellulose have a calorific value of 18,600 kJ/kg, while lignin has a calorific value of (23,260–26,580) kJ/kg.

There have been developed equations to estimate the calorific value of fuels from lignocellulosic materials and of various vegetable oils based on their chemical composition. There are many attempts to correlate calorific value with composition. The high calorific value (HHV) for dry, ash-free sample can be calculated based on the chemical composition as follows:

$$HHV = 0.889 \cdot LC + 16821.8 \ [\text{kJ/kg}] \tag{7}$$

where:

LC—content of lignin of dry, ash-free sample, %.

The *biomass density* depends on the presentation form (Fig. 10). Due to the lower heating value (LHV) of biomass feedstock, it requires mechanical processing in order to increase the energy density by compacting, pressing, and briquetting. Therefore, the size of solid biofuels can vary from a few millimeters for sawdust up to a maximum of 80 cm for wood logs or straw bales.

The pellets are produced by grinding the wood chips, shavings, bark, or straw/stalks followed by the pressing of the obtained dust. The heat generated by friction forces is sufficient to soften the lignin. During the cooling of pellet or briquette, the lignin becomes rigid and it acts as a binder. The pellets have cylindrical or spherical diameter of less than 25 mm. The briquettes have rectangular or cylindrical form and are obtained by pressing together the sawdust, wood chips, shavings, or bark in a press with piston or screw. The energy content of the pellets and briquettes is about 17 GJ/ton, and the moisture content is 10 %, and density of about 600–700 kg/m^3.

The *ash content* and inorganic substances of solid biofuels depend on the species of the plant and the contamination of the soil in which the plant has grown, ranging from 0.4–0.5 % (dry basis) for wood to 22.3 % for rice husks. The ash can be classified into:

- *bottom ash* which is produced in the primary combustion zone and which falls below the combustion grate in case of fixed bed or remains in bed, in case of fluidized bed combustion, due to the high density and tendency to coalescence.

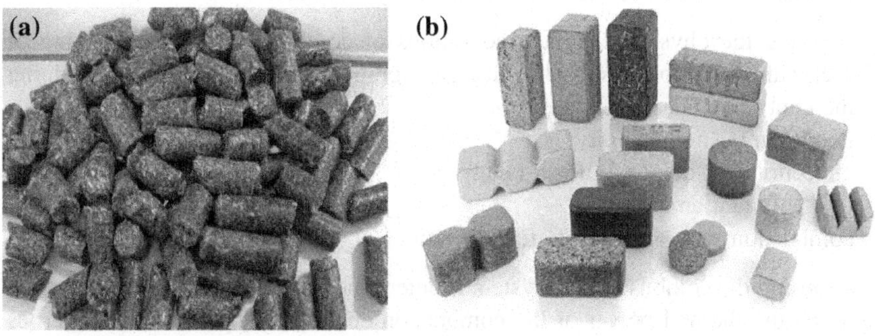

Fig. 10 Compacted solid biofuels, **a** Pelletes, **b** Briquettes

In fixed-bed combustion plants there is the risk of ash melting at high temperatures. Bottom ash represents about 60–90 % of the total ash produced in this system;
- *cyclone-ash* is the ash transported by exhaust gases and collected in cyclone. It accounts for about 10–35 % of the total ash produced in the fixed bed of the combustion plant;
- *filter fly-ash* is very fine ash which can be collected by electrostatic precipitators, wet filter, or bag filter. In the absence of advanced filtration systems, that ash is discharged directly into the atmosphere. In fixed-bed combustion boilers, this ash accounts for 2–10 % of the total ash produced.

The *moisture content* varies from one plant to another. It is lower for stem and higher for roots and crown. Higher moisture content leads to decreasing in calorific value and adiabatic flame temperature. Higher moisture content means a larger volume of flue gas and consequently larger combustion chamber. The *moisture content* influences thermal behavior during pyrolysis affecting the physical properties and quality of the pyrolysis oil. Also high moisture content reduces the efficiency of biomass combustion plant.

3.2 Biofuel Combustion

The combustion process is affected by composition of biomass, especially by moisture content and calorific value and by operational parameters: temperature, residence time, stoichiometry, and turbulence.

The efficiency of biomass combustion unit depends on the fuel type, excess of combustion air, and the temperature of the exhaust flue gas.

Combustion systems consist of a combustion chamber (furnace), where the flame and combustion gases are developed; burners or grates, with the associated fuel supply subsystem; combustion air supply subsystem; gas exhaust system and exhaust subsystem of combustion solid products (ash/slag), in the case of solid fuel combustion.

The main functions of the combustion subsystems are:

- ensuring the physical-chemical conditions of combustion (intimate mixing of fuel with air, proper disposal of flue gas, ignition temperature, and maintaining the continuous burning);
- good combustion stability;
- collection and disposal of ash (slag);
- low pollutant emission;
- combustion with appropriate fuel/air ratio.

Choosing the combustion subsystem is determined by the nature of the fuel to be used and the thermal power of the combustion chamber. The manifold properties and features of the fuel determine the existence of various combustion systems. The possibility of fuel and ash storage also influence on the selection of burning subsystem.

The air supplied into combustion chamber is divided into:

- primary air, when air is introduced with the fuel in a homogeneous mixture (combustion chambers with burners) or when passing through the fuel layer (burning on grate);
- secondary air, introduced into the flame developed inside the combustion chamber to ensure complete combustion and maintain the combustion temperature;
- tertiary air, introduced into certain areas of the combustion chamber, either to burn the fuel components incompletely burnt, or to swirl the combustion gases and ensure uniform temperature in different parts of the combustion chamber and combustion gas path.

The air is supplied into the combustion chamber either with a fan and/or natural circulation. The fans used to introduce the air into the combustion chamber and to extract the combustion gases from the combustion chamber may be radial (centrifugal) and axial. The flow rates of the air fan and exhaustor are determined by the calculation of the fuel combustion with air excess, considering an over sizing of 10 % due to the ash deposit on the heat exchange surfaces, which cause the increase of gas dynamic resistance. The power of the fans is established according to the flow rate of the fluid to be circulated and the gas dynamic resistances that they must overcome. Adjustment of air and flue gas flow rates is achieved by changing the speed of the fan driving engine.

To increase the combustion thermal efficiency and stability, the combustion air is preheated by means of the waste heat contained in flue gas before being exhausted in atmosphere.

The characteristics of the fuel supply system depend on the fuel type. This system is more complex when solid and liquid fuels are used. The solid fuel supply system consists of the storage and handling facility, the supply facility to the bunker of the combustion chamber (conveyor belt, scrapers unit, screw feeder, vacuum unit) and the fuel preparation system (drying and grinding) for pulverized fuel combustion. Liquid fuel supply systems consist of a tank, circulation pumps, filters, and preheater, when high viscosity liquid fuels are used. The simplest system is that using gas fuels.

Except for the case when the direct combustion of fuel is suitable, as is the case of gaseous fuels, fuels require special preparation for combustion. When solid fuels are burnt on grate, it is necessary to dry or compact them as bales, pellets, and briquettes. To burn pulverized solid fuels, they have to be dried and ground in the dust preparation system. Liquid fuels require preheating before combustion to reduce viscosity for handling and spraying fine drops. A gas fuel supply system includes the supply pipeline on which there are fitted manual and electrical valves, filters, pressure gauge, pressure controller, valves, venting pipe, and gas meter.

All combustion facilities should have protection systems which must comply with the following minimum technical conditions:

- burner ignition and extinguishing are safe, by correlating these operations with the functions of the control loops of the system they belong to;
- ignition and flame monitoring and protection during operation;
- protection against reaching minimum fuel pressure at which burner operation is no longer safe;
- protection against air supply cut off;
- protection against power supply interruption.

Combustion ignition is usually achieved by means of ignition devices, mobile or fixed, also called igniters such as a pilot burner used to ignite the main flame both at starting and after the burner was activated, operating continuously in parallel with the main flame, thus providing stability. Igniters operate with gas fuels (natural gas, liquefied petroleum gas as propane and butane) or liquid fuels (diesel and light fuel oil). Ignition of pilot burners can be performed nonelectrically or by electrical sparking.

Normal operation of combustion systems should comply with the material and energy balances (steady operation). During the operation, fuel intake and its parameters, along with the parameters of both air and flue gas are closely controlled. For this purpose, on the fuel side, the fuel flow rate, pressure, and temperature are measured, when appropriate. On the flue gas side, pressure, temperature downstream the combustion chamber till the chimney inlet are measured, and the chemical composition is analyzed, both to control combustion and to watch the pollutant emissions.

Under optimum operation conditions of combustion system, the following factors are important: thermal output, plant safety, and reduced pollutant emissions. In abnormal circumstances, such as when the temperature of the combustion chamber metal exceeds the upper limit, the boiler control system should focus on it as being the most important thing, sacrificing the other two factors.

In most cases, the combustion system operates at variable parameters as a result of heat consumer's request. In the transitional modes, the energy and material balance is no longer met, and therefore variations of the combustion gas parameters are reported. This leads to the need to provide an automatic control system able to provide optimal combustion conditions, i.e., to maintain a proper ratio between fuel and air flow rate, and the other factors that influence the combustion process (primary air- secondary air distribution, the pressure in the combustion chamber).

The automatic control system of the combustion plant must achieve the following functions:

- maintain the optimum flow rate ratio combustion air–fuel;
- ensure the development of a heat flow inside the combustion chamber as required by the consumer;
- maintain the pressure in furnace to a fixed value, slightly less than the atmospheric pressure to avoid gas leakage.

The three functions are achieved by means of the burner fuel supply system equipped with the necessary regulating devices, with the fan that supplies air to the furnace and/or the fan drawing the flue gas from the furnace, fitted with its own regulating devices.

It is obvious that the first two functions must be exercised in an absolute synchronization, so that adjustment of the heat developed by combustion is conducted with an optimal ratio between the air and fuel flow rates, a ratio which usually depends on the load of the combustion chamber. The third function is designed to be independent of the first two, considering the combustion chamber large enough, which makes it independent of the fluid flow that passes through it. This means that, by maintaining constant the flow rates of fuel and air entering the combustion chamber, pressure inside the furnace can be changed by changing the flow of gases extracted, by changing the speed of the flue gas fan driving engine.

The main indices of the furnaces allowing comparison of both combustion chambers and their sizing are:

- volumetric thermal load:

$$q_v = \frac{B \cdot LHV}{V_f} \quad [\mathrm{kW/m^3}] \tag{8}$$

- cross-section thermal load:

$$q_S = \frac{B \cdot LHV}{S} \quad [\mathrm{kW/m^2}] \tag{9}$$

where: *B*—fuel flow rate, kg/s or Nm^3/s; *LHV*—fuel low heating value, kJ/kg or kJ/Nm^3; V_f—furnace volume, m^3; *S*—furnace cross section, m^2.

The combustion process is an exothermic chemical reaction, a reaction that releases energy as it occurs. The amount of heat generated by complete combustion of a unit weight of a fuel is a constant for any given fuel (combination of combustible elements and compounds) and is not affected by the manner in which the combustion occurs. This constant is called “heat of combustion”, “calorific value”, or “heating value”. The heat of combustion is determined by direct measurement in a calorimeter of heat evolved during combustion of a certain amount of fuel. Combustion products inside the calorimeter are cooled to the initial temperature and the heat transferred to the cooling medium is measured to determine the *higher heating value* (HHV) as the water vapors in flue gas are transformed into liquid water (by cooling) loosing their heat of vaporization. When the flue gas contains water vapors, the heating value is called LHV. In combustion practice the lower heating value is used.

A very important issue in combustion technology is to ensure the proper amount of air that minimizes pollutant emissions and fuel consumption (energy efficiency).

In practice are used two terms: the air to fuel ratio, *AF* (kg air/kg fuel), and the excess air factor defined as:

$$\lambda = \frac{AF}{AF_{\mathrm{stoich}}} = \frac{\frac{m_{\mathrm{air}}}{m_{\mathrm{fuel}}}}{\frac{m^0_{\mathrm{air}}}{m_{\mathrm{fuel}}}} = \frac{m_{\mathrm{air}}}{m^0_{\mathrm{air}}} = \frac{V_{\mathrm{air}}}{V^0_{\mathrm{air}}} \tag{10}$$

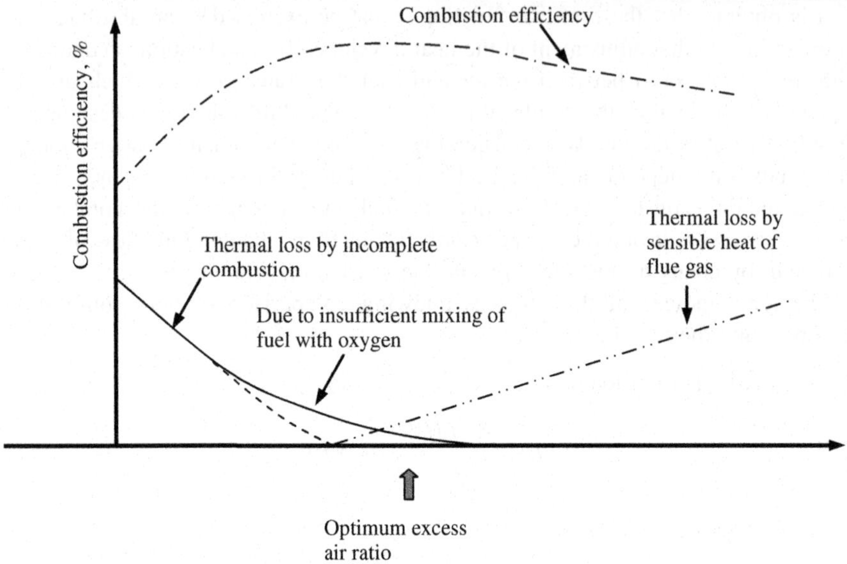

Fig. 11 Combustion efficiency versus excess air factor (λ)

The *AF* is a property of each fuel and can be calculated from the ultimate chemical composition of the fuel.

In Fig. 11, the variation of combustion efficiency with excess air factor (λ) is shown. Unnecessary high excess air volumes besides the reduction of combustion temperature increases the thermal loss with hot flue gas stream exhausted to the atmosphere through the stack.

If a burner is operated with stoichiometric air or a deficiency of air, some combustible components of the fuel remain unburned and carbon monoxide (CO) and hydrogen (H_2) will appear in the products of combustion. The unburned fuel means a thermal loss due to its chemical energy content and therefore reduced combustion efficiency. The soot, smoke, and carbon monoxide exhaust create additional pollutant emissions and surface fouling. The surface fouling may lead to plant damage. So, the combustion chamber should be fed with optimal amount of excess air to prevent combustion problems associated with less excess air and to avoid decrease of combustion efficiency due to higher excess air.

Considering that the mixing of fuel with air is well performed, the amount of excess air should be controlled. A combustion control device should be programmed to adjust the heat production to the heat demand (load control) and to optimize the combustion process with respect to maximizing thermal efficiency and minimizing pollutant emissions.

4 Stirling Engine as Cogeneration Unit

Nelu Cazacu

This engine makes use of the property of gases to expand strongly when heated and conversely to contract as they cool. Two pistons run in a hermetically sealed cylinder filled with an operating gas (helium usually). One end of the cylinder is heated by a burner while the other is cooled by water from the heating circuit in the building.

Two types of Stirling engines show potential for residential cogeneration—kinematic Stirling and free-piston Stirling.

The Free-Piston Stirling Engine (FPSE) is a recent innovation that combines Stirling principle with a modern linear alternator (Fig. 12). The result is a device that generates electricity from heat which is derived from an external source such as gas or biomass.

For FPSE, one of the two pistons—known as the displacer piston—alternately displaces the operating gas from the cold side to the hot side and vice versa. This alternation between heating and cooling produces a pressure difference which moves the second piston—the power piston. The power piston forms part of a generator which converts the piston movement into electricity.

The kinematic Stirling engine—a single-acting 90° model with two cylinders in V formation—is so-called alpha type. The Stirling module comprises the components shown in Fig. 13. The special feature of this engine is that the process gas is pushed back and forth between two cylinders without leaving the engine. The heating energy which fed to the process gas (helium or hydrogen) comes from outside via a heat exchanger, hereafter referred to as the heater, and this is converted into motion energy by the crankshaft.

As with a conventional combustion engine the Stirling also has two cylinders. The heater, the regenerator, and the gas cooler are situated between the two cylinders.

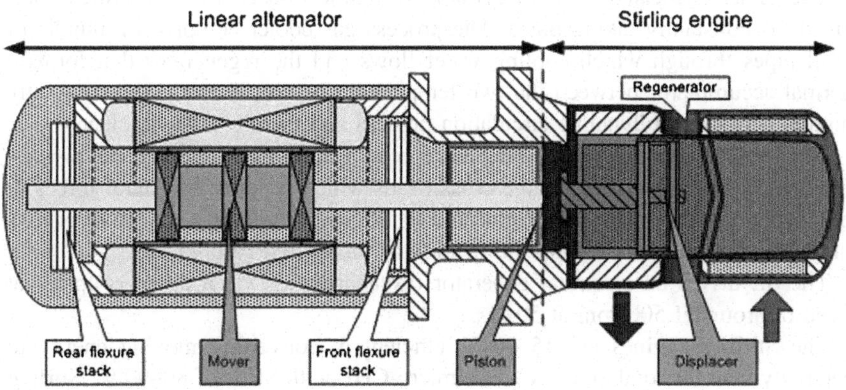

Fig. 12 Principle diagram of a free-piston Stirling engine [15]

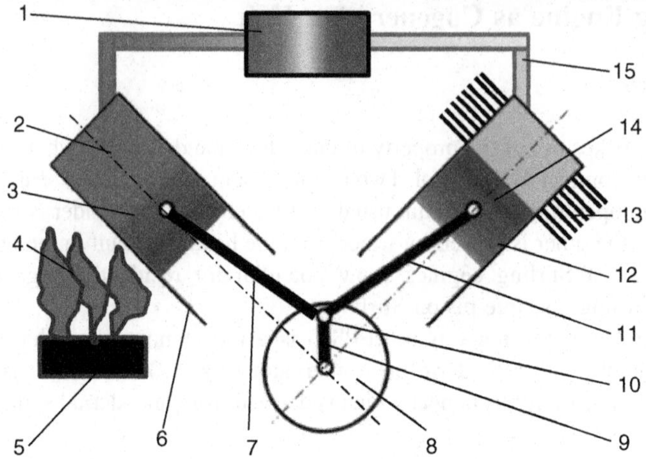

Fig. 13 Alpha Stirling engine, *1* regenerator, *2* expansion piston, *3* hot piston, *4* source of heat, *5* fuel, *6,12* cylinder jacket, *7, 11* connecting rod, *8* flywheel, *9* crankshaft, *10* hinge, *13* radiator, *14* compression piston, *15* gas cooler

As the process gas is compressed at lower temperatures and expands after being heated, it pushes the pistons back and forth between both cylinders. One cylinder is therefore termed the expansion cylinder, while the other is the compression cylinder.

The working gas is located between the upper ends of the two pistons in the actual Stirling process, in the heater, regenerator, and gas cooler. This is where the circulation process, with constant changes of pressure and temperature, takes place. As the process gas remains in the engine, it can be put under higher pressure which offers improved and/or variable performance by varying the pressure. One of the supply tank with working gas is the bottle. Here, (in bottle) the pressure level is always higher than the average cycle pressure. Helium may flow into the engine or be pumped back.

The heater is located inside a cylindrical combustion chamber in which combustion is constantly taking place. The process gas cooler comprises a bundle of small pipes through which cooling water flows and the regenerator that forms a thermal accumulator between the two temperature levels. The regenerator greatly improves the thermodynamic circulation process and therefore the efficiency.

A crankshaft with friction bearing and forced feed lubrication drives the pistons. The con rods are connected to the cross heads which support the lateral forces in their cylinders. The pistons are connected to the cross heads by a rod. Both pistons run inside the cylinders without lubrication as this would carbonize.

The SE drives an electrical generator (asynchronous) via a direct coupling at speed of around 1,500 rpm at 50 Hz.

The Stirling engines are 15–30 % efficient in converting moving energy to electricity, and the total efficiency of a micro-CHP with SE is of 80–90 %. Stirling engines are more efficient on account of their thermal balance, as the difference

Table 1 Technical data of commercial Stirling unit

Electrical power	1–10 kW
Total operating efficiency (without use of the exhaust heat/gross calorific value technique)	>90%
Electrical efficiency (electrical output/combustible)	20–30 %
Thermal power (Stirling process)	3–30 kW (power to heat ratio = 0.3)
Working gas	Helium
Mean pressure	35–150 bar
Upper process temperature	650–720 °C

between the heat extracted by the spent gases and the heat extracted by the coolant liquid.

The commercial cogeneration units with SE (Table 1) have the following technical data:

5 Boilers

Lizica Paraschiv

A boiler is a device which produces hot water or steam using the heat from burning a fuel or from an electric resistor. In the first case, producing hot water or steam is achieved through two energetic processes:

(a) the transformation of the fuel's chemical energy into heat, in the burning process, resulting in high temperature burning gases;
(b) the transfer of heat from the gases to the water or steam through the boiler's metallic surfaces. From the functional point of view, the boiler may be considered in this case as a heat exchanger which transfers heat from one fluid to another.

Boilers are built in a wide range of sizes and configurations, depending on the characteristics of the fuel used, on the thermal power demand and on the requirements regarding the control of polluting emissions. Some boilers may only produce hot water, and others are conceived to produce steam.

Warm water boilers are those which heat water up to a maximum of 115 °C (the water pressure is between 2 and 10 bars). *Hot water* boilers are those which heat water at temperatures exceeding 115 °C (water pressure is usually between 6 and 15 bar). *Steam* boilers are those which produce steam at a higher pressure than the atmospheric one.

The thermal efficiency of a boiler, η, defined as the ratio between the useful thermal power and the consumed thermal power, is calculated as follows:

$$\eta = \frac{Q_u}{B \cdot LHV} \quad [\%] \tag{11}$$

where: Q_u—heat output [kW]; B—fuel flow rate [kg/s] or [Nm^3/s]; LHV—lower heating value of the fuel [kJ/kg] or [kJ/Nm^3].

The thermal efficiency of boilers has values of over 90 % when functioning with liquid and gaseous fuels and at temperatures of the burning gases exhausted in the funnel of under 150–180 °C.

For heating boilers, the notion of seasonal operating efficiency is defined. It is expressed as the ratio between the total heat produced during the heating season and the total heat developed through the burning of fuel during the heating season. This efficiency depends on the efficiency of the boiler in stationary mode, on the losses in standby and on cyclical losses:

$$\text{Seasonal efficiency} = \frac{\text{Total useful seasonal output}}{\text{Total seasonal input}}$$

Boilers are equipped with boiler fittings (machine components that are fitted on the body of the boiler itself for the safety of the boiler and for complete control of the process of steam generation) and boiler accessories (components which are installed either inside or outside the boiler to increase the efficiency of the plant and to aid the proper working of the plant). The most frequently used boiler fittings are: two safety valves; two water level indicators; pressure gauge; fusible plug; steam stop valve; feed check valve; blow off cock; and man and mud holes. The most frequently used boiler accessories are: air preheater; economizer; steam superheated; feed pump; burner. Most boilers meant for heating operate at a maximum efficiency when they produce maximum power. In most cases, the operation of boilers at maximum power represents only approximately 60 % of the duration of a heating season. Consequently, the seasonal efficiency of a boiler remains at a low level, and the primary energy resources are wasted.

To increase the seasonal efficiency, the boiler is dimensioned for a certain level of heat demand, corresponding to the external temperature specific to the application and to the standard interior temperature.

For domestic installations there are three common types of boiler:

- *Wood pellet boilers*. Most wood pellet boilers are fully automatic in operation with feed mechanism (Fig. 14). Fuel handling is convenient and requires a relatively small storage volume due to the high energy density of the pellets. Storage pellets can be adjacent to the boiler itself or in separate fuel store.
- *Wood chip boilers*. Wood chip boilers are very similar to wood pellet boilers. Similar to pellet boilers, a small amount of ash is generated that needs to be removed periodically. The main advantage of chips over pellets is that the fuel and running costs are cheaper.

Combination boilers are also available which can burn one or more of the above fuels. Typical wood fuel storage solutions include:

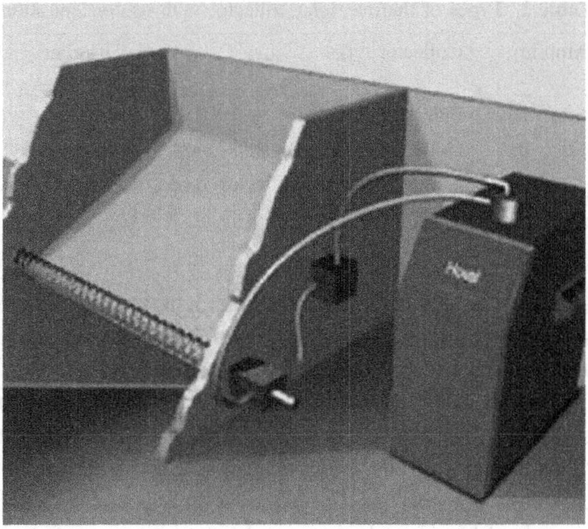

Fig. 14 Wood pellet feed mechanism

- Wood pellets—concrete store, prefabricated storage tanks (metal or plastic), fabric stores supported by metal frames, bagged pellets (can be stored practically anywhere).
- Wood chips—concrete stores next to boiler room, storage tanks (metal or plastic, above/below ground).

6 Solar Thermal Collectors

Spiru Paraschiv

Solar thermal collectors represent the main element of most solar energetic systems. Collectors absorb the solar irradiation and transform it into heat. This heat is then transferred to a fluid (which may be water, air, or oil) which flows through the collector. The heat conveyed by the fluid may be used to heat buildings, water, to produce electric energy, to dry wood and agricultural products, to prepare food, or it may be stored in an accumulation tank.

The efficiency of collecting solar energy depends of the type of collector used. Each collector is designed to absorb light with smaller wave lengths, received from the Sun (0.3 μm in length) and to prevent the loss of heat with wave lengths of 2–10 μm.

Solar thermal collectors may be classified in keeping with their mobility and their operating temperature (Table 2).

There are two types of basic solar collectors without concentration and with concentration. Collectors without concentration have the same surface for interception and for absorption of the solar irradiation, while collectors with a system of following the Sun and with concentration usually have reflecting concave surfaces

Table 2 Types of thermal solar collector with its concentration ratio and temperature

Mobility	Collector type	Absorber type	Concentration ratio	Temperature range (°C)
Stationary	Flat plate collector	Flat level	1	30–80
	Collector with evacuated tubes	Flat level	1	50–200
	Composed level collector made of parabolic grooves	Tubular	1–5	60–240

to intercept and focalized the incident solar irradiation on a smaller absorbing surface, thus increasing the radiant flux. The ratio between the incidental surface and the absorber surface is called concentration ratio.

Besides the types of collectors enumerated above, there is also special type of collectors which simultaneously convert solar energy into electricity and heat. These hybrid collectors are called thermophotovoltaic collectors. Flat plat solar thermal collectors are only used in places with warm and sunny weather. Their efficiency is considerably reduced when conditions become unfavorable (when it is cloudy, cold, or windy). Moreover, the condensation of air humidity on the collector interior metallic surfaces leads to their deterioration and to the decrease in efficiencies.

These drawbacks may be reduced in collectors with evacuated tubes. These are made of two coaxial tubes, the one on the outside made of glass and the one on the inside (the absorber) made of metal, between which there is void. Water circulates through the tube on the inside. Due to the vacuum, the heat losses through conduction and convection are eliminated, and the negative effects of the air humidity are eliminated also, since there is no air between the absorber and the transparent surface. The construction of these collectors is difficult because it necessitates perfect sealing to maintain the void.

Collectors with evacuated tubes reach higher temperatures and have higher efficiencies than the flat plate ones, but they are more expensive.

Concentration collectors use mirror surfaces to concentrate solar energy on an absorber called receiver. They reach higher temperatures than flat plate collectors, even if they only concentrate on direct solar irradiation. Their efficiency is lower on days without sunshine. They are used more frequently in areas with great irradiation, close to the Equator. The concentrators function best when they are directly oriented toward the Sun. For this, these systems use their tracking devices to move the collector during the day, so that it may always be oriented toward the Sun. The tracking devices with a single axis move from the east to the west, and those with two axes move from the east to the west and from the north to the south.

6.1 Construction and Operation

A *flat plate solar collector* (Fig. 15) is made up of a thermally insulated box which has a transparent cover and a black absorbent plate. Flat plate collectors may be

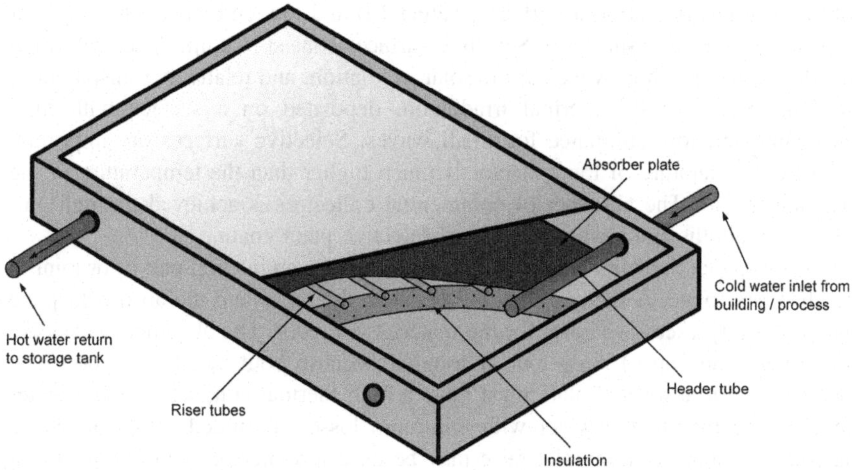

Fig. 15 Flat plate solar collector [17]

with liquid or with air. In a collector with liquid, the solar energy heats the liquid flowing through the tubes attached to the absorbent plate. The tubes may be arranged in parallel using entrance and exit collectors or in coils. The coils ensure a uniform flow, but raise problems when emptied, since they cannot be totally emptied, in order to avoid the freezing of the water.

The cover has the role of reducing losses through convection from the absorbent plate through the air stratum in between the two surfaces, as well as losses through irradiation from the collector, the cover being transparent for the short waves of the irradiation received from the Sun and almost opaque for the long waves of the thermal irradiation emitted by the absorbent plate. The cover may be made of transparent or translucent glass or of plastic materials. Translucent glass with a low content of iron is the most widely used material because it transmits a high percentage of the solar irradiation (approximately 0.85–0.90 of the normal incident irradiation on the glass surface) and is almost opaque for the thermal irradiation emitted by the absorber.

The sheets or plates made of plastic have a good transmittance for the short waves of the irradiation received from the Sun, but a high transmittance (of approximately 0.4) for the long waves of the thermal irradiation. The use of plastic materials is limited by the temperature they are resistant at without deteriorating or modifying dimensions. Few plastic materials resist for a long time to the action of the ultraviolet rays of the Sun. The advantages of the plastic covers consist in flexibility, reduced mass, and the fact that they do not brake when hail falls or stones hit. Antireflexive covers and textured surfaces improve transmittance.

The absorbance of the collector surface for the short waves of the solar irradiation depends on the nature and color of the cover and on the angle of incidence. Since almost all black paints reflect approximately 10 % of the incident irradiation, the absorbent plates undergo electrolytic or chemical treatments to obtain surfaces with high solar irradiation absorbance values (α) and low emittance values (ε) for long waves (selective surfaces). Selective surfaces consist in a thin layer with high absorbance for the short waves of the solar irradiation, and relatively transparent to the long waves of the thermal irradiation, deposited on a surface with high reflectance and low emittance for small waves. Selective surfaces are important when the temperature of the collector is much higher than the temperature of the environment air. The absorber of commercial collectors is achieved through galvanisation, anodisation, evaporation, and selective paint coating.

The tubes that the work fluid flows through must be an integral part of or joint to the absorbent plate. A thermal welding between the tubes and the absorbent plate must not imply excessive costs for the materials or work. The absorbent plates are most often made up of copper or aluminum, because both metals are good heat conductors. An absorbent plate must have a high thermal conductivity to transfer the collected heat to the water with minimum losses. Extruded plates of plastic materials resistant to ultraviolet raise may be used in collectors which function at low temperatures, if the entire surface is in contact with the fluid which transfers the heat.

In practice, most collectors are fixed and that is why they must be so oriented as to collect a maximum of solar irradiation during 1 day. This is the reason why fixed collectors are arranged in a leaning position, facing the Equator. Usually, for small latitudes, the leaning angle of the collector is equal to the angle of the latitude, but it increases with 10° at latitudes above 40°. The leaning angle of the collector depends on the latitude and the day of the year. If the leaning angle is equal to the latitude, then the solar raise will be perpendicular to the surface of the collector in mid March and September. To maximize the irradiation collected during summer, the leaning angle of the collector must be smaller, and to maximize the irradiation collected during winter, the leaning angle must be large.

In a flat plate solar collector, the radiant energy incident on the surface cannot be increased and that is why the surface needs to absorb as much of the incident irradiation as possible, and the energetic losses of the collector must be as small as possible.

The performances of the conventional flat plate collector are much reduced during cloudy weather, cold weather, or windy weather. Moreover, condensation and humidity deteriorates the material inside the collector, having as a result a decrease in performances and eventually the destruction of the system.

Solar collectors with evacuated tubes function differently from the flat plate collectors. The evacuated tube is made of borosilicate transparent glass, and the absorber in the shape of a plate or a tube—of selective cover copper.

These collectors were developed in many constructive variants namely heat pipe, direct flow, U-tube, etc. according to the method used to extract heat from the evacuated tube, but all of them use vacuum to insulate the absorber. The vacuum envelope reduces the heat losses through convection and conduction, so that higher temperature may be reached than in the flat plate collector. These collector types are formed from an array of evacuated tubes joined to a manifold through which the heat transfer liquid (water or water/glycol) flows.

In evacuated tube collectors (ETCs), the heat can either be gathered by means of a solar collector fluid flowing through the absorber as in flat plate collectors or it can be collected by means of the heat pipe principle.

The heat pipe is the method of heat transfer from inside the tube to the manifold, where a water glycol mixture flows past the condenser heat at the top of the heat pipe and pulls the energy away. A heat pipe evacuated tube collector uses heat pipes to transfer the collected solar heat from the tube into the fluid in the manifold. Heat pipes are made up of copper tubes which contain a very small amount of water in a partial vacuum (Fig. 16).

The heat pipe is encased in the inner glass tube. The energy transfer takes place in four steps:

1. This fluid is evaporated by the solar radiation.
2. The vapor rises to the top where it meets a (colder) pipe where a liquid flows through.
3. The vapor is condensed thus transferring the latent heat to the liquid in the top pipe.
4. The condensed fluid in the evacuated tubes runs back to the bottom of the tube where the process can start again.

The difference between heat pipes and direct flow ETCs can be seen in Fig. 17.

Heat pipe evacuated tube collectors are not suitable for horizontal installation, as inclination should be at least 25° to function. Instead of coaxial direct flow, ETCs

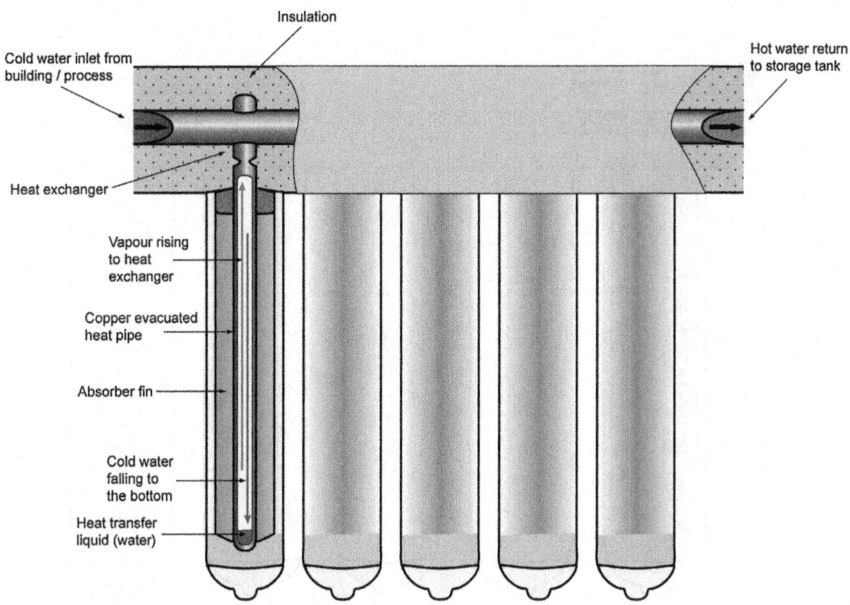

Fig. 16 The heat pipe principle [17]

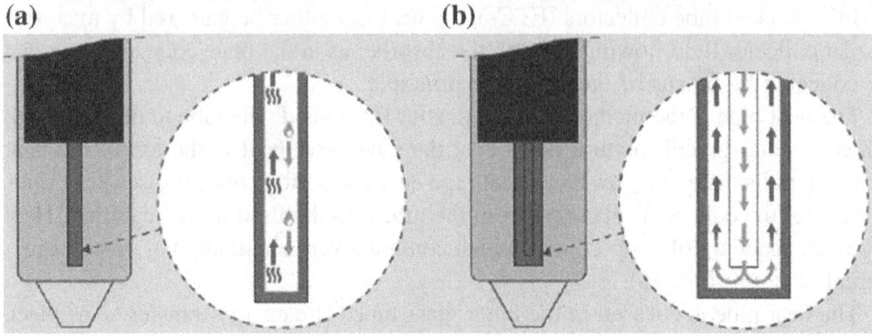

Fig. 17 Heat pipe principle and direct flow ETC principle [16], **a** Heat pipe, **b** Direct flow

can also be made as "U-pipe" where the solar collector fluid runs through one single pipe shaped as U (Fig. 18).

Collectors with evacuated tubes may be equipped with concentrating surfaces which may be situated on the inside or on the outside of the tubes.

Concentrators may be of the reflecting or the refringent type, cylinder or parabolic, continuous or segmented. Receivers may be convex, level, cylinder, or concave, covered in glass or not.

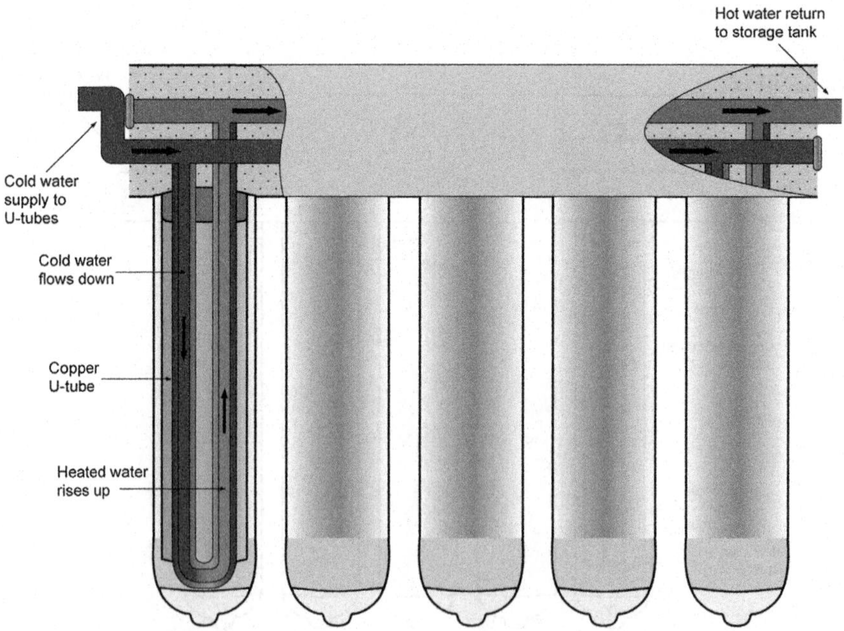

Fig. 18 Evacuated U-tube [17]

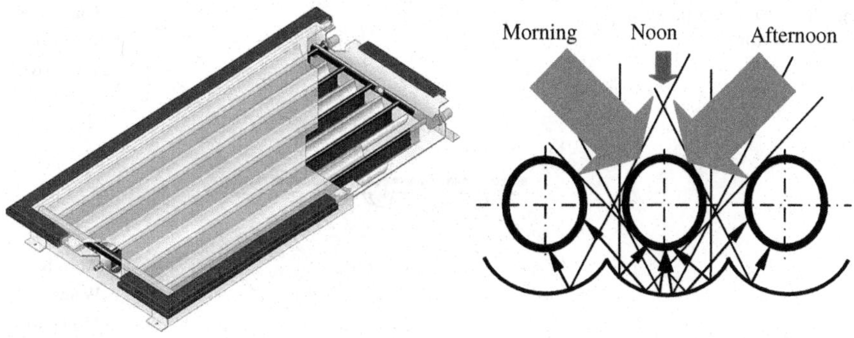

Fig. 19 Compound parabolic concentrator [18]

The simplest type of solar collector is represented by the flat plate collector equipped with reflectors which may increase the quantity of direct radiation reaching the collector. Another type of concentration collector is the namely compound parabolic concentrator (Fig. 19). It has the capacity to reflect all the incident radiation toward the absorber.

Increasing the temperature of the work fluid in the solar collector may also be achieved through reducing the surface that the heat losses occur on. Higher temperatures than those obtained with flat plate solar collectors may be reached if a large quantity of solar radiation is concentrated on a relatively small surface. This may be achieved by interposing an optic device between the source of radiation and the absorbing surface.

As compared to flat plate collectors, the concentration collectors have the following advantages:

- the work fluid may reach higher temperatures, which means that the thermodynamic efficiency is higher;
- the thermal efficiency is higher due to the fact that heat losses are lower as compared to the radiation receiving surface;
- the reflecting surfaces necessitate less material and are structurally simpler.

The disadvantages of these collectors are:

- they collect less diffuse radiation, in relation to the concentration ratio;
- they necessitate systems which orient the collector toward the Sun;
- the reflecting surfaces may lose their reflectance in time, which necessitates their periodic cleaning and shining.

Hybrid thermophotovoltaic collectors are devices which simultaneously convert solar energy into electricity and heat. A hybrid thermophotovoltaic collector consists in a photovoltaic module, on the back of which a level absorber is fitted, so as to cool it down, thus increasing its electric efficiency and collecting heat (which would otherwise have been lost) to be used for heating domestic water (Fig. 20). The photovoltaic module converts the solar radiation into electricity with a

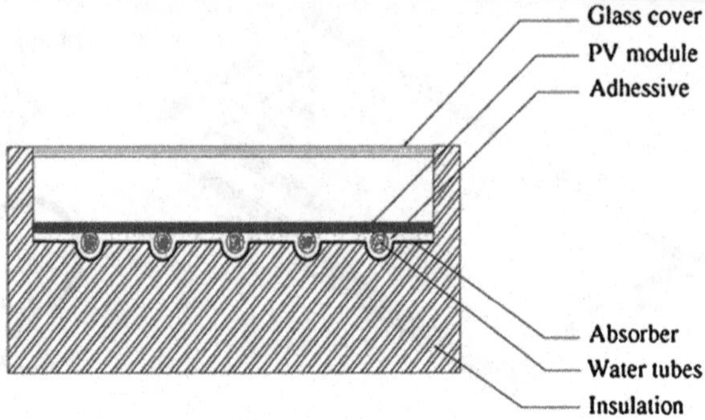

Fig. 20 Cross section of a PV/T collector

maximum efficiency of 5–20 %. The electric and thermal performance of thermo-photovoltaic collectors is smaller than that of photovoltaic panels and separate classic solar thermal collectors, but two thermophotovoltaic collectors together produce more energy on the surface unit than a photovoltaic panel and a solar thermal collector situated close to each other [19]. This is important when the available surface is important. Thermophotovoltaic collectors may be level plan or with concentration, and may work with water or with air.

6.2 Performances of Thermal Solar Collectors

The thermal efficiency of the collector depends on the instantaneous thermal efficiency and on the collector time constant, which depends on the thermal capacity of the collector and is shown by equation:

$$\eta = \eta_0 - k_1 \frac{T_i - T_a}{H} - k_2 \frac{(T_i - T_a)^2}{H} \tag{12}$$

were:

T_i is the temperature of the inlet fluid;
T_a is the ambient temperature;
H is the irradiance [W/m^2];
η_0 is the optical efficiency. The optical efficiency is the efficiency of the collector at the point where the average collector temperature is equal to the ambient temperature.
k_1 and k_2 are the heat loss coefficients.

The efficiency of the collector, as defined by the EN 12975 on the aperture area of the collector shall meet, or exceed the following conditions: optical efficiency at least 0.75, k_1 heat loss not more than 1.18 W/m^{2o}C, k_2 heat loss not more than 0.010 W/m^{2o}C^2.

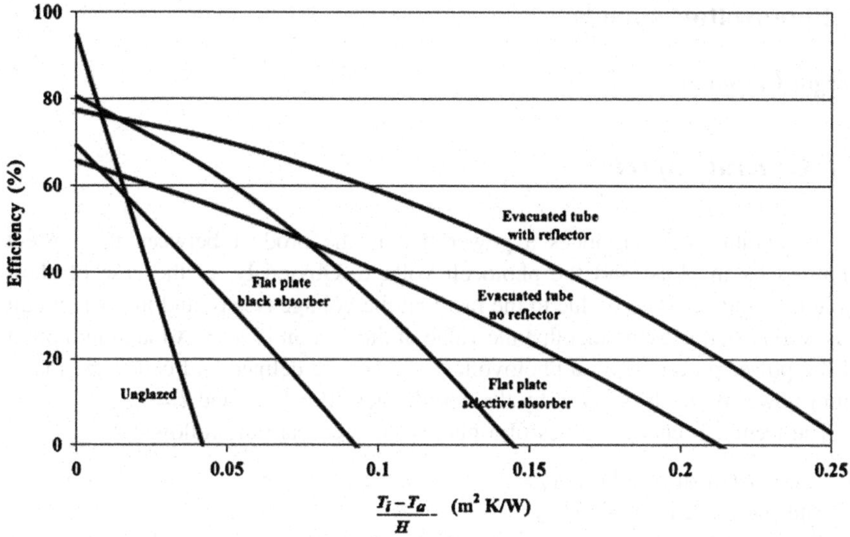

Fig. 21 Efficiency curves for various types of solar collectors [17]

The optic efficiency of the collector depends on the properties of the materials and on the cover used (absorbance, transmittance, and emittance), and on the incident angle modifier. This angle is important both for stationary collectors and for those with a solar tracking system. Typical efficiency characteristics for the solar collectors are shown in Fig. 21.

The collector power output can then be calculated as:

$$\dot{Q}_{ST} = A_{ST} \cdot \eta_{ST} \cdot H \tag{13}$$

where:

η_{ST}—collector efficiency [%];
$\dot{Q}_{ST}$—power output from the collectors [W];
A_{ST}—collector area [m^2];
H—irradiance on the collector surface [W/m^2].

To compare the heat output of different solar collectors, the product $(A_{ST} \cdot \eta_{ST})$ of aperture area and efficiency at the operating condition considered should be compared rather than efficiency alone. The choice of collector type depends on several factors such as price, efficiency, operating temperature, and location (available solar radiation, ambient temperatures).

7 Photovoltaic Panels

Sergiu Caraman

7.1 General Aspects

A photovoltaic cell produces a power that is the product between the current imposed by the load and the photocell voltage. Obviously, in the case of short circuit, the power is zero, due to the fact that the voltage is zero and in open circuit it is zero also, due to the fact that the value of the current is zero. A maximum point of the power produced by a photovoltaic cell can be defined and called the maximum power point. It is given by the coordinates: $V = V_{\max}$ and $I = I_{\max}$.

The technical characteristics of a photovoltaic cell are the following:

- The tension of backlash, V_{OC};
- The short circuit current, I_{SC};
- The tension in the optimum point of operation, $V_{\max}$;
- The current in the point of maximum power $I_{\max}$;
- The maximum power estimated $P_{\max}$;
- The filling factor FF;
- Coefficient of modifying power with cell temperature.

The efficiency of the conversion is defined as the ratio between the maximum electric power and the continuous power in the total incidental light (the power received from the Sun by the photovoltaic cell—P_{in}), so:

$$\eta = \frac{V_{\max} I_{\max}}{P_{in}} = FF \frac{V_{OC} I_{SC}}{H \cdot A} \tag{14}$$

where V_{OC}—the tension given by the solar cell in the open circuit, I_{SC}—the short circuit current, FF—the filling factor, I—the irradiance [kW/m^2], and A—the surface of the solar cell. The filling factor, FF, is defined as the ration between the maximum electric power and the product $V_{OC} I_{SC}$:

$$FF = \frac{V_{\max} I_{\max}}{V_{OC} I_{SC}} \tag{15}$$

The technical characteristics of photovoltaic cells are defined in keeping with their operation in standard test conditions, which are:

- Irradiance (Luminous intensity) of 1,000 W/m^2 in the panel area;
- The temperature of the solar cell—constant 25 °C;
- The spectrum of light AM 1.5 global.

It can be noticed that, especially at noon, during summer, the temperature of the solar cells (depending on position, wind conditions etc.) may reach 30 up to 60 °C,

Table 3 The voltage variation of the solar cell in open circuit

Temperature [°C]	27	35	40	45	50	55	60
V_{OC} [mV]	567.5	550	538.4	526.9	515.3	503.8	492.9
$\Delta V_{OC}/\Delta T$ [mV/°C]		−2.18	−2.3	−2.3	−2.31	−2.31	−2.18

Table 4 Main characteristics of the solar cells

Material	Efficiency (AM 1,5) (%)	Life cycle (years)	Costs (Euro/W)
Amorphous silicon	5–10	<20	
Polycrystalline silicon	10–15	25–30	5
Monocrystalline silicon	15–20	25–30	10
Gallium arsenure (monolayer)	15–20		
Gallium arsenure (two layers)	20		
Gallium arsenure (three layers)	25 (30 % la AM0)	>20	20–100

which leads to a low efficiency. This is the reason why another parameter, P_{NOCT}, which indicates the power at the normal operating cell temperature, should be taken into consideration. For example, Table 3 presents the voltage given by the solar cell in open circuit, as well as the voltage variation, depending on temperature, which also implies a decrease of its efficiency.

Solar cells may supply an electric power on the surface unit of up to 300 W/m^2. Included in the module (photovoltaic system), it will be lower, since there is a certain distance between the cells and the edge of the module. To calculate the electric power, a radiance value of 1,000 W/m^2, in optimum conditions of natural illumination (clear sky, temperature between 25–40 °C, total exposure to solar rays, etc.) may be taken into consideration. The irradiance may decrease up to the value 0 during the night. Otherwise, intermediary values of 200, 400, 600, 800 W/m^2, depending on conditions of illumination, exposure, etc. may be taken into consideration. Table 4 presents the efficiency, life cycle, and the cost for different types of solar cells.

7.2 Connection to Load

Photovoltaic systems are designed to fuel both the electric loads in continuous current, and those in alternative current. In both cases, they may operate in insular mode (operating as an independent generator)—Figs. 22 and 23, or connected through the electric grid, injecting electric energy into it—Figs. 24 and 25.

From Figs. 22, 23, 24, and 25 what may be observed is the fact that the electric accumulator appears in all the schemes, as an element with the role of taking up

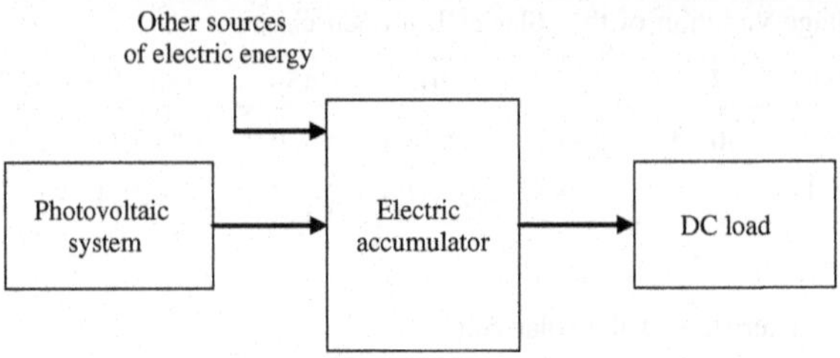

Fig. 22 System of insular supply with photovoltaic panels for continuous electric loads

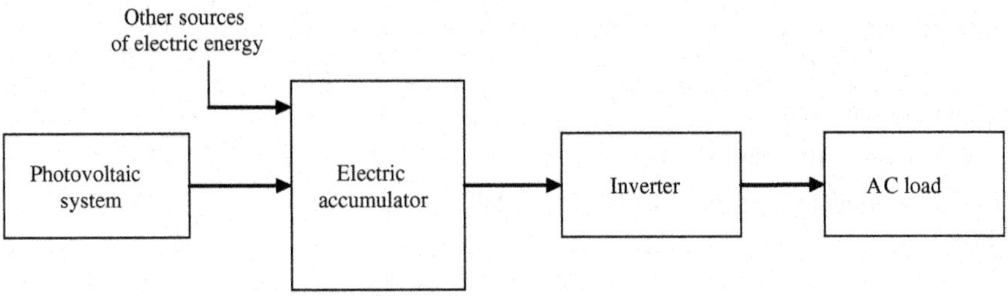

Fig. 23 System of insular supply with photovoltaic panels for electric loads in alternative current

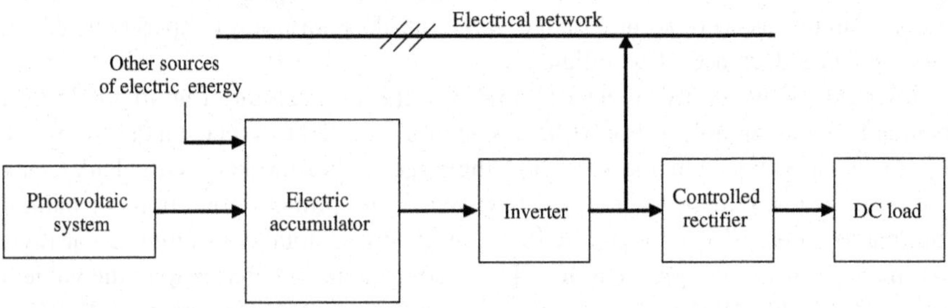

Fig. 24 System of supply with photovoltaic panels connected to the grid for continuous electric loads

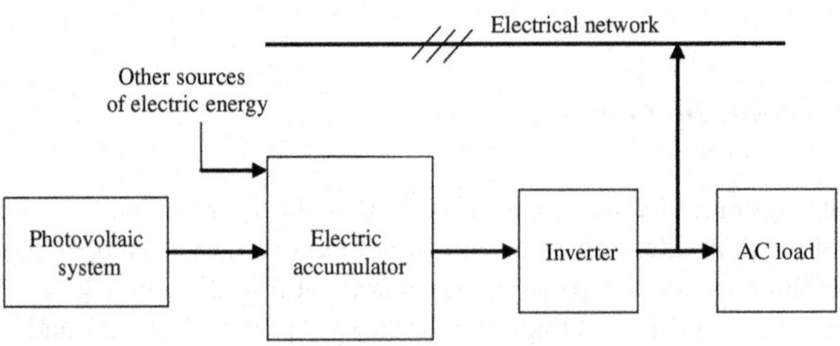

Fig. 25 System of supply with photovoltaic panels connected to the grid for electric lads in alternative current

the electric loads, as well as the other DC-AC conversion elements, depending on the type of electric load or on whether the supplying system is connected to the grid or not.

References

1. Directive 2010/31/EU Directive on the Energy Performance of Buildings OJ L 153, pp 13–35
2. EN 15603 Energy performance of buildings. Overall energy use and definition of energy ratings
3. Socal L (2013) High energy performance buildings: design and evaluation methodologies new features in EN 15603, overall energy use and definition of energy ratings Bruxelles
4. http://www.sciencedaily.com/articles/r/renewable_energy.htm
5. Twidell J, Weir AD (2006) Renewable energy resources. Taylor& Francis, Oxford
6. Report of the World Commission on Environment and Development: Our Common Future (1987) Oxford University Press
7. http://re.jrc.ec.europa.eu/pvgis/solres/solmod1.htm
8. PVGIS: The Photovoltaic Geographical Information System is an online-tool with interactive maps for PV performance calculations, powered by the European Commission. REC, 2012
9. European Photovoltaic Industry Association [EPIA] (2009) p 16)
10. Wachenfeld V (2012) Intelligent integration of decentralized storage systems in low voltage networks, SMA AG
11. EPIA Connecting the sun- solar photovoltaics on the road to large/scale grid integration—Report 2012
12. Schossig P (2008) Thermal energy storage. In: 3rd international renewable energy storage conference, Berlin/Germany, 24–25 Nov 2008
13. European Environment Agency Report No 6/2013, EU bioenergy potential from a resource-efficiency perspective (eea.europa.eu/enquiries). ISSN 1725-9177 © European Environment Agency, 2013
14. Elbersen B, Boywer C, Kretschmer B (2012) Atlas of EU biomass potentials: summary for policy makers, Institute for European Environmental Policy. http://www.biomassfutures.eu
15. ENATEC Free Piston Stirling Engine. http://www.groenebrandstof.nl/?p=1002
16. Technical guide—Solar thermal systems, Viessmann GmbH (2009). http://www.viessmann.pt/content/dam/internet_pt/pdf_documents/brochures/universal/technical_guide_-solar.pdf
17. Master Plumbers and Mechanical Services Association of Australia (MPMSAA) and Sustainability Victoria (SV). Large scale solar thermal systems design handbook. www.sustainability.vic.gov.au/…/small%20scale
18. Power Sol project-AO Sol Portugal. https://www.psa.es/webeng/projects/joomla/powersol/index.php?option=com_content&task=view&id=2&Itemid=7
19. Kalogirou S (2006) HybridPV/T solar water heaters-TEI Patras

Structural Design of the mCCHP-RES System

Nicolae Badea and Alexandru Epureanu

Abstract In order to be shown in detail, the design process algorithm was divided into two stages, namely *structural design* and *functional design*. In this chapter is presented the first stage and in Chap. Functional Design of the mCCHP-RES System the second. Then, in Chap. Experimental Case Study it is shown a concrete example. Here, the structural design is divided into seven steps and of each step has been devoted a paragraph. Regarding the contents of steps, they are as follows The first three steps (described in Sects. 1, 2 and 3) are preparatory and consist of establishment of both *the conceptual framework* (i.e., delimitation of the space where will be searched solutions for the problems arising during the design process) and *the manufacturer business plan* (on which is based the choice of design approach that can be open-ended or closed-ended), followed by *the initial data collection* (namely the data on which is based the design calculations, such as the residence building features, the customer needs and requirements, the functional needs of residence, and the residence energetic environment). The fourth step (described in Sect. 4) consists in building *the general structural model* of the mCCHP system, based on the data and information gathered in the previous steps, and then in identifying a set of *potential structural models* by customizing the general model. The last three steps (described in Sects. 5, 6 and 7) consist in both *the consumption estimation* (for each month of the year and each of the consumers incorporated in the system), and *the load estimation* (for each month of the year and each energy supplier incorporated in the system), followed by *the structural model selection,* based on the evaluation and improving the performance of each potential structural model identified in the fourth step. The output of this first stage is the set of acceptable structural models, these representing the input in the second stage, called *functional design* and presented in Chap. Functional Design of the mCCHP-RES System.

N. Badea (✉) · A. Epureanu
"Dunarea de Jos" University of Galati, Galati, Romania
e-mail: nicolae.badea@ugal.ro

A. Epureanu
e-mail: alexandru.epureanu@ugal.ro

N. Badea (ed.), *Design for Micro-Combined Cooling, Heating and Power Systems*,
Green Energy and Technology, DOI 10.1007/978-1-4471-6254-4_5

1 Conceptual Framework

The step with which the design process starts is establishment of its conceptual framework. Practically, this consists in delimitation of the space where will be searched solutions for problems that occur during the design process. This space will be called *the search space*. Each such type of problem represents *a dimension* of this space. Further, each restriction that must be satisfied when solving such type of problem represents *a limitation* of the respective dimension. *All these limitations delineate the search space*. On the other hand, since the search space represents a conceptual framework, the problems should be *strategic*.

An example of strategic problem is the architecture of the mCCHP system. A solution to this problem could be the generic structural scheme shown in Fig. 1. In solving this problem, a restriction imposed was that the stored energy can only be power or heat. This restriction delineates the search space excluding the alternative in which the stored energy could not only be electrical or thermal, but also chemical, for example.

Another example. Residence means a couple of building and mCCHP system. A strategic problem, which could be one dimension of the search space, is whether, in order to improve the performance, it may or may not alter the building, and if so, what are the restrictions imposed. Can we change the thermal insulation, or even other parts of this building?

The above examples show that: (a) the dimensions *define* the search space, while the restrictions *show* its shape, (b) the reduction of the search space *reduces* both *uncertainty* and *cost* of design process, and (c) the search space can have many dimensions, but should be considered *only those that appear to be important* as they give either conceptual clarification or information impact.

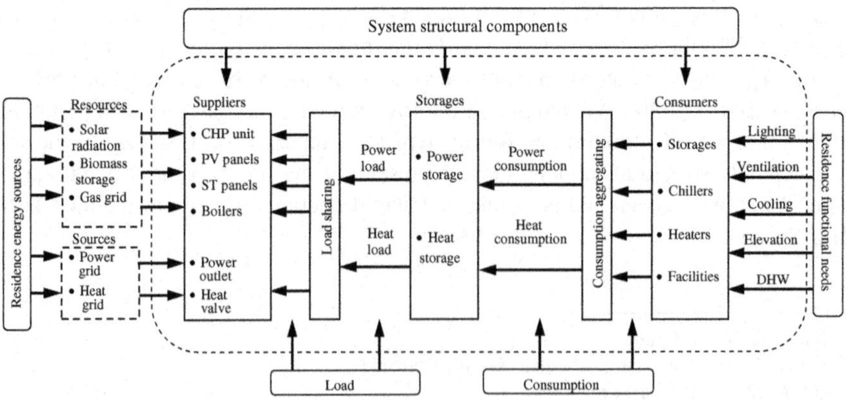

Fig. 1 Generic structural scheme of the mCCHP system: residence functional needs and residence energy sources; system structural components: consumers, suppliers, and storages; consumption aggregating and load sharing

Here will be considered only four such dimensions, namely: (a) general aspects regarding the approach of the design problems, (b) system architecture that leads to the system conceptual scheme, (c) typical actions of the design process, and (d) structuring the design process in stages and steps.

Below, these dimensions of the search space will be discussed in more detail.

1.1 General Aspects

Design is a creative process by which new technical structures and processes (such as devices, machines, installations, manufacturing processes, etc.) are developed to satisfy specific needs. Design is an essential element in engineering practice. It may be the design of an individual component, or of a system (an internal combustion engine, for example) that consists of several components interacting with each other. Perception of design ranges from the creation of a new device or process to the routine calculation and presentation of specifications of the different items that make up a system. Design must incorporate some elements of creativity and innovation, in terms of a new and different approach in order to solve an existing engineering problem that has been solved by other methods or a problem not solved before.

Design process involves the following four general aspects: synthesis versus analysis actions; closed-ended versus open-ended approaches; optimal versus acceptable solutions; concept, model, and prototype stages.

(a) Synthesis and analysis are two typical design techniques that may be applied. *Synthesis* means selection of several components with appropriate features, which are brought together to yield the characteristics of the overall system. Establishing the properties and characteristics of various components that may be incorporated represents an essential design action, because the components selection is a very important factor in obtaining an acceptable or optimal system.

Analysis means using of input data and information derived from basic areas such as statics, thermodynamics, fluid mechanics, and heat transfer in order to evaluate and to compare the operating parameters and the performance indicators of system.

Frequently, synthesis and analysis are employed together in the development of a system, by selection of the components and analysis of the system. Standard items such as valves, control sensors, heaters, flow meters, and storage tanks are usually selected from catalogs of available equipment. Similarly, pumps, compressors, fans, and condensers may be selected, rather than designed, for a given application. Selection of components needed for a system will be considered only as a first step in the design process. The second step is analysis of the system as a whole.

(b) In solving every problem, be it of design or business, there are two approaches, namely closed-ended approach and open-ended approach.

Closed-ended approach means that all the appropriate entries needed are usually given and the results are generally unique and well-defined, so that the solution to the given problem may be carried out to completion, thus obtaining the final result that satisfies the provided inputs and requirements.

Open-ended approach means that the entries may be vague or incomplete, making it necessary to seek additional information or to accept approximations and assumptions. Also, the results are not well-known or well-defined at the beginning. A unique solution is generally not obtained and one may have to choose from a range of acceptable solutions. In addition, a solution that satisfies all the requirements may not be obtained, and it may be necessary to relax some of the requirements to obtain an acceptable solution. Therefore, the trade-offs generally form a necessary part of the solving process, and individual judgment based on available information is needed to decide on the final solution.

For both problems of design and business, the open-ended approach is more frequent than closed-ended.

(c) It is crucial *to optimize* the designed system or process, so that a chosen quantity, known as the objective function, is maximized or minimized.

In a mCCHP system, it is necessary to seek an optimal design that will, for instance, consume the least amount of energy per unit. This measure is closely linked with the overall efficiency. In addition, by reducing the energy consumed for removing a unit of thermal energy, a unique system is not obtained and the design may vary over wide ranges, given in terms of the hardware as well as the operating conditions. All these designs may be termed as *acceptable* or workable because they satisfy the given requirements and constraints. But only one of them is considered as *optimal*.

(d) Design process starts with a *basic concept*, and continues with modeling and *evaluating* different alternatives, for *selecting* a final one that meets the requirements and constraints.

Before going into production, the system may be *tested* by carrying out a *prototype*. This way, the design is directed at creating a new process or system.

In recent years, we have seen a tremendous growth in the development and use of thermal systems in which fluid flow and transport of energy play a dominant role. These systems arise in many diverse engineering fields such as those related to manufacturing, power generation, pollution, air conditioning. The survival and growth of most industries today are strongly dependent on the design and optimization of the relevant systems. It has become important to apply design and optimization methods that traditionally have been applied to thermal systems and processes. The design process must consider many arrangements and determine the relevant characteristic parameters for all cases.

1.2 System Conceptual Scheme

The mCCHP system (Fig. 1) has as *input* the so-called *residence energetic environment,* which has two sets of characteristics. The first set refers to the structure and potential of the residence energy sources that are available at the place where the residence is located. These could be the electrical and heat grid—which are secondary energy sources—or the sun, wood, thermal water, gas grid, etc.,—which are primary energy sources (in particular renewable or nonrenewable they can be). The second set refers to the potential energy exchange, which may occur naturally between the residence and their environment.

As *output*, the system has the residence functional needs, which leads in a set of residential functions (lighting, air conditioning, space heating, or food preparing, for example) of that residents in using it to obtain residency-desired comfort. These functions can be considered as characteristics of *the residents living environment.*

The mCCHP system is defined as *energetic interface* between the environment in which the residence is located and the environment in which the occupants of residence are living. In short, *it uses the residence environment to create the living environment.* Above, the system is viewed as unitary item. If we look at it as a structured element, three groups of components can be distinguished, as follows.

Firstly, the satisfaction of the residence functional needs requires energy in various forms such as mechanical energy for lift, light energy for lighting, thermal energy for DHW, cold for air conditioning, and so on. Under this forms, the energy is obtained from the domestic equipments, which in terms of energy are *consumers.* They provide various forms of energy, consuming energy in only two forms, namely power and heat. The amount of energy that the consumers utilize them during a time period (for instance 1 month) is called *consumption.*

Secondly, to power the group of consumers, the system should incorporate another group of structural elements, called *suppliers*, which (a) by utilizing the energy resources from the environment generates energy, and/or (b) by connecting system to the energy grids exchanges energy (taking or giving up energy). The amount of useful energy transmitted by suppliers, be it generators (prime movers) or grids, during a period of time (for instance 1 month) in one way or another (i.e., to or from power and heat storages) is called *load.* It should notice that the suppliers offer energy in only two forms, namely power and heat.

Thirdly, in every moment, the consumption and load should be equal. Because, practically, this does not happen naturally, the consumption and load are continually monitored, and when their imbalance exceeds certain limits the control subsystem intervenes to mitigate them. Due to the delayed reaction of the system components, the mitigation becomes difficult (or even impossible) when the imbalances are minor (i.e., duration and/or magnitude of imbalances is small). This is why, to mitigate minor imbalances, the system must incorporate two additional structural elements, namely *the power* and *heat storages*. The higher the capacity of these storages is, the longer can be the duration of imbalance between consumption and load, and the greater may be the magnitude of imbalance (i.e., the minor imbalances that they mitigate may be higher). Obviously, the storages may mitigate only minor imbalances, and this makes it without difficulty; the major imbalances remain to be mitigated by the control subsystem.

On the other hand, if we look the mCCHP system not as an ensemble of components (as above), but as a system of links, then we can distinguish two types of relations, namely physical relations and causal relations.

Physical relation refers to the fact that energy, which is the output of a component, is transmitted as input of another component. In this way, there is an *energy flow,* which starts from residence energy sources and arrives at the residence functional needs. Operation follows the energy flow.

Causal relation refers to the fact that one of the two linked components determines the amount and form of energy transmitted and the other must satisfy this demand. Thus, there is a *causality flow* departing from the residence functional needs and reaching the residence energy sources. In Fig. 1 the arrows show the sense of the causality flow. Design process follows the causality flow. So, the use of the residence functions determines both the amount and form of energy used by the consumers, i.e., their consumption. By *aggregating consumptions* is determined the power and heat consumptions of the whole system. The storages immediately cover these consumptions. On the other hand, the filling level of storages determines the level of system load so as not to get in trouble. By *load sharing* is determined how much each supplier must contribute to fill the storages. This contribution represents the suppliers load.

1.3 Typical Actions

In the case of a mCCHP system intended for a given residence, the design process includes the following typical actions:

(1) *Problem formulation*

In this action, the hypothesis, conclusion, and variables of the problem are established concretely, so that the latter can be solved during the design process.

- Establishing the hypothesis means: (a) estimating the energetic needs of the residence, (b) identifying the energy sources available in the area where the residence is located as well as establishing (from the structural and quantitative point of view) how they may become primary or secondary energy sources of the mCCHP system.
- Establishing the conclusion of problem means identifying the requirements regarding the qualitative and quantitative performances, which the system should meet.
- Establishing the variables of problem means: (a) identifying those characteristics of the system, which influence decisively the link between hypothesis and conclusion, and their consideration as *independent variables* of an *objective function*, as well as (b) identifying those characteristics of the

system, which may highlight essential aspects of its exploitation, and considering them as *dependent variables* of this objective function. The objective function thus defined is then used to optimize the design solutions. Frequently, in designing a mCCHP system, the structures of the three subsystems, which make up the system (the thermal subsystem, the electric subsystem, and the control subsystem) are adopted as independent variables, and one or more of the established performance indicators are adopted as dependent variables (that is the objective function or optimization criterion). Thus, the optimization is a structural one, with a combinatory basis. The first step of optimization consists in the elaboration of a set of potential structural models, which may satisfy the functional needs of residence at the requested performance level. The second step is the selection of that model, which best satisfies the requirements included in conclusion.

(2) *Elaboration of the general structural model of system*
Using the information resulting from the previous stage, the system's ensemble is conceptually structured (made up of the two energetic subsystems—the thermal subsystem and the electric subsystem only). What results is the general structural model of this ensemble which, through adequate particularization and dimensioning of its components, may cover the energetic needs of the residence.

(3) *Identifying the structural models of system*
Firstly, the energy balances of the general structural model are formulated mathematically. The balance equations thus obtained are then used in the comparative analysis of the consumption diagrams, on the one hand, and of the load diagrams of the system as a whole, on the other. Finally, through particularizing the general structural model and imposing the condition of satisfying the balance equations, a set of structural models are selected.

(4) *Elaborating the functional models of system*
This means that, for every structural model, the structure of the energetic ensemble is completed with auxiliary components (fluid recirculation pumps or ventilators, for instance), and that the components are dimensioned so that the balance equations should be satisfied. Thus, each structural model becomes a functional model, whose components features are known as regards their nominal values. Then, the functional models, which are viable from the point of view of static performances, are selected. These form the set of functional models.

(5) *Establishing the control models of system*
The principle according to which the potential control models are elaborated is that of using both the electric potential of the electric energy accumulator (evaluated through the battery voltage) and the thermal potential of the heat storage tank (evaluated through the temperature of the thermal agent) as control variables (because they are sensitive to the unbalance between the energy produced and the energy consumed).
If the energy produced is inferior to the energy consumed, then the electric/thermal potential decreases and vice versa. The control system must maintain

these control variables at their reference values, through adjusting the energy produced. Thus, the balance between production and consumption is achieved.

(6) *Modeling and simulation of the system's operation in dynamic mode*
To this end, the system's global mathematical model is built, on the basis of the energy balances, where the main accumulations of energy existing in the system are taken into consideration:

- Accumulation of electric energy in the battery, by means of which the electric power transfers are achieved,
- Accumulation of thermal energy in the heat storage tank, by means of which the heat is transferred from sources to consumers.

The dynamic components of this model are generated by:

(a) energy balance equations, at level of the two energy accumulators,
(b) models of the control subsystem components,
(c) simplified models of dynamics of the energy sources, which have their own control system (like the Stirling engine or the boiler).

The purpose of this analysis is twofold:

- to check if the nominal values of the component outputs (which have been established in basic action 4 and have been considered satisfactory from the point of view of the system's operating in stationary mode) are satisfactory in the dynamic mode also, and to adjust these values if not;
- to elaborate the global functional model of the whole system, based on the comparative evaluation of its global performances when a certain functional model of the two energetic subsystems ensemble is applied (namely the thermal subsystem and the electric subsystem), associated with a certain structural model of the control subsystem.

1.4 Design Process

In what follows, the design process for a mCCHP system will be divided into two stages:

- *Structural design*, which includes the first seven steps of the design process and is finalized by a number of structural models for the mCCHP system, based on which the shift to the system dynamics analysis is possible;
- *Functional design*, which includes the next five steps of the design process and is finalized by the documentation based on which the mCCHP system may be actually manufactured.

The flowchart of the design process, which includes these twelve steps, is given in Fig. 2.

(a) *Structural design*

The first step consists in *Establishing the conceptual framework* in order to simplify the design process and to reduce the design costs. The conceptual framework is established by restricting the alternatives that could be considered when we look for solutions to problems occurring in the design process.

On the one hand, the typology of these problems is extensive and the number of alternatives is large. This leads to complicated activities required in this step. On the other hand, drastically limiting conceptual framework simplifies and makes it easy the design process. Therefore, a reasonable compromise should be found by considering a *small* number of *important* problems.

The second step in achieving a mCCHP system is the analysis of *Manufacturer business plan* in order to select the appropriate approach. We have considered the case in which, for a given residence (be it new or old), the design of a dedicated mCCHP system is required, using non-dedicated (or general use) components, available on the market today. Obviously, in this case, the open-ended approach will be applied. In the other cases, the closed-ended approach will be applied. The last, being very similar to the algorithm currently used in industry, it will not be detailed in what follows.

The third step is *Initial data collection*, in which the designer selects, analyses, and collects the starting data, in keeping with the specificity of the residence and the customer's objectives. The result is several specific data, grouped into four categories, regarding:

- residence building features,
- residence functional needs,
- customer requirements,
- residence energetic environment.

The fourth step is *System structural modeling*, whereby the general structural model of the system is built firstly, being defined as that structural model of the system which: (a) presupposes the use of all the types of sources available in the residence energetic environment, and (b) considers all the alternatives by means of which these sources may be used to support the residence functional needs (which define the residence living environment). Then, by particularization of the general structural model, the set of potential structural models is established.

The fifth step is *Consumption estimation* and consists in estimating, as little approximately as possible, the energy consumption occasioned by every function that the residence is equipped with. The most frequent functions of the residence are space heating, hot water, air conditioning, interior and exterior illumination, water supply, sewage, charging various appliances, and

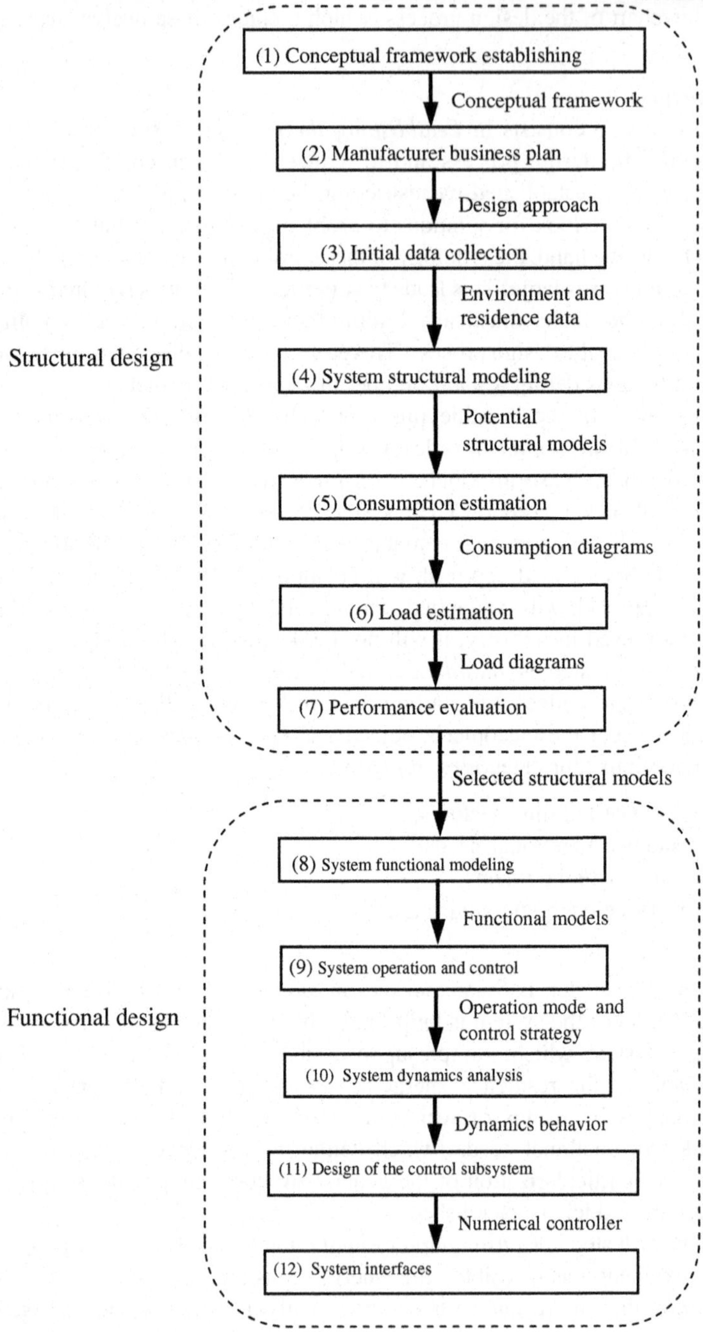

Fig. 2 Flowchart of the design process

food preparation. Moreover, complementary functions like lifts, garden, or garage equipment may occur.

Estimation is done under two aspects: (a) the monthly consumption for fulfilling each function, (b) the form of energy consumed to support each function. Further, by consumptions aggregating, the diagrams of the power and heat consumption at system level are obtained.

The sixth step is *Load estimation* and consists in (a) construction of the load diagrams at system level based on the consumption diagrams at the same level (which have been obtained in the previous step), and (b) sharing this load on each supplier, based on a load sharing strategy, which was previously established.

Finally, the load diagram of every system component is obtained.

The seventh step is *Performance evaluation* and consists in (a) evaluation of the performance indicators for each potential structural model established in the fourth step, and (b) elimination of the less efficient models. Thus was made up the set of structural models, which will be fully analyzed in the functional design stage. One of these will be actually implemented.

(b) *Functional design*

This stage begins with *the eighth* step of the design process, namely *System functional modeling*. Two actions take place within this step: (a) building the potential functional models, and (b) sizing the system components.

The first action consists in the transformation of the system's structural models into functional models, by completing them with the auxiliary components (work fluid recirculation pumps, ventilators, ventilator-convectors, heat and power accumulators, heat exchangers, pipes, etc.).

The second action consists in determining the nominal values of the components output which, on the one hand, should cover the components load (estimated in the sixth step) and, on the other hand, should satisfy all the balance equations regarding energy and mass flows.

The ninth step consists in elaborating the *System operation and control* strategies. Starting point in this step is the observation that the momentary values of the residence consumption modify permanently, depending on the momentary necessities of the residence. In essence, these strategies regard the manner in which the load of the system's components will be modified, so that it might supply the residence with the necessary energy at all times.

The tenth step is *System dynamics analysis* and consists of the numerical modeling of the mCCHP system and then of the simulation of the dynamic operation for the electric and thermal subsystems in summer and winter modes. The modeling and simulation are carried out for each functional model, in view of comparatively analyzing the results obtained. These results describe the behavior of the system in each particular case and allow a good technical characterization.

The eleventh step is *Design of the control subsystem* and consists in carrying out two actions. The first action is the design of the monitoring and control system, which includes choosing the field equipment (transducers, actuators, etc.), designing the device for connecting this equipment to the system of data acquisition, and designing the software for monitoring the whole device. The second action is the design of the numerical controllers in the electric and thermal subsystems.

The twelfth step is *System interface* and consists in designing the hardware and software components, which support the system's communication with the exterior and its reaction to foreseeable dangerous disturbances.

Thus, the designer has at his disposal a number of solutions, which are well characterized from the technical point of view, and may therefore make a decision with regard to the best functional model.

2 Manufacturer Business Plan

The approach applied in designing a mCCHP system is fundamentally influenced by the way in which the business is conceived. In principle, there are two possible business plans.

The first plan is based on the manufacturing and commercializing of mCCHP systems for general use. With such a system, certain general technical characteristics (referring to the nominal input-output values) are defined so that the system may be selected in view of being used in a specific application. When the client selects such a system, the technical criteria used are linked to the need for the system to cover the useful energy demand of his residence.

The second plan is based on the manufacturing and commercializing of several general use components and their fitting in a system dedicated to the client's residence. According to the two business plans, the design approach might be closed-ended or open-ended, respectively.

Closed-ended approach. In this approach, the hypothesis stipulates both the primary sources (few in number, even one single source) and the system's structure. Moreover, the exit requirements are well established and refer to the flow and the structure of the useful energy that the system transmits to the residence. The variables of the design problem are only the individual characteristics of the components. That is why the design consists in their optimal dimensioning. Achieving such a system consists in designing, manufacturing, and fitting the components. This approach is applied to the general use system, which is made and commercialized as distinct complete units.

Open-ended approach. In this approach, the hypothesis only stipulates the area where the residence is located and its characteristics. The output requirements are established only in as far as the residence functions are concerned, functions that the system must cover energetically (for instance: heating, cooling, charging of home appliances or other utilities such as lifts, fountains, etc.). This approach is applied to that system, which is made and commercialized as dedicated unit to a given residence.

Table 1 presents the differences between the two approaches that appear in connection with the main features of the design process.

Table 1 Comparison between the design approaches

Main feature of the design process	Closed-ended approach	Open-ended approach
Initial data	Initial data represent the system's input (that is the primary energy sources) and the system's output (that is the heat, cold, and electricity produced). These data are clearly specified and imposed	Initial data represent only the characteristics of the residence and the area in which it is located. These data are vaguely specified
Output requirements	The imposed output requirements are the useful nominal power distributed between the components (heat, cold, and electricity), which the system must transmit to the residence	The imposed output requirements are the functions that the system must cover permanently, as thermal comfort winter-summer or charging of home appliances and other utilities of the residence
Primary energy sources	Primary energy sources are imposed and clearly specified while the area in which the residence is located is neither imposed, nor clearly specified	Primary energy sources are not imposed, but must be from among the ones available in the area in which the residence is located, while this area is imposed and clearly specified
System structure	The system structure is clearly specified and imposed	The system structure is not specified, being established during design process
Main system components	The technical characteristics of the main system components are determined during the functional design and represent the input data in their manufacturing	The technical characteristics of the main system components are determined during the functional design and are ensured through selection from those available on the market
Optimization variables	Optimization is continuous and achieved during technical design of the main component by convenient adjustment of their construction	Optimization is discrete (combinatorial) and is achieved during structural design of the system, by adjusting the structure and selecting the components
System-residence compliance	The system-residence compliance is obtained by selection of a general use system from those available on the market, so that it might cover the energetic demand of the given residence	The system-residence compliance is obtained by conceiving a system that is dedicated to the given residence; the system components are of general use and are selected from those available on the market

3 Initial Data Collection

After the first two steps in achieving the mCCHP system, namely *Conceptual framework* and *Manufacturer business plan*, have been completed, the design should be continued with the collection of all initial data that form the hypothesis of the design problem. Initial data refers to (a) the residence building features, (b) the customer needs and requirements, (c) the residence functional needs, and (d) the residence energetic environment.

3.1 Residence Building Features

The data, which must be collected, refer to those geometrical and functional characteristics of the residence, which influence its energy consumption. These are the following: (a) the residence location expressed by latitude and longitude and the location of the residence (building) within the built-up area of the city or not, and (b) information about the characteristic elements of the architecture such as:

- ground floor area (GFA) and total floor area (TFA);
- exterior wall structure, surface, and material;
- windows surface, type, orientation, and structure;
- heated surface;
- heated space average height;
- perimeter;
- height state of the residence, and building shading.

3.2 Customer Needs and Requirements

In designing a mCCHP system, regardless of whether it is carried out according to open-ended or closed-ended approach, the customer needs and requirements must be known in as much detail as possible, since they decisively influence the solutions that the designer adopts during the designing process. The customer needs and requirements mainly refer to the functions of the residence, to the space taken by the mCCHP system, and to the costs related to system purchase and operation. Due to various reasons, it is difficult to formulate these needs and requirements precisely enough from the very beginning. That is why, if the open-ended approach is used, it is possible for certain decisions to be taken by customer with the support of designer so that an optimum compliance between the customer and the designed system should be attained.

3.3 Residence Functional Needs

Energy is used in buildings for various purposes as heating and cooling, ventilation, lighting, and preparation of domestic hot water.

Space heating is by far the most important energy user in the residential building. Space heating fuel shares vary significantly from country to country. In many countries the natural gas is the main fuel for space heating, followed by oil (including oil products such as liquefied petroleum gas). Recently, in most countries, both the oil and renewable sources has fallen, and has been replaced by natural gas or electricity. The main drivers behind higher natural gas consumption include increased energy use for space heating by gas boilers in decentralized system. A boiler is the heating plant used to create *domestic hot water* and to supply with hot water in radiant heating systems. Hot water boilers can be small, compact, energy efficient, and low maintenance, and can use a variety of fuels including natural gas, propane, oil, or electricity.

To room air conditioning there are various cooling systems from portable appliances to mini-split systems. At present all these systems are multifunction units that combine air conditioning with dehumidification or heating and an air circulating fan, all in one. Their operation requires power supply.

Electricity is used for space heating, in particular, in Nordic countries that are, generally, large producers of electricity from renewable sources (wind and hydropower). In many countries, for example Denmark, Finland, Sweden, and partially Romania, district heating represents the most important energy commodity for space heating.

Strong growth in electricity consumption was largely driven by increases in the ownership and use of electrical appliances and, in particular, a wide range of smaller appliances and cooling systems to air conditioners.

A house owner plays an important role in choice of heating, cooling, and electricity supply systems. Occupant behavior plays an important role in determining the total energy use in the residence. However, other factors, such as the household incomes and the cost of energy, are also important. For instance, higher incomes allow people to purchase more and larger energy-consuming equipment. Higher energy costs are likely to restrain household energy consumption to some extent by encouraging occupants to purchase energy efficient equipment.

The home owner in order to reduce the energy bill can opt for a nonconventional system that includes solar thermal panels, photovoltaic panels, and cogeneration unit.

The solar thermal panels will contribute to obtaining hot domestic water and to cooling the residence during summer (by supplying the adsorption chiller with hot thermal agent). The electricity of the building can be obtained from photovoltaic panels and cogeneration unit.

Distribution of thermal energy can be made either with two distinct circuits (one for heating and other for cooling) or with single circuit, namely a turn-return circuit for the heating/cooling, using ventilator-convectors, running in the cold season with a thermal agent—hot water of 80/60 °C for heating the residence, and running in the hot season with thermal agent—cold water of 5/15 °C for cooling the residence.

3.4 Residence Energetic Environment

3.4.1 Local Climate

The energetic consumption of a building depends on external and internal factors. The external factors are linked to climate aspects, specific to the area where the residence is geographically located, namely air temperature, wind speed, sunlight, and air humidity. The internal factors are linked to the features of the residence building.

Supplying the space with heat should compensate all heat losses, the most important of which are those transferred through the walls, the roof, and the ventilation system. As far as the local climate is concerned, the outside temperature is the most important variable, which influences both the daily magnitude and the seasonal/annual variation of the thermal energy consumption.

On the other hand, the quantity of thermal energy necessary for heating is influenced by the caloric input given by the solar radiation, and by the wind. The caloric input is mainly generated by the solar radiation through the windows, and is enhanced by the natural ventilation due to the pressure of the wind, which surrounds the buildings, especially on very windy (stormy) days. This is why information regarding the local climate in the residence location is needed. The information regarding the local climate in the residence location is available at EU Commission [4].

This information consists in the average statistic values of the climate parameters corresponding to a certain period of time (day, month, season, and year), namely:

- average annual temperature outdoors,
- average daily exterior temperature,
- average daily interior temperature,
- average temperature during the heating season,
- solar radiation,
- number of heating/cooling degree days,
- relative humidity,
- wind speed.

Average annual temperature outdoors, in Europe, varies from 2 to 19 °C, offering very different local conditions for heating the living space. The variation of the average annual temperature outdoors in Europe is presented in Fig. 3. The map does not take into consideration the mountain areas, where the values of the temperature outdoors are lower.

The average daily exterior temperatures $\vartheta_{e,i}$ are useful to calculate the length of the heating season. Also, these are used to determine the average temperature during the heating period ($\vartheta_{e,hs}$) and the number of heating degrees-days (HDD). The length of the heating season d_{hs} expressed in days (Fig. 4) is determined in accordance with relation:

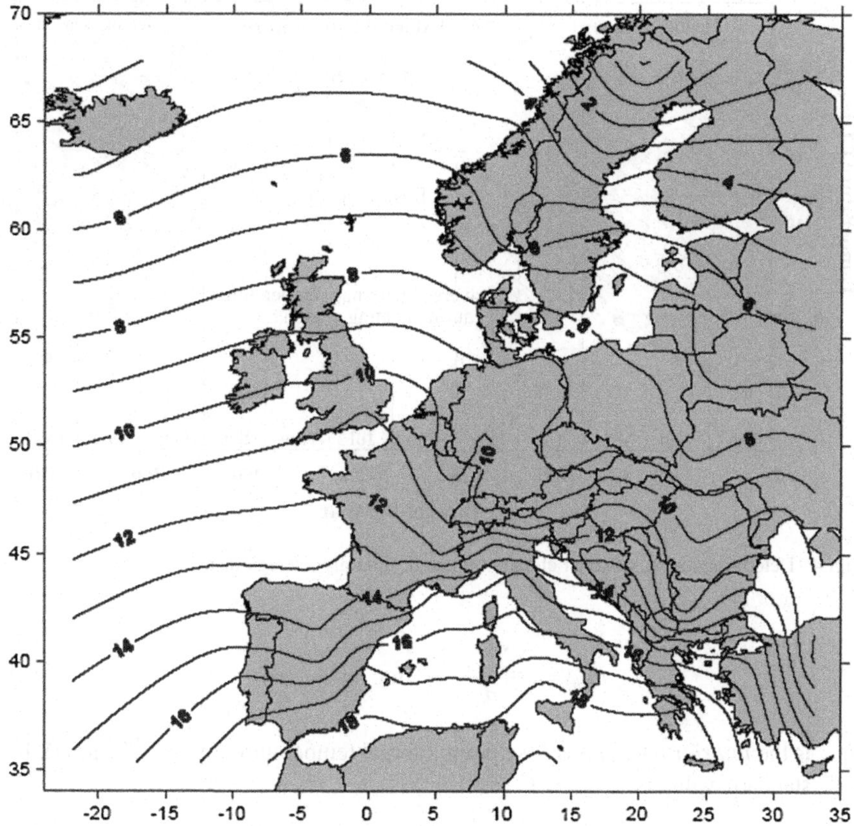

Fig. 3 Variation of the average annual temperature outdoors in Europe [1]

$$d_{hs} = \sum_{i} d_i \quad \text{for} \quad \vartheta_{e,i} \leq \vartheta_b \tag{1}$$

where:

- d_i—is number of days;
- $\vartheta_{e,i}$—temperature of the external environment, average value for the respective day;
- ϑ_b—is the base temperature ($\vartheta_b = 12°C$) sometimes called the balance point temperature.

The average temperature during the heating season can be calculated with relation:

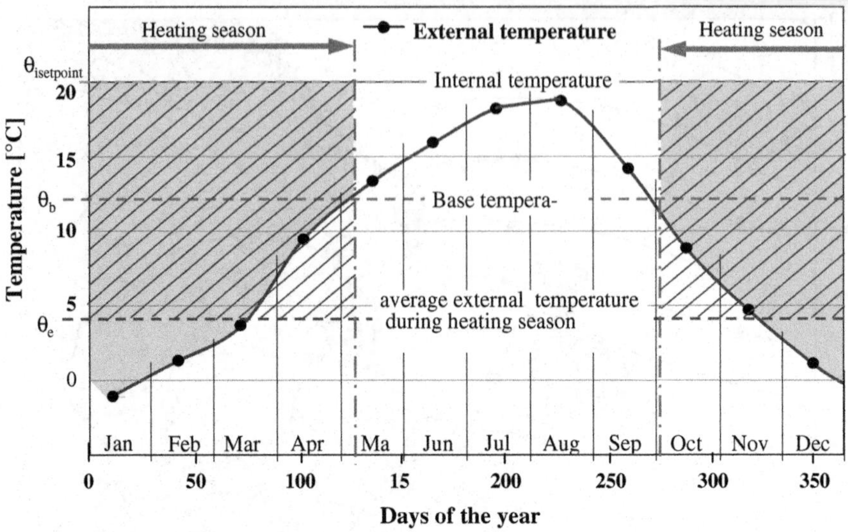

Fig. 4 Determination of seasonal climate by use of monthly average data

$$\vartheta_{e,hs} = \frac{\sum_{1}^{365} \vartheta_{e,i}}{d_{hs}} \quad \text{for} \quad \vartheta_{e,i} \leq \vartheta_b \tag{2}$$

ϑ_i is the *interior temperature* (setpoint internal temperature for space heating). Its standard value is $\vartheta_i = 20°\mathrm{C}$

Solar radiation is the radiant energy received from the Sun. It is the intensity of sunrays falling per time unit and per area unit, and is usually expressed in Watts per square meter (W/m^2). The radiation incident on a surface varies from moment to moment depending on its geographic location (latitude and longitude of the place), orientation, season, time of day, and atmospheric conditions. Solar radiation is the most important weather variable that determines whether a place experiences high temperatures or is predominantly cold. The instruments used for measuring of solar radiation are the pyranometer and the pyrheliometer. The duration of sunshine is measured using a sunshine recorder.

The solar irradiation H_{solar} during the heating season is given by equation:

$$H_{solar} = \frac{\sum_i H_{solar,i}}{d_{hs}} \quad \text{for} \quad \vartheta_{e,i} \leq \vartheta_b \tag{3}$$

where:

$H_{solar,i}$ is daily solar irradiation on 1 m^2 surface in heating season

The climate parameter called "Degree-Days" is a measure of the severity and duration of cold weather or warm weather. They are, in essence, a summation over time, of the difference between a reference or "base" temperature and the outside temperature. The basic definition of degree-days is the difference between the base temperature and the mean daily outdoor temperature.

Two values may be calculated for "Degree-Days". The first value namely Heating Degree-Days (HDD) are a measure of the severity and duration of cold weather and the second value, Cooling Degree-Days (CDD), are a measure of the severity and duration of warm weather. The colder the weather in a given month is the larger the cooling degree-day value for that month is obtained.

In a heated building during cold weather, the heat is lost to the external environment. Some of this heat is replaced by casual heat gains to the space—from people, lights, machines, and solar gains—while the rest is supplied by the heating system. Since the casual gains provide a contribution to the heating within the

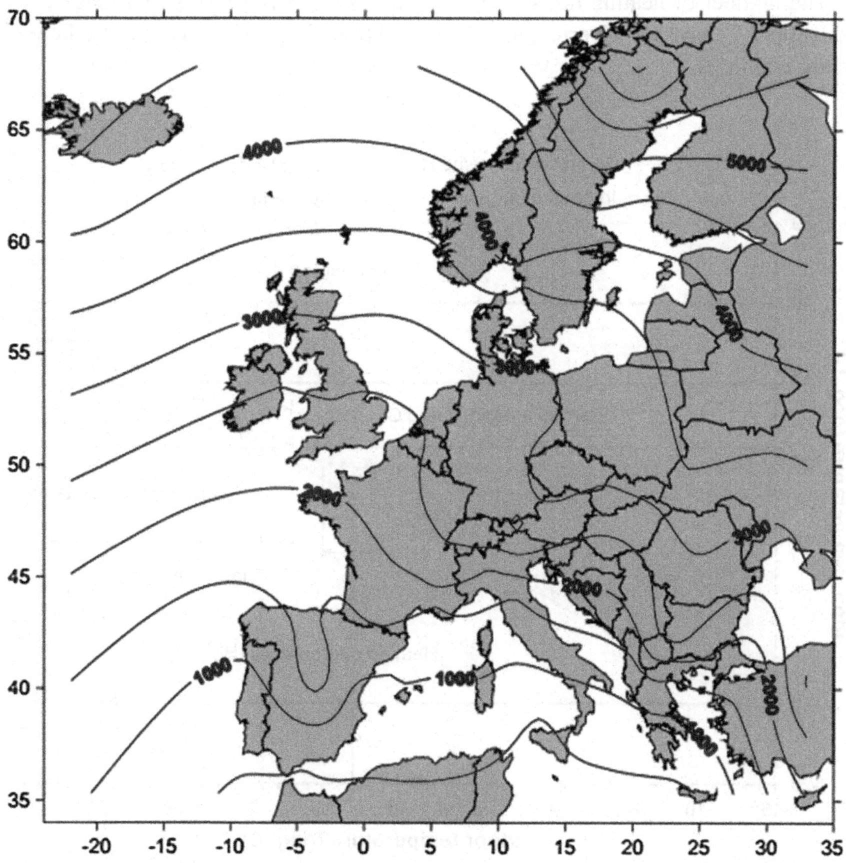

Fig. 5 Map of Europe with number of annual "heating degrees-day" [1]

building, there will be some outdoor temperature, below the occupied set point temperature, at which the heating system will not need to run. At this point the casual gains equal the heat loss. This temperature will be the base temperature for the building. If the outdoor temperature is below the base temperature, then it is required heat from the heating system. Heating degree-days are a measure of the amount of time when the outside temperature falls below the base temperature. They are the sum of the differences between outside and base temperature whenever the outside temperature falls below the base temperature. The number of degree days for a heating season is determined with relation:

$$HDD = \sum_{day} (\vartheta_b - \vartheta_{e,i}) \text{ for } \vartheta_b \geq \vartheta_{e,i} \quad [°\text{C day}] \tag{4}$$

Heating degree-days for a geographical site between given dates correspond to a time integral of $\Delta\vartheta_{bal} = \vartheta_b - \vartheta_{e,i}$ in which only positive values of the difference are counted.

The number of heating degree-days in EU countries (Fig. 5) is in a range from 700–800 [°C day] for Cyprus and Malta to 4,000–5,000 [°C day] in Nordic and Baltic countries.

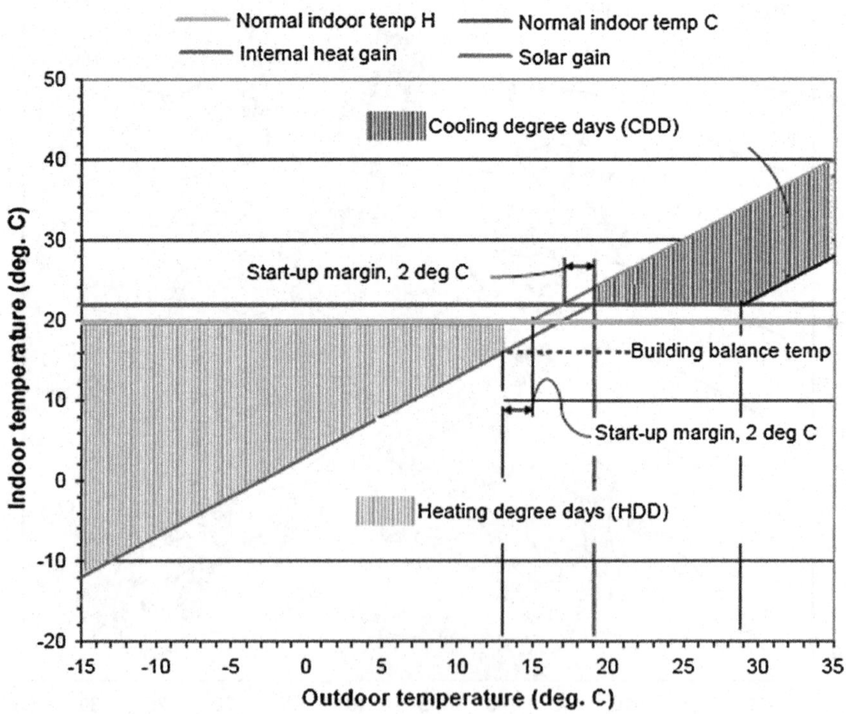

Fig. 6 Correlation between temperatures and degree-days [2]

HDD is useful to calculate the annual heat necessary for a building and the fuel necessary for heating. Thus, the heating energy demand (kWh) = overall heat loss coefficient (kW/°C) · HDD (°C day) · 24 (h/day). The 24 is included to convert from days to hours.

Cooling degree-days are defined in the same manner. It is used to evaluate the annual energy consumption for cooling. For an actively cooled building the base temperature is the outdoor temperature at which the cooling plant need not run, and is again related to the casual heat gains to the space (which now add to the cooling load). In this case cooling degree-days are related to temperature differences above this base.

The cooling systems are designed to maintain an indoor temperature of 22 °C only when the outdoor temperature is below 29 °C. When the outdoor temperature exceeds this limit, the indoor temperature will start to slide at a constant difference of 7 °C below the outdoor temperature.

Correlation between indoor and outdoor temperature and the Heating degree-days and Cooling degree-days, respectively, is shown in Fig. 6.

Illustrative exercise on the climate data collection, for the city of Galati, Romania, is shown in Table 2.

3.4.2 Local Energy Sources and Resources

Primary energy sources (meaning energy is created directly from the actual resource) can be classified in two groups: nonrenewable or renewable. *Nonrenewable energy sources* are energy extracted from the ground, either in the form of gas, liquid, or solid. They cannot be replenished, or made again, in a short period of time. Oil, natural gas, and coal are called fossil fuels because they have been formed from the organic remains of prehistoric plants and animals.

Renewable energy sources are the energy that comes from a source that is constantly renewed, and can be replenished naturally in a short period of time. Examples include solar, wind, biomass, and hydropower. In 2020 at EU level, 20 % of total electricity used will come from renewable resources. Secondary energy sources are derived from primary sources. Secondary sources of energy are used to store, move, and deliver energy in an easily usable form. Examples include electricity and hydrogen. More than 85 % of the world's current energy needs are met through fossil fuels whose price is rising due to resource depletion.

Energy sustainability is about finding the balance between a growing economy, the need for environmental protection, and the social responsibilities in order to provide an improved quality of life for current and future generations. In short, it is meeting the needs of the present without compromising the needs of the future. Energy sustainability can inspire technical innovation with an environmentally conscious mindset. Renewable resources such as sunlight, wind, and biomass provide a source of sustainable energy. This includes biofuels like ethanol, which is created from crops like corn or sugarcane.

Table 2 The climate data for the city of Galati (illustrative exercise) [3]

Month	Air temperature °C	Relative humidity %	Horizontal solar daily irradiation $kWh/m^2/day$	Wind speed m/s	Earth temperature °C	Heating degrees days °C days	Cooling degrees days °C days
January	−1.7	87.0	1.47	5.7	−1.5	611	0
February	−0.6	84.0	2.31	5.7	0.0	521	0
March	5.0	79.0	3.44	5.7	5.9	403	0
April	11.1	74.0	4.94	6.2	13.6	207	33
May	16.7	73.5	5.94	4.6	20.0	40	208
June	20.0	75.0	6.69	4.6	23.4	0	300
July	21.7	72.5	6.56	4.1	26.2	0	363
August	21.7	71.5	5.75	4.6	26.1	0	363
September	17.8	73.5	4.39	4.6	20.6	6	234
October	11.7	77.0	2.97	4.6	13.7	195	53
November	5.0	86.0	1.64	5.1	5.3	390	0
December	0.6	89.0	1.19	5.7	−0.4	539	0
Annually	*10.8*	*78.5*	*3.95*	*5.1*	*12.8*	*2,912*	*1,554*

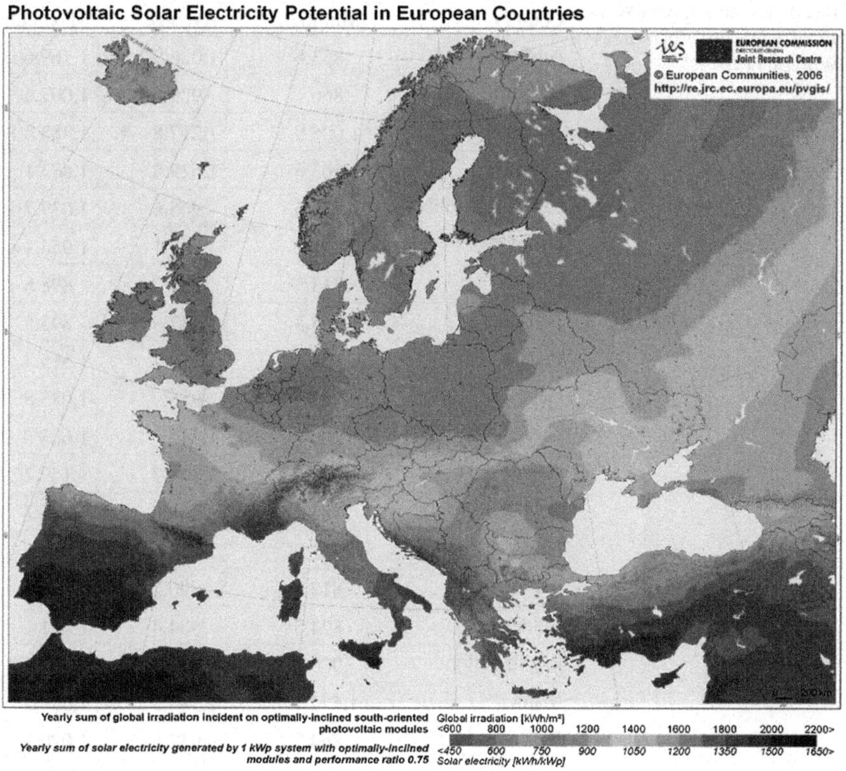

Fig. 7 Solar energy potential in Europe [4]

Since the solar radiation is the main source of renewable energy and, moreover, using solar energy is suitable for microCCHP, this source will be evaluated further. On the other hand, the PV panels and ST panels can be placed directly on the ground, without any restriction, or on the building, when several limitations (of physical and architectural nature) occur. That is why we evaluate below separately the solar energy potential of the territory and the solar energy potential of the building.

A. Solar energy potential of the European territory

Under territory of EU member states, on a horizontal surface of 1 m^2, it is possible to capture an annual quantity of solar energy ranging between 800 and 1,650 kWh/m^2/year, according to the geographical location (Fig. 7). Annual total potential in NW Europe is approx. 100 kWh/m^2/year (approx. 50 % direct and 50 % diffuse radiation) and in southern Europe is approx. 2,000 kWh/m^2/year.

The resource distribution was taken into account to estimate the PV production for each of the EU 27 countries according to their potential based on the irradiation level (Table 3).

Table 3 PV output per kW at optimal angle in urban areas (kWh/year) [5]

Country	Minimum	Average	Maximum
Austria	853.6	1,026.9	1,169.6
Belgium	866	929.6	1,007.6
Bulgaria	1,005.9	1,217.8	1,388.4
Cyprus	1,563.6	1,629.8	1,683.1
Czech Republic	838.6	945.6	1,039.7
Denmark	841.2	945.1	1,054.1
Estonia	813.5	867.7	898.8
Finland	765.3	837.9	895.5
France	858	1,116.7	1,515.3
Germany	825.5	936	1,085.8
Greece	1,200	1,445	1,667
Hungary	991.6	1,104.7	1,159.2
Ireland	789.5	908.6	1,066.7
Italy	772.9	1,326	1,624
Latvia	817.8	890.2	992.6
Lithuania	824.5	884.4	1,011
Luxembourg	900.3	939.6	967.5
Malta	1,572.1	1,584.2	1,599.2
Netherlands	864.7	932.6	1,020.7
Poland	833.6	937.2	979.7
Portugal	1,270.7	1,494	1,648.6
Romania	891.1	1,132.7	1,278.3
Slovakia	845.5	1,020.7	1,116.8
Slovenia	931.6	1,085.2	1,249.7
Spain	968.2	1,470.7	1,664
Sweden	639.1	826	1,050.8
United Kingdom	710.8	920.2	1,121

The optimum inclination is the angle at which a south-facing surface receives the largest amount of total yearly global irradiation.

The optimum tilt (or inclination angle) depends on the latitude and varies, e.g. between 48° for a northern city like Oslo/Norway (59°55′ North,) and 23° for city like Athens/Greece (36°57′ North). The optimum inclination angle for each location at EU level is shown in Fig. 8.

With insolation power (see Table 3), a free standing crystalline silicon system with an installed peak power of 1 kWp will harvest electricity in a range of

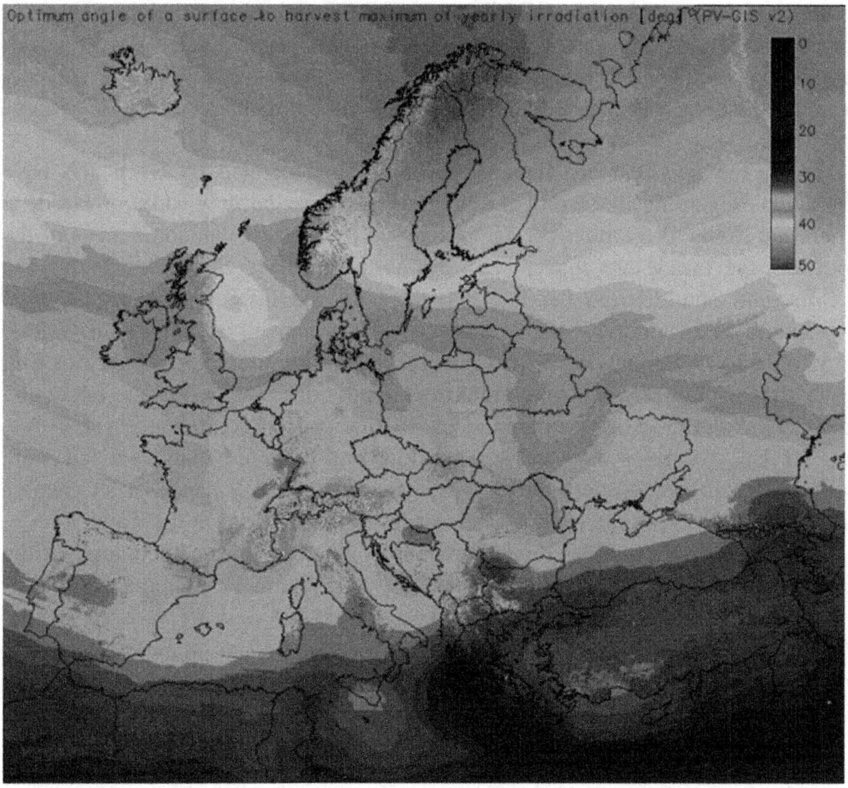

Fig. 8 Optimum inclination angle of a surface to harvest maximum irradiation over a year [5]

826 kWhel/year (Sweden) to 1,630 kWhel/year in the extreme (Cyprus). The scale unit "Kilowatt peak" (kWp) describes a PV module's nominal power output of one Kilowatt at standard test conditions of 1,000 W/m^2 irradiance at 25 °C module temperature.

B. Solar energy potential of the building

In general, a photovoltaic installation is composed of various modules of solar cells and the balance-of-system including typically an inverter (given that the device is connected to the household), cables, and the mounting installation. These may be constructed either without a connection to the electricity grid (off-grid installation) or on grid connected installations.

Besides the different types of solar system (PV or ST panels), there are different options of mounting such installation. PV/ST panels may be mounted on top of roofs, placed directly at ground level, or integrated into buildings. In the latter case, the PV/ST panels can be placed on the roof or integrated into the façade. Moreover, if they are mounted on or integrated into the building's roof or facade they will require no additional land use. While most of the free-field installations tend to be larger installations of centralized character, the building integrated PV/ST installations can be characterized as decentralized installations systems. They are ideally positioned where are unshaded throughout the year and with an orientation towards the South.

Unlike the free-field installations, the building integrated PV/ST panels have limitations regarding the use of the building surfaces (be they roofs or façades). These are limits to the physical nature (caused by the fact that some areas of the roof and the facade are shaded, i.e., the north facade of the building) and the architectural nature (from the fact that the roof and facades have architectural elements such as windows that do not allow the placement of the panels).

In order to assess the solar energy potential either of building or of panels as well as to express the physical or architectural limitations, we define *the utilization factor, k,* as the ratio between area of panels that convert solar energy (be it A_{PV}, A_{ST}, or $A_{panels} = A_{PV} + A_{ST}$) and ground floor area (A_{GFA}). Notations k and A are completed with indices, depending on the type of panels (PV panels or ST panels), and their location [on the roof, (r), or façades, (f)], as shown in the Table 4).

The physical and architectural limitation is given by the methodology of the International Energy Agency Photovoltaic Power Systems Programme and can be synthesized with a simple rule of thumb: *for every* m^2 *of building area ground floor, there is on average 0.4* m^2 *of rooftop area and 0.15* m^2 *of façade area with good solar potential.* With the notations mentioned above, the limitations are expressed as follows:

$$k^r \leq 0.4 \quad k^f \leq 0.15 \tag{5}$$

Both limits do not contain restrictions on (a) the type of panels placed (i.e. PV panels or panels ST) considering that $k = k_{PV} + k_{ST}$, or (b) the type of surface on which they are placed (i.e., roofs or facades), considering that $k = k^r + k^f$.

On the other hand, the solar suitability takes into account the relative amount of irradiation for the surfaces depending on their orientation, inclination, and location as well as the potential performance of the photovoltaic system integrated in the building. A "good" solar yield (Y) for building is understood as 80 % of the maximum local annual solar input, separately defined for slope roofs and façades and individually for each location.

Table 4 Utilization factor for buildings

Panels area	Roof	Façade	Roof and façade
PV panels, $A_{PV} = A_{PV}^{(r)} + A_{PV}^{(f)}$	$k_{PV}^r = A_{PV}^r / A_{GFA}$	$k_{PV}^f = A_{PV}^f / A_{GFA}$	$k_{PV} = k_{PV}^r + k_{PV}^f$
ST panels, $A_{ST} = A_{PV}^{(r)} + A_{PV}^{(f)}$	$k_{ST}^r = A_{ST}^r / A_{GFA}$	$k_{ST}^f = A_{ST}^f / A_{GFA}$	$k_{ST} = k_{ST}^r + k_{ST}^f$
PV and ST panels, $A_{pannels} = A_{PV} + A_{ST}$	$k^r = k_{PV}^r + k_{ST}^r$	$k^f = k_{PV}^f + k_{ST}^f$	$k = k_{PV} + k_{ST}$ or $k = k^r + k^f$

Below, we evaluate the solar energy potential of building based on the following methodology.

Firstly, in the structural design are allowed the following: (a) the basic unit of time is the month, and (b) within a month, the value of any variable does not change. Therefore, the daily value of a variable is obtained by dividing its monthly value by the number of days per month, while the annual value is obtained by summing the monthly values corresponding to the months of the year. If any documentary source gives variable values expressed in other units of time, then these values must be converted into monthly values.

Secondly, once the building and panels features are known,

- the monthly global power production E and heat production Q of the building,
- the monthly specific power production e and heat production q of the panels relative to the *ground floor area*, as well as
- the monthly specific power production e_{PV} and heat production q_{ST} of the panels relative to the *total floor area*

may be obtained with the following relations.

Regarding the building:

$$E = E_r + E_f; \quad Q = Q_r + Q_f \quad [\mathrm{kWh/month}] \tag{6}$$

Regarding the rooftops in relation to ground floor area:

$$\begin{aligned}
E_r &= k_{PV}^r \cdot A_{GFA} \cdot Y_r \cdot \eta_e \cdot H \cdot N_{days} = e_r A_{GFA} \quad [\mathrm{kWh/month}] \\
Q_r &= k_{ST}^r \cdot A_{GFA} \cdot Y_r \cdot \eta_t \cdot H \cdot N_{days} = q_r A_{GFA} \quad (\mathrm{kWh/month}) \\
e_r &= k_{PV}^r . Y_r \cdot \eta_e \cdot H \cdot N_{days} \quad [\mathrm{kWh/m^2_{GFA}/month}] \\
q_r &= k_{ST}^r \cdot Y_r \cdot \eta_t \cdot H \cdot N_{days} \quad [\mathrm{kWh/m^2_{GFA}/month}]
\end{aligned}$$

Regarding the façades in relation to ground floor area:

$$\begin{aligned}
E_f &= k_{PV}^f \cdot A_{GFA} \cdot Y_f \cdot \eta_e \cdot H \cdot N_{days} = e_f A_{GFA} \quad [\mathrm{kWh/month}] \\
Q_f &= k_{ST}^f \cdot A_{GFA} \cdot Y_f \cdot \eta_t \cdot H \cdot N_{days} = q_f A_{GFA} \quad [\mathrm{kWh/month}] \\
e_f &= k_{PV}^f . Y_f \cdot \eta_e \cdot H \cdot N_{days} \quad [\mathrm{kWh/m^2_{GFA}/month}] \\
q_f &= k_{ST}^f \cdot Y_f \cdot \eta_t \cdot H \cdot N_{days} \quad [\mathrm{kWh/m^2_{GFA}/month}]
\end{aligned}$$

The monthly specific power and heat production of the panels relative to *the ground floor area are:*

$$e = e_r + e_f; \quad q = q_r + q_f \quad [\mathrm{KWh/m^2_{GFA}/month}] \tag{7}$$

Regarding the rooftops in relation *to panel area:*

$$\begin{aligned} e^r_{PV} &= Y_r \cdot \eta_e \cdot H \cdot N_{days} \quad [\mathrm{kWh/m^2_{PV}/month}] \\ q^r_{ST} &= Y_r \cdot \eta_t \cdot H \cdot N_{days} \quad [\mathrm{kWh/m^2_{ST}/month}] \end{aligned} \tag{8a}$$

Regarding the façades in relation to panel area:

$$\begin{aligned} e^f_{PV} &= Y_f \cdot \eta_e \cdot H \cdot N_{days} \quad [\mathrm{kWh/m^2_{PV}/month}] \\ q^f_{ST} &= Y_f \cdot \eta_t \cdot H \cdot N_{days} \quad [\mathrm{kWh/m^2_{ST}/month}] \end{aligned} \tag{8b}$$

In addition, between e_{PV} and q_{ST} components of the specific production of the panels relative to *the total floor area*, on the one hand, and the components e^r_{PV}, e^f_{PV}, q^r_{ST}, and q^f_{ST} of the specific production of panels, on the other hand, there are the following relations:

$$e_{PV} = e^r_{PV} \cdot \frac{A^r_{PV}}{A_{TFA}} + e^f_{PV} \cdot \frac{A^f_{PV}}{A_{TFA}} \tag{9a}$$

$$q_{ST} = q^r_{ST} \cdot \frac{A^r_{ST}}{A_{TFA}} + q^f_{ST} \cdot \frac{A^f_{ST}}{A_{TFA}} \tag{9b}$$

where:

- A^r_{PV}, A^f_{PV}—is area of the roof and façade PV panels [m^2];
- A^r_{ST}, A^f_{ST}—is area of the roof and façade ST panels [m^2];
- A_{GFA}—is the ground floor area [m^2],
- A_{TFA}—is the total floor area [m^2],
- η_e, η_t—are the overall PV/ST panels efficiency,
- H is the irradiation received in each monthly, by a fixed surface of 1 m^2, [kWh/m^2],
- N_{days}—number of days in month,
- Y_r,Y_f—are the solar yield for rooftops/façades, defined as the average over all rooftop/façade surface orientations,
- e_r, e_f—are the monthly specific power production of rooftops/façades, $[\mathrm{kWh/m^2_{GFA}}]$,
- q_r, q_f—are the monthly specific heat production of rooftops/façades, $[\mathrm{kWh/m^2_{GFA}}]$,
- e_{PV}, q_{ST}—are the monthly specific power production, and heat production of the panels, $[\mathrm{kWh/m^2_{TFA}/month}]$,
- e^r_{PV}, q^r_{ST}—are the monthly specific power production, and heat production of the panels installed on the roof, $[\mathrm{kWh/m^2_{PV}/month}]$, $[\mathrm{kWh/m^2_{ST}/month}]$,
- e^f_{PV}, q^f_{ST}—are the monthly specific power production, and heat production of the panels installed on the facade, $[\mathrm{kWh/m^2_{PV}/month}]$, $[\mathrm{kWh/m^2_{ST}/month}]$.

Thirdly, if the values of *the utilization factors* k, $k^{(r)}$, and $k^{(f)}$ are the maximum and satisfy the limitations (5) the global and specific production, given by the relations (6), (7), and (8a, b) represents *the energy potential of the building*. If the values of k, $k^{(r)}$, and $k^{(f)}$ satisfy the limitations (5) and correspond to the panels actually installed, the global and specific production given by the relations (6), (7) and (8a, b) represent *the full load of the panels*.

For example, to Galati city the PVGIS estimates of the long-term monthly averages, are given in the Table 5:

As illustrative exercise, let us consider a residential building with ground floor area A_{GFA} = 100 m^2, located in Galati city. We installed, on the roof, both the photovoltaic panels with area $A^r_{PV} = 22.7$ m^2 and efficiency $\eta_e = 0.15$, and the solar thermal panels with area $A^r_{ST} = 16.3$ m^2 and efficiency $\eta_t = 0.7$, for temperature difference 60 °C. On the roof, the solar yield is $Y_r = 0.8$.

The result is $k^r_{PV} = 0.227$ and $k^r_{ST} = 0.163$. The limitations (5) are satisfied because:

$$k^{(r)} = k^{(r)}_{PV} + k^{(r)}_{ST} = 0.39 < 0.4$$
$$k^{(f)} = 0 < 0.15$$

Table 5 Monthly global irradiation for Galati city [4]

Month	H_h	H_{opt}	I_{opt}	T_D	T_{24h}
January	1,470	1,830	62	−1.0	−1.4
February	2,310	3,050	57	2.0	1.4
March	3,440	4,410	46	5.9	5.2
April	4,940	5,410	31	12.2	11.3
May	5,940	6,260	17	18.4	17.2
June	6,690	6,260	12	22.0	21.1
July	6,560	6,500	16	24.5	23.4
August	5,750	6,310	27	23.8	22.6
September	4,390	5,140	41	18.3	17.3
October	2,970	3,960	55	13.1	12.1
November	1,640	2,360	63	7.0	6.3
December	1,190	1,850	63	0.7	0.3
Year	3,950	4,450	35	12.2	11.4

where

H_h Irradiation on horizontal plane [Wh/m^2/day]
H_{opt} Irradiation on optimally inclined plane [Wh/m^2/day]
I_{opt} Optimal inclination [degree]
T_D Average daytime temperature [°C]
T_{24h} 24 h average of temperature [°C]

Table 6 Monthly specific power and heat production of the panels and building (illustrative exercise)

Month	Horizontal solar daily radiation, H_h, ($kWh/m^2/day$)	Monthly specific power and heat production of the panels		Monthly specific power and heat production of the building	
		e^r_{PV} (kWh/m^2_{panel})	q^r_{ST} (kWh/m^2_{panel})	e_{PV} (kWh/m^2_{TFA})	q_{ST} (kWh/m^2_{TFA})
January	1.47	5.47	25.52	1.24	4.14
February	2.31	7.76	36.22	1.76	5.88
March	3.44	12.80	59.72	2.91	9.69
April	4.94	17.78	82.99	4.04	13.47
May	5.94	22.10	103.12	5.02	16.73
June	6.69	24.08	112.39	5.47	18.24
July	6.56	24.40	113.88	5.55	18.48
August	5.75	21.39	99.82	4.86	16.20
September	4.39	15.80	73.75	3.59	11.97
October	2.97	11.05	51.56	2.51	8.37
November	1.64	5.90	27.55	1.34	4.47
December	1.19	4.43	20.66	1.01	3.35
Annually	*3.95*	*172.97*	*807.18*	*39.31*	*130.98*

Table 6 presents the monthly specific power and heat production of both the panels and building, calculated with relations (6)–(9a, b).

Table 7 Monthly global AC energy production of the building (illustrative exercise)

Month	No. days	Solar radiation, I ($kWh/m^2/day$)	Specific DC energy, e ($kWh/m^2_{TFA}/month$)	Global AC energy $E = 0.8 \cdot e \cdot A$ (kWh/month)
January	31	1.47	1.24	99
February	28	2.31	1.76	141
March	31	3.44	2.91	233
April	30	4.94	4.04	323
May	31	5.94	5.02	402
June	30	6.69	5.47	438
July	31	6.56	5.55	444
August	31	5.75	4.86	389
September	30	4.39	3.59	287
October	31	2.97	2.51	201
November	30	1.64	1.34	107
December	31	1.19	1.01	81
Annually	*365*	*3.95*	*39.31*	*3,144*

In addition, based on a value of 80 % for the PV system performance to convert DC in AC electricity we can determinate the AC energy produced from solar energy. It is shown in Table 7.

As example, choosing PV panels, with nominal power of 250 W/m^2_{panel} for roof area of 22.7 m^2, will result 14 panels with total installed capacity of 3.5 kWp.

4 System Structural Modeling

4.1 Building the General Structural Model

In Sects. 1, 2 and 3 were described the first three steps of the design process. As a result, we now have complete information on the context in which will be realized the mCCHP system. Thus: (a) we know that the system will operate conceptually in accord with the scheme shown in Fig. 1, (b) we have relevant data and information on the building that will become a residence because the mCCHP system will be incorporated into it, (c) we know what are the functional needs of residence, which the system must satisfy, (d) we know which are the energy sources of residence (both renewable and nonrenewable), that could be considered in designing the system, and finally (e) we know that, according to the business plan of the manufacturer, the design approach is open-ended.

In this paragraph will be described the fourth step of the design process, in which the designer must perform two actions that lead to the development of a set of possible structural models of the system.

The first action consists in construction of the general structural model of system, through customizing the generic structural scheme (shown in Fig. 1). Customizing is done in agreement with the data and information collected in the previous steps.

The second action is to define a set of potential structural models of the system. This action starts from the observation that any structural model is defined by two features of the system, namely: (a) the system structure, (b) the system integration in its energetic environment. Each such feature has several alternatives available.

Any combination of the alternatives that identify the two features of the system represents one of the potential structural models of the system.

The most promising combinations make up what we call the set of potential structural models of the system.

Regarding the general structural model, a typical example is shown in Fig. 9. Here, the functional needs of residence are those shown in Fig. 1, i.e., lighting, ventilation, cooling, heating, elevation, and domestic hot water.

The needs will be met by incorporating into the system of several consumers as home appliance, chiller with mechanical or thermal compression (MCC or TCC), radiator, heater, and domestic facility. These components of the system will be

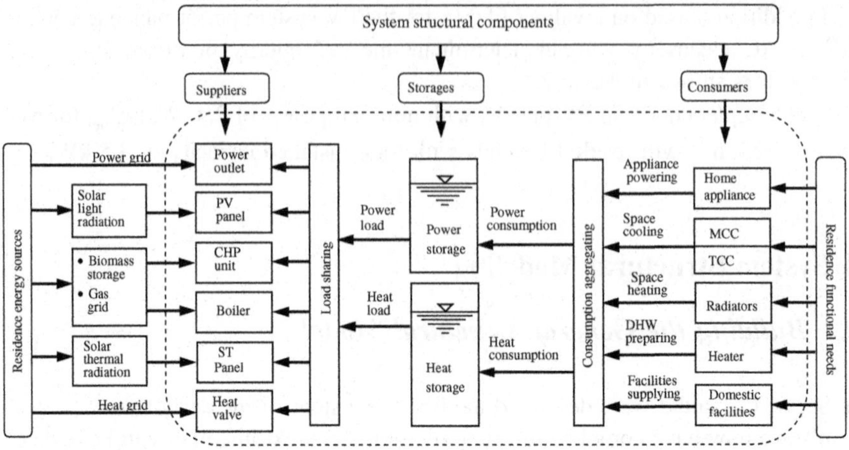

Fig. 9 General structural model of the CCHP system: residence functional needs and energy sources; system structural components: consumers, storages, and suppliers; consumption aggregating and load sharing

powered either by electricity (voltage 220 V) or heat (temperature of thermal agent for 60–90 °C). In this example, the other parameter values of energy or the other forms of energy were excluded when it was determined the search space that defines the conceptual framework.

Consumptions occasioned by the system operation will be aggregated, resulting in heat and power consumption at the system level. To cover these consumptions will be incorporated only two storages, power storage and heat storage, respectively.

On the other hand, each of the two storages will be supplied with energy when the amount of energy contained in the respective storage is below a level considered acceptable. When one of the storages is supplied, the energy will come from some suppliers. What are these suppliers is established through load sharing. In this example it is considered that the suppliers will only be power grid, PV panel, CHP unit (for instance Stirling engine), boiler, ST panel, and heat grid (in agreement with the data and information collected into the previous step) (Table 8).

In their turn, the system suppliers will be supplied from the residence energy sources, which we consider that (in agreement with the data and information obtained in the previous step) could be just these: power, heat, and gas grids, solar irradiation, and biomass (wood pellets).

In this way, using the data and information obtained in the previous step, the generic structural scheme was customized, becoming the general structural model of the system. The aim of the general structural model is that of its being used as base for the elaboration of a set of potential structural models.

Table 8 Set of potential structural models

System feature and feature's alternatives			Number of the structural model						
			1	2	3	4	5	6	7
Structuring strategy	Prime mover	Stirling engine	X	X	X	X	X	X	X
	Additional sources	PVpanels (K_{PV})	X	–	–	X	X	X	–
		ST panels (K_{ST})	X	X	–	–	–	X	–
		Boiler with pellets	X	X	X	X	X	X	–
		Boiler with gas	–	–	–	–	–	–	X
	Storages	Power storage	X	X	X	X	X	X	X
		Heat storage	X	X	X	X	X	X	X
	Consumers	MCC	–	–	–	–	X	X	–
		TCC	X	X	X	X	–	–	X
Integration strategy	Electric grid	Off-grid	X	X	X	X	X	X	–
		On-grid	–	–	–	–	–	–	X
	Thermal grid	Off-grid	X	X	X	X	X	X	–
		On-grid	–	–	–	–	–	–	X
	Gas grid	Off-grid	–	–	–	–	–	–	X
		On-grid	X	X	X	X	X	X	–

4.2 Identifying the Set of Potential Structural Models

For a given general structural model, establishing the potential structural models depends on *the strategy of system structuring and integration.*

- *Strategy of system structuring* refers to the type of the components embedded. Here will be considered only Stirling engine as CHP unit, PV panels, ST panels, boilers, electrical batteries, hot water tank, and two types of chiller, namely with *mechanical compression* and with *thermal compression.*
- *Strategy of system integration* in its energetic environment refers to the system's ability for connecting to the existing energy networks, as well as to the primary energy resources. Depending on the fact that the system receives or supplies energy from or to the energy grids, a large number of cases can occur. Of these, two are typical, and will be considered. These are *the off-grid strategy* to which between system and grids there is no transfer of energy, and *the on grid strategy* to which the electricity can be transferred from the system to the network, and vice versa. As regards the primary energy resources, here will consider that only the solar radiation, gas, and wood pellets are available.

For example, Tab. 8 presents a set of potential structural models, which was established on the basis of the above.

In the next steps, these structural models will be concretized and comparatively analyzed. The ultimate goal of analysis is to obtain an *optimal trade-off* between the *performance indicators* and *investment cost*. Finally, one of these models will be selected to be applied practically.

5 Consumption Estimation

Consumption means the form and quantity of energy consumed in a time unit. Depending on the time unit, the consumption may be annual, monthly, or daily.

Firstly, the consumption may be estimated either at consumer (end-use) level, or at residence level.

At residence level, consumption may be global or specific. Consumption is global when it refers to the whole residence. If the global consumption is divided to the total floor area (TFA) of the residence building, then it becomes specific.

Moreover, the consumption is estimated for each form of energy consumed and each function of the residence covered. For example, the forms of energy considered could be heat and power, while the residence functions covered could be heating, air conditioning, hot water, and domestic facilities. Evaluating of the residence consumption based on the energy form is called consumption aggregating.

Understanding energy uses in the buildings sector is complex because of lack of reliable data on the energy consumption by end-use (e.g., heating, cooling, lighting, IT equipment and appliances), and because of the large variety of building categories in this sector. Subdivision of energy consumption can be particularly difficult in the cases of electricity, where air-conditioners, appliances, lights, pumps, and heating installations all draw electricity and often from the same metering. In residential buildings, the installed equipment and appliances require electrical energy, as do removable devices like mobile phone chargers, laptops, and electric cars in the future. However, identification of fixed and fluctuating demand for electrical energy rarely appears in a building's consumption metric, as single measurement consider only the total amount consumed by the whole building. Natural gas, too, can serve several end uses at once, including heating, cooking, and the provision of domestic hot water. The typically amount of energy required in the residential buildings and the percent of total energy need is illustrated in Fig. 10.

Given the difficulty in subdividing buildings' energy requirements and the use of different fuel types, most analysis examines energy use in building as defined by end-use: space heating, cooling, cooking, etc. The split in use of energy will be less accurate due to uncertainties, and it will vary with different types of buildings and also with the age and use of the buildings.

The importance of consumption for space heating and cooling varies by country and region depending on climate. Most energy in the building sector is used for space and water heating, while the energy consumption for cooling is generally modest. In hot countries, with little or no space heating needs, cooling is much more important.

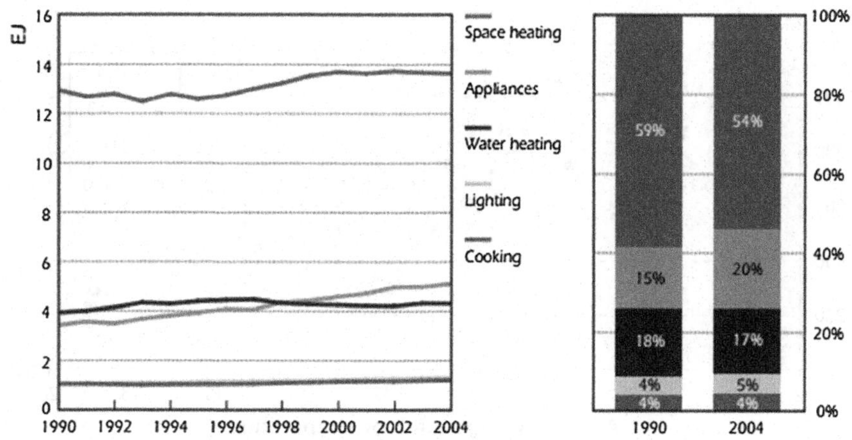

Fig. 10 Energy use in residential buildings [6]

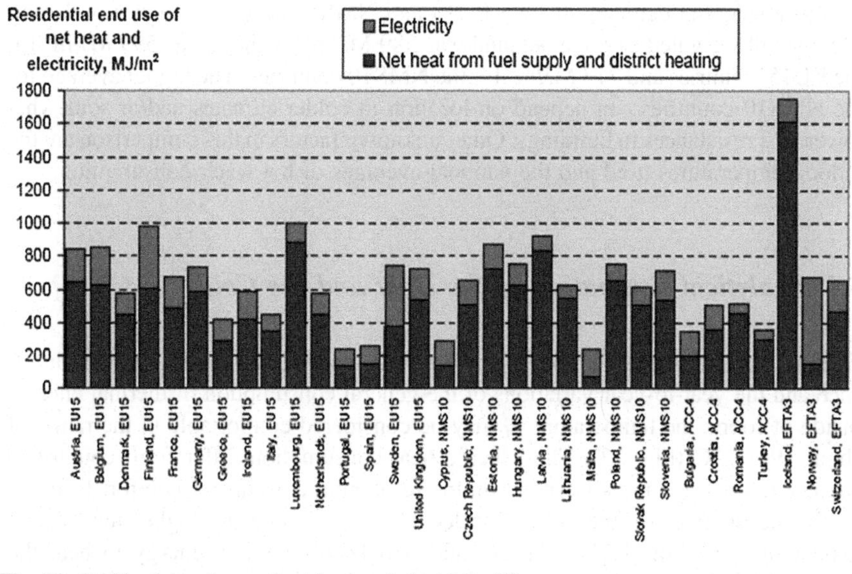

Fig. 11 Residential end use of net heat and electricity [1]

In order to compensate for varying national use of floor areas, the specific use of net heat and electricity per square meter has been estimated for the residential buildings. The total values with division between electricity and net heat are presented in Fig. 11 for the residential sector at EU level.

The corresponding correlations with the national climates are found in Fig. 12. These total values include all end users for space heating, hot water preparation, ventilation, cooling, cooking, lighting, etc.

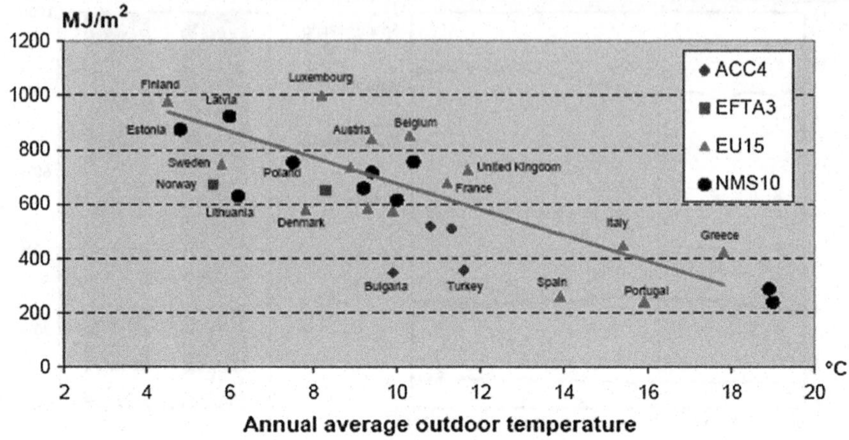

Fig. 12 Residential end use of net heat and electricity versus outdoor temperature [1]

The residential demands for the EU15 and NMS10 countries align now better. The overall weighted average demand was 589 MJ/m^2, while it was 599 MJ/m^2 for the EU15 countries and 713 MJ/m^2 for the NMS10 countries. The higher average for the NMS10 countries can depend on location in colder climates and/or somewhat lower heat resistances in buildings. Other unknown factors in this comparison are the indoor temperatures used and the national averages of hot water consumption.

5.1 Analytical Estimation of the Heat and the Cold

External temperature is the most important variable for explaining the daily influence and the year-to-year variations of the general consumption of thermal energy. Inside of room, the temperature ϑ_i may be considered comfortable in the range of about 19–21 °C (Fig. 13). The specific heat consumption differ from country to country (climate), but also depend on the residential consumers' comfort level.

On the other hand, the colder the local climate is and the higher the interior temperatures are, the higher the demand will be for thermal energy to heat the space. A comfortable air temperature depends on the humidity, the received radiation flux, wind speed, clothing and that person's activity, metabolism, and lifestyle.

The internal built environment should be at such a "comfort temperature", using the minimum artificial heating or cooling, even when the external (ambient) temperature ϑ_e is well outside the comfort range.

The level of residential energy consumption for space heating, regardless of the location of the building, *depends of construction and materials used* in achieving

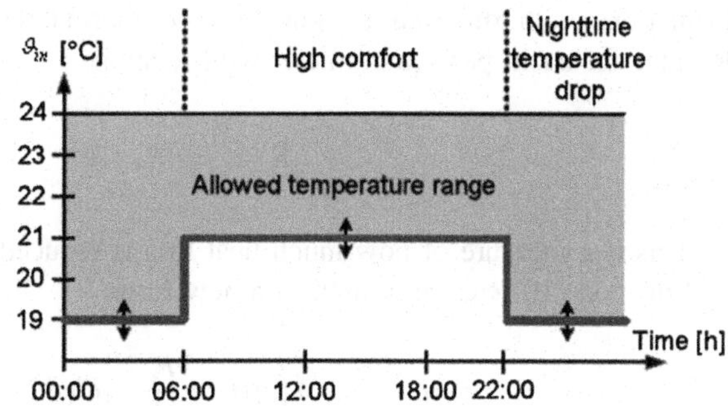

Fig. 13 Room temperature

them. The heat balance in the building inside taking into account the solar input is described by the following equation:

$$mc\frac{d\vartheta_i}{dt} = \alpha\tau GA + P_a - \frac{\vartheta_i - \vartheta_e}{R_w} \tag{10}$$

where:

m lumped building mass (interior walls and floors) (kg);
c specific capacity of the wall material (Wh/kg °C);
ϑ_i room temperature (°C);
α building absorbance;
τ building transmittance;
G– irradiance (W/m^2);
A– surface (m^2);
P_a power heating sources (W);
ϑ_e external temperature (°C);
$R_w = \frac{1}{\lambda}\frac{\delta}{A}$ is the wall resistance to heat loss (°C/W);

where λ is the thermal conductivity and δ is the material thickness.

Thermal mass within the building can absorb some of the heat and release it at night. Internal thermal mass reduces temperature swings within a space because the thermal mass absorbs solar energy during the day, preventing the building from overheating, and releases the energy at night. Thermal mass is most effective when it can gain energy directly from the Sun. An ideal thermal mass for solar heating has high heat capacity, moderate conductance, moderate density, and high emissivity.

Thermal conductivity (known as Lambda, λ, or as k-value) is a standardized measure of how easily heat flows through an actual plate regardless of its material and thickness. The lower thermal conductivity indicates the better thermal performance. It provides a quick way to easily compare to thermal performance of different insulation. Units are Watts per meter Kelvin W/mK. To compare two insulations with different thicknesses and thermal conductivities, calculate the R-value for each, per square meter with relation:

$$R = \frac{\delta}{\lambda} = R_w \cdot A \tag{11}$$

This is a measure of how much heat loss is reduced.

Equation (10) can be written in a new form:

$$mc\frac{d\vartheta_i}{dt} = A\left(\alpha\tau G + \frac{P_a}{A} - U(\vartheta_i - \vartheta_e)\right) \tag{12}$$

where

$U = \frac{1}{R_w \cdot A} = \frac{\lambda}{\delta}$ is the 'overall heat loss coefficient or U -value' (W/°C m^2).

U-value is a measure of how much heat is lost through a given thickness of any specific material, which includes conduction, convection, and radiation. The U-value of a material (or several materials in series, e.g., brick and insulation in a wall) is calculated by taking the reciprocal of the R-value (i.e., 1/R-value), and adding convection and radiation heat losses. The lower U-value indicates the better thermal performance.

The thermal capacity of the building's elements delays the heat transfer to the interior of the building, by soaking up excessive heat for several hours. During the night, when the external temperature is lower, the stored heat is slowly expelled to the environment by radiation and by convection. The time delay due to the thermal mass is known as time lag. The thicker the walls are and more resistive the material is, the longer it will take for heat waves to pass through building walls. The reduction in cyclical temperature on the inside surface compared to the outside surface is knows as the decrement factor.

Time lag and decrement factor are very important characteristics to determine the heat storage capabilities of any material. The time it takes for the heat wave to propagate from the outer surface to the inner surface is named as "time lag" and the decreasing ratio of its amplitude is named as "decrement factor". The schematics of time lag and decrement factor are shown in Fig. 14. At the cross-section of the outer wall of a building, there are different temperature profiles during any instant of a 1-day period. These profiles are function of inside temperature, outside temperature, and materials of the wall layers.

During this transient process, a heat wave flows through the wall from outside to inside and the amplitude of this wave shows the temperature magnitudes and wavelength shows the time. The amplitude of the heat wave on the outer surface of the wall is based on solar radiation and convection in between the outer surface of the wall and ambient air. During the propagation of this heat wave through the wall, its amplitude will decrease depending on the material and the thickness of the wall

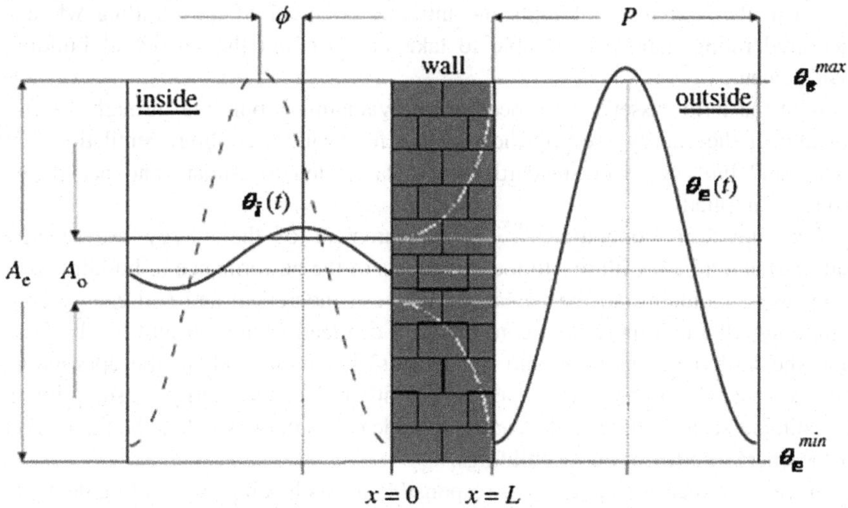

Fig. 14 Explanation of time lag and decrement factor on the heat transfer [7]

It follows that reducing the need for heating/cooling of a residential house can be done by increasing the thickness of insulation and the use of materials with low U-value.

5.1.1 Analytical Estimation Based on the Heat Transfer Coefficient of Building

The thermal performance of a building refers to the process of modeling the energy transfer between the building and its surroundings. For a conditioned building, it estimates the heating and cooling load, and hence the sizing and selection of HVAC equipment can be correctly made. For a non-conditioned building, it calculates temperature variation inside the building over a specified time, and helps one to estimate the duration of uncomfortable periods. These quantifications enable one to determine the effectiveness of the design of a building, and help in evolving improved designs for realizing energy efficient buildings with comfortable indoor conditions. The lack of proper quantification is one of the reasons why the passive solar architecture is not popular among architects. Occupants would like to know how much energy might be saved, or the temperature reduced to justify any additional expense or design change. Architects too need to know the relative performance of buildings to choose a suitable alternative. Thus, knowledge of the methods of estimating the performance of buildings is essential to the design of buildings. EN 15217 provides methods that are required to express the energy performance of buildings. The certificate can be based on either the 'measured' (or "operational") or the 'calculated' (or "asset") rating.

A calculated rating highlights the intrinsic potential of the building while a measured rating makes it possible to take into account the impact of building management.

With the term "asset", the procedure of evaluating a building through the calculation of the energy used by the building for heating, cooling, ventilation, hot water, and lighting, with standard input data related to climate and occupancy modes is implied.

With the term "operational", the measurement of the energy use (energy auditing) is implied, with in situ measurements of the performance of building as is being used, including all the deviations between theoretical and real energy consumption. Although more time consuming, it can lead to more useful feedback on both the building conditions and the tenants' behavior, and on the appropriate measures that should be taken, so as to efficiently reduce the energy consumption in the building sector (and thus the carbon dioxide emissions), which is the final target of the energy certification of buildings.

Since the building's envelope is responsible for its heating and cooling demand, the estimation of the heat transfer coefficient of building is a very important method for evaluating a building's energy performance.

A. Finding the calculated (asset) overall heat transfer coefficient of the building
The building heat loss coefficient (UA) is found by identifying every route of heat loss from a building and adding these together. The routes for heat transfer between the interior and exterior are through the building envelope, air exchange between inside and outside (infiltration), and sometimes through the ground (perimeter), depending on how the building meets the ground (Fig. 15).

Fuelling the heating space has to compensate for the losses of heat transmitted through walls and the roof, and also the losses given by the heated air from the mechanical or natural ventilation systems.

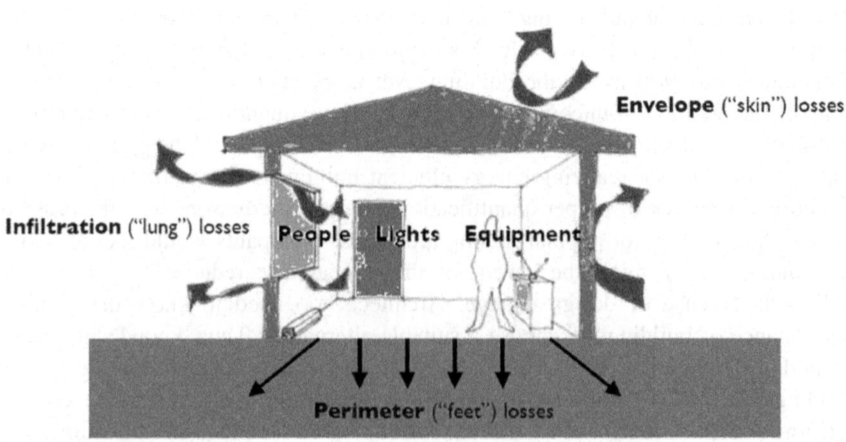

Fig. 15 Building heat transfer

Temperature ϑ_b (or the base temperature) is determined from relation:

$$\dot{Q}_{\text{losses}} = UA_{\text{ref}} \cdot (\vartheta_i - \vartheta_b) = \dot{Q}_{\text{gain}} \quad (13)$$

where

- UA_{ref} is the overall heat transfer coefficient of the building in W/°C;
- Standard value $\vartheta_i = 20\,°\text{C}$;
- $\dot{Q}_{\text{losses}}$ is the heat loss of building and is given by the equation:

$$\dot{Q}_{\text{losses}} = \dot{Q}_{\text{envelope}} + \dot{Q}_{\text{airchanges}} + \dot{Q}_{\text{perimeter}} \quad (14)$$

The terms of equation are:

$$\dot{Q}_{\text{envelope}} = \text{UA}_{\text{envelope}}(\vartheta_{\text{i}} - \vartheta_{\text{e}}) - \text{Envelope loss ("skin")} \quad (15)$$

$$\dot{Q}_{\text{Air Changes}} = \text{UA}_{\text{Air Changes}}(\vartheta_{\text{i}} - \vartheta_{\text{e}}) - \text{Infiltration loss ("lungs")} \quad (16)$$

$$\dot{Q}_{\text{perimeter}} = \text{UA}_{\text{perimeter}}(\vartheta_{\text{i}} - \vartheta_{\text{e}}) - \text{Perimeter loss ("feet")} \quad (17)$$

- $\dot{Q}_{gain}$ is the heat generated in the building and is given by the equation:

$$\dot{Q}_{\text{gain}} = \dot{Q}_{\text{people}} + \dot{Q}_{\text{lights}} + \dot{Q}_{\text{equipments}} \quad (18)$$

This results in balance point temperature:

$$\vartheta_b = \vartheta_i - \frac{\dot{Q}_{\text{losses}}}{\text{UA}_{\text{ref}}} \quad (19)$$

There is considerable uncertainty associated with the determination of the balance point temperature, but based on the observations of typical buildings, it is usually taken to be 19–21 °C in Europe.

The rate of energy consumption of the heating system is

$$\dot{Q}_h = \frac{\text{UA}_{\text{ref}}}{\eta_h} \cdot (\vartheta_b - \vartheta_i) \quad (20)$$

Expressing the design energy consumption of a building for heating as:

$$\dot{Q}_{\text{design}} = \frac{\text{UA}_{\text{ref}}}{\eta_h} \cdot (\vartheta_i - \vartheta_e)_{\text{design}} \tag{21}$$

Comparing it to the annual energy consumption gives the following relation between energy consumption at design conditions and the annual energy consumption:

$$\frac{\dot{Q}_{\text{heating,year}}}{\dot{Q}_{\text{design}}} = \frac{\text{HDD}}{(\vartheta_i - \vartheta_e)_{\text{design}}} \tag{22}$$

where
$(\vartheta_i - \vartheta_e)_{\text{design}}$ is the design indoor–outdoor temperature difference.

The annual energy need for heating can be determined by integration (or by summation over daily or hourly averages) as follows:

$$Q_{\text{heating,year}} = \text{UA}_{\text{ref}} \cdot \int (\vartheta_b - \vartheta_i)\text{d}t \approx 0.024 \cdot \text{UA}_{\text{ref}} \cdot \text{HDD}\ [\text{kWh/year}] \tag{23}$$

Similarly calculated is the annual cold consumption, Q_C, the difference resides only in that the formula for calculation becomes the following:

$$Q_C = 0.024 \cdot \text{UA}_{\text{ref}} \cdot \text{CDD} \quad [\text{kWh/year}] \tag{24}$$

The total building heat loss coefficient UA_{ref} is given by relation:

$$\text{UA}_{\text{ref}} = \text{UA}_{\text{envelope}} + \text{UA}_{\text{AirChange}} + \text{UA}_{\text{perimeter}} \tag{25}$$

The conductive heat transfer through all elements of the building envelope from one side to the other (walls, floor, roof, windows, doors, etc.) is given by the following relationship:

$$\text{UA}_{\text{envelope}} = U[W/(m^2K)] \cdot A[m^2] \tag{26}$$

To calculate the UA value for whole building, may find in Table 9 the U-value for each element of the building envelope, multiply this value by the area of element, and then summing the UA values thus obtained.

Infiltration is the exchange through leaks, cracks, and ventilation. This represents a loss of energy through the replacement of conditioned interior air with (colder)

Table 9 References U value required for residential buildings [8]

Envelope units	U′max [W/(m² K)] maximum thermal transmittances corrected with the influence of the thermal bridges[a]		U′max [W/(m² K)] maximum thermal transmittances corrected with the influence of the thermal bridges[b]
	Newly built	Existing to be renovated	Newly built after 1998
External walls	0.57	0.71	0.71
External windows	1.30	2.50	2.00
Terraces	0.20	0.33	0.33
Floors of unheated basements	0.35	0.60	0.60
Ground floors (no basements)	0.22	0.33	0.22
Floors of heated basements	0.21	0.24	0.20
External walls of heated basements	0.35	0.50	0.41

[a] C107/2010
[b] C107/1997

outside air. Overall heat transfer coefficient for infiltration can be determined with equation:

$$\mathrm{UA}_{\mathrm{AirChange}n} = \mathrm{VHC}_{\mathrm{air}}\left[Wh/m^3K\right] \cdot V\left(m^3\right) \cdot \mathrm{ACH}[no/h] \tag{27}$$

where:

- V is the volume of enclosed space in building (m^3);
- ACH is the Air Changes per Hour usually estimated by assessing the relative tightness of the building (Table 10);
- VHCis the Volumetric Heat Capacity of air, which is equivalent to the specific heat, or ability of air to store energy (like thermal mass). The volumetric heat capacity is defined as having SI units of J/m³ K.

Perimeter heat transfer is applicable[1]to that house, which has direct contact with the ground via a slab-on-grade. The "UA" is a function of the length of the building's perimeter [m] and F, which is an empirically derived constant (Tables 11and 12). Overall heat transfer coefficient for perimeter can be determined with equation:

$$\mathrm{UA}_{\mathrm{perimeter}} = F(W/mK) \cdot P(m) \tag{28}$$

[1] Is not applicable when the house/apartments are not built with the floor in contact with the earth.

Table 10 Estimated infiltration rates (ACH) for small buildings

Part B. Design infiltration rates (ACH) for winter: indoors 68 °F (20 °C)*; wind speed* = 15 mph (6.7 m/s)

Construction type	Winter outdoor design temperature									
	°F:50	*40*	*30*	*20*	*10*	*0*	*−10*	*−20*	*−30*	*−40*
	°C:10	*4*	*−1*	*−7*	*−12*	*−18*	*−23*	*−29*	*−34*	*−40*
Tight	0.41	0.43	0.45	0.47	0.49	0.51	0.53	0.55	0.57	0.59
Medium	0.69	0.73	0.77	0.81	0.85	0.89	0.93	0.97	1.00	1.05
Loose	1.11	1.15	1.20	1.23	1.27	1.30	1.35	1.40	1.43	1.47

Part C. Design infiltration rates (ACH) for summer: indoors 75 °F (24 °C)*; wind speed* = 7.5 mph (3.4 m/s)

Construction type	Summer outdoor design temperature					
	°F:85	*90*	*95*	*100*	*105*	*110*
	°C:29	*32*	*35*	*38*	*41*	*43*
Tight	0.33	0.34	0.35	0.36	0.37	0.38
Medium	0.46	0.48	0.50	0.52	0.54	0.56
Loose	0.68	0.70	0.72	0.74	0.76	0.78

Part D. Infiltration rates per unit floor area

Ceiling height	Air flow	Air changes per hour																	
		0.3	0.4	0.5	0.6	0.7	0.8	0.9	1.0	1.1	1.2	1.3	1.4	1.5	1.6	1.7	1.8	1.9	2.0
2.3 m	L/s m^2	0.20	0.25	0.31	0.41	0.46	0.51	0.56	0.66	0.71	0.76	0.81	0.91	0.97	1.02	1.07	1.17	1.22	1.27
2.4 m	L/s m^2	0.20	0.25	0.36	0.41	0.46	0.56	0.61	0.66	0.76	0.81	0.86	0.97	1.02	1.07	1.17	1.22	1.32	1.37
2.6 m	L/s m^2	0.20	0.31	0.36	0.46	0.51	0.56	0.66	0.71	0.81	0.86	0.91	1.02	1.07	1.17	1.22	1.32	1.37	1.42
2.7 m	L/s m^2	0.25	0.31	0.41	0.46	0.56	0.61	0.71	0.76	0.86	0.91	1.02	1.07	1.17	1.22	1.32	1.37	1.47	1.52

Table 11 Empirically constant [9]

R-Value, position, and width (or depth) of insulation	F2	
	Btu/h ft °F	W/m K
Uninsulated slab	0.73	1.26
R-5 (SI:R-0.88) horizontal insulation, 2 ft (0.6 m), no thermal break	0.70	1.21
R-10 (SI:R-1.76) horizontal insulation, 2 ft (0.6 m), no thermal break	0.70	1.21
R-15 (SI:R-2.64) horizontal insulation, 2 ft (0.6 m), no thermal break	0.69	1.19
R-5 (SI:R-0.88) horizontal insulation, 4 ft (1.2 m), no thermal break	0.67	1.16
R-10 (SI:R-1.76) horizontal insulation, 4 ft (1.2 m), no thermal break	0.64	1.11
R-15 (SI:R-2.64) horizontal insulation, 4 ft (1.2 m), no thermal break	0.63	1.09
R-5 (SI:R-0.88) vertical insulation, 2 ft (0.6 m)	0.58	1.00
R-10 (SI:R-1.76) vertical insulation, 2 ft (0.6 m)	0.54	0.93
R-15 (SI:R-2.64) vertical insulation, 2 ft (0.6 m)	0.52	0.90
R-5 (SI:R-0.88) vertical insulation, 4 ft (1.2 m)	0.54	0.93
R-10 (SI:R-1.76) vertical insulation, 4 ft (1.2 m)	0.48	0.83
R-15 (SI:R-2.65) vertical insulation, 4 ft (1.2 m)	0.45	0.78
R-10 (SI:R-1.76) fully insulated slab	0.36	0.62

Table 12 Empirically constant perimeter [9]

Construction	Insulation	Btu/h °F ft perimeter degree days (65 °F base)			W/K m perimeter degree days (18 °C base)		
		2,950	5,350	7,433	1,640	2,970	4,130
(a) Block wall, 8 in. (200 mm), brick facing	Uninsulated	0.62	0.68	0.72	1.07	1.17	1.24
	Insulated from edge to footer: (SI: R-0.95 m^2K/W)	0.48	0.50	0.56	0.83	0.86	0.97
(b) Block wall, 4 in. (100 mm), brick facing	Uninsulated	0.80	0.84	0.93	1.38	1.45	1.61
	Insulated from edge to footer: (SI: R-0.95 m^2K/W)	0.47	0.49	0.54	0.81	0.85	0.93
(c) Metal stud wall, stucco	Uninsulated	1.15	1.20	1.34	1.99	2.07	2.32
	Insulated from edge to footer (SI: R-0.95 m^2K/W)	0.51	0.53	0.58	0.88	0.92	1.00
(d) Poured concrete wall with duct near perimeter	Uninsulated	1.84	2.12	2.73	3.18	3.67	4.72
	Insulated from edge to footer: (SI: R-0.95 m^2K/W)	0.64	0.72	0.90	1.11	1.24	1.56

Table 13 Steps in building heat loss coefficient

Step 1: *envelope heat transfer coefficient* (*UA conduction*)					
Building element		U-value	Area	U × area	Description
Orientation	Element	kW/ m^2 °C	m^2	kW/°C	
South	Walls				
	Window				
East	Walls				
	Window				
North	Walls				
	Window				
West	Walls				
	Window				
	Ceiling/roof				
UA_{envelope} (kW/°C) = $U \times A$ (kW/°C)					
Step 2: *Infiltration heat loss* (*"UA" infiltration*)					
Volumetric heat capacity (air)		kWh/m^3 °C			
Building volume		m^3			
Air changes per hour (ACH)		1/h			
UA infiltration = VHC × Volume × ACH [kW/°C]					
Step 3: *perimeter heat loss* (*"UA" perimeter*)					
Heat loss coefficient (F)		kW /m °C			
Length of building perimeter (P) [m]					
UA perimeter = F × P			kW/ °C		
UAref			kW/ °C		

The steps in building heat loss coefficient (UA) are shown in Table 13.

B. Finding the measured overall heat transfer coefficient of the building
Thermographic survey techniques, based on infrared cameras, are not a recent development and are "operational" rating. There are really two essential elements for achieving a good infrared audit, namely, the high performance instrument and the technical skill of the infrared auditor. Analysis is very useful for evaluating the building's energy performance, both for its envelope and its facilities. In fact it could lead to identification of many energy problems such as bad final design, construction, installation, or building malfunctions. For example, the knowledge of energy performance of existing buildings could render feasible their respective retrofitting activities, while in new buildings it is possible to check the accuracy of the "as-built" construction compared to the project details. It is also used to verify the presence of air seepage, moisture, or water leaks.

In order to correctly define the performance of a building envelope, it is very important to know, in addition to the characteristics of the building materials, the reflected ambient temperature, as well as the weather conditions (outside temperature trend, relative humidity, rainfall intensity, wind direction and velocity, solar radiation, etc.).

Thermography can detect surface temperature variations as small as 0.08 K and graphic images can be produced that visibly illustrate the distribution of temperature on building surfaces. Variations in the thermal properties of building structures, such as poorly fitted or missing sections of insulation, cause variations in surface temperature on both sides of the structure. They are therefore visible to the thermograph. However, many other factors, such as local heat sources or reflections and air leakage, can also cause surface temperature variations. The professional judgment of the thermographer is usually required to differentiate between real faults and other sources of temperature variation. Increasingly, thermographers are asked to justify their assessment of building structures and, in the absence of adequate guidance, it can be difficult to set definite levels for acceptable or unacceptable variation in temperature.

In order to calculate the density heat flux Φ ([W/m^2]) between two environments separated by a wall, in the steady state conditions, it is possible to use equation:

$$\Phi = U_w \cdot (\theta_i - \theta_e) \tag{29}$$

where:

U_w (W/m^2 K) is the U-value of the wall, and

θ_i and θ_e, expressed in K are respectively the internal and the external air temperature.

The density heat flux Φ can be calculated also as a function of the external (convective) heat transfer coefficient h_{we} (W/m^2 K) and the difference between the external surface temperature of the wall θ_{we} and the external air temperature θ_e.

$$\Phi = h_{we} \cdot (\theta_{we} - \theta_e) \tag{30}$$

Combining Eqs. (29) and (30), it is possible to estimate the U-value of the wall U_w as a function of the coefficient h_{we}, the internal air temperature θ_i, the external air temperature θ_e, and the external surface temperature of the wall θ_{we} with relation:

$$U_w = h_{we}\frac{(\theta_{we} - \theta_e)}{(\theta_i - \theta_e)} \tag{31}$$

Knowing the values of the external surface temperature of the wall, derived indirectly from the thermal map, the value of the internal air temperature and the value of the external air temperature, it would then be possible to obtain, in a simple way, the U_w-value of the wall in question.

This evaluation is only theoretical as the heat transmission in reality does not take place in a steady-state condition, and the external surface temperature of the wall could be affected by the phenomena of thermal transition due to the thermal inertia. Other factors could render unreliable the values of the thermal map generated by the thermographic survey, for example, the presence of moisture in the wall, radiant heat exchange, or the effect of solar radiation.

The effect due to the presence of moisture is controlled by identifying, in the thermal mapping provided by the infrared camera, a homogeneous area of the wall in which to evaluate the average value of the surface temperature, while the effect due to solar radiation is controlled by choosing walls with an orientation that, in the winter season, is not subject to the sunshine. The effect of the non-steady state heat flow condition can be partially reduced by identifying periods in which to perform the measurements with external climatic stability.

The main factors affecting the thermographic study relevance during a building envelope thermal behavior inspection [10] are shown below:

- *Climatic conditions*: insulation, wind, ambient temperature, relative humidity, greenhouse gases concentration (water vapors, CO_2).
- *Pattern characteristics*: emissivity/reflectivity, roughness or unevenness, stains and color of wall surface; construction of wall finish (e.g., extremely thick finish).
- *Environmental conditions and deficiencies*: angle of vision and survey distance, orientation of building to the path of sunshine during the survey, existence of any heat generating plants or machines inside the building; screening objects (e. g., trees, shade of eaves, or adjacent building).

Climatic conditions. At the interface between air and wall heat flux by conduction is equal to the convective heat flux. From Eq. (30) it is possible to observe the significant influence of the convective coefficient h_{we}. Its value, in fact, has an effect proportionally to the U_w value of the U-value.

In the standard calculations of the external heat transfer coefficient, pre-calculated values or simple equations are provided by the technical standards, such as the ISO 6946 standard. The convective coefficient is used to calculate the heat loss during the design phase of the building envelope and have ISO value, equal to 25. In reality, the convective coefficient is function of the wind speed. For calculate the convective coefficient, the Jurges equation [11] is given as follows:

$$h_{\mathrm{we}} = 5.8 + 3.805 \cdot v_w \text{ with } v_w < 5\,\mathrm{m/s} \tag{32}$$

where:

h_{we} is convective heat transfer coefficient;
v_w is wind velocity near the building element

For three different values [12] of wind sped the h_{we} coefficient have the following values:

Table 14 Emissivity coefficients of the most common construction materials

Material	ε
Concrete	0.94
Sand	0.76
Brick	0.75–0.80
Limestone	0.95
Plaster	0.90–0.96
Glass	0.90–0.96
Wood	0.80–0.90
Roofing felt	0.93
Gypsum	0.80–0.90
Paints (all colors')	0.90–0.96
Clay	0.95
Brick earth	0.93

$$h_{we} = \begin{cases} = 5.8 & \text{for} \cdot v_w = 0\,\text{m/s} \\ = 9.68 & \text{for} \cdot v_w = 1\,\text{m/s} \\ = 24.8 & \text{for} \cdot v_w = 5\,\text{m/s} \end{cases} \tag{33}$$

We observe that if the h_{we} value is high, the temperature differences are the smallest obtainable, and the order of magnitude of these values is comparable to the margin of error of the measurement instruments, thus not detectable even with high-level instrumentation. On the other hand if the coefficient stands at lower values, the temperature difference between the surfaces may be higher, and the possibility of error is reduced. In addition to this we must consider the error introduced by not knowing the exact value of emissivity, as well as the mean radiant temperature, air temperature, distance, humidity, all of which are parameters necessary for the equipment to make the correct conversion of the radiance measured in temperature value.

Pattern characteristics. Different tests [13] using four distinct values of emissivity (0.62, 0.85, 0.91 and 0.95) showed that emissivity (Table 14) variation induced changes in the thermal images, both during the absorption and drying periods.

Thermal images obtained with an emissivity of 0.62 were different from the other images, but differences among the other thermograms (emissivity's 0.85, 0.91 and 0.95) were not very significant.

Environmental conditions. To achieve best results from a thermographic survey it is important to consider the environmental conditions and to use the most appropriate thermographic technique for the task.

Thermal anomalies will only present themselves to the thermographer where temperature differences exist and environmental phenomena are accounted for. Each type of thermographic survey has its own specific environmental requirements. For example, insulation continuity and thermal bridging surveys require temperature difference across the building fabric to be greater than 10 °C to be maintained. In addition to temperature, there are other environmental conditions that should also be taken into account when planning a thermographic building survey. External inspections for example, may be influenced by radiation emissions and reflections from adjacent buildings or a cold clear sky, and even more significantly the heating effect that the Sun may have on surface.

For calculation of the overall heat transfer coefficient (U-value) in building envelopes with infrared thermography the following conditions must be met [14]:

- The minimum test duration is 72 h if the temperature is stable around the flow meter, and to be sure that a sufficient heat flow is present.
- The minimum difference between temperatures has to be of the order of 10–15 °C.
- For at least the 24 h before the beginning of the infrared audit and during its execution the following relationship must be respected:

$$(\theta_i - \theta_e) > 3/\mathrm{U} \tag{34}$$

where: U is the value (W/m^2 K) of the element being considered.

However, $\Delta\theta = (\theta_i - \theta_e)$ has to be at least 5 °C.

- For at least the 12 h before the beginning of the infrared audit and anyway during its execution, the analyzed building surfaces should not be exposed to direct solar radiation.
- During the thermographic analysis the exterior and the interior temperature must not vary by more than ± 5 °C and ± 2 °C, respectively.

5.1.2 Analytical Estimation Based on the Global Coefficient of Building's Thermal Isolation

For heating, the annual consumption can be calculated with the formula:

$$Q = 0.024 \frac{G \cdot V}{\eta_{\text{heating}}} \text{HDD} \quad (\text{kWh/y}) \tag{35}$$

where:

η_{heating} efficiency of the heating device;

V heated volume of the building (m^3);

HDD annual heating degree-days;

G global coefficient of the building's thermal isolation (which represents the sum of heat losses resulting from direct transmission through the area of the building's envelope, for a difference in temperature of 1 °K between the interior and the exterior, related to the volume of the building, to which the heat losses due to refreshing the interior air and to the

Table 15 Number of hourly air exchanges n [1/h] in residences

Building category		Sheltering class	Permeability class		
			High	Medium	Low
Individual buildings (one family houses, duplex buildings or series of buildings, etc.)		Non-sheltered	1.5	0.8	0.5
		Moderately sheltered	1.1	0.6	0.5
		Sheltered	0.7	0.5	0.5
Buildings with more apartments, hostels etc.	Double exposure	Non-sheltered	1.2	0.7	0.5
		Moderately sheltered	0.9	0.6	0.5
		Sheltered	0.6	0.5	0.5
	Simple exposure	Non-sheltered	1.0	0.6	0.5
		Moderately sheltered	0.7	0.5	0.5
		Sheltered	0.5	0.5	0.5

The notions in the table have the following significations:
Sheltering class
non-sheltered very tall buildings/in the outskirts of towns /in squares
moderately sheltered buildings inside the towns
sheltered buildings at the center of towns/in woods
Permeability class
high buildings with exterior woodwork without insulation measures
medium buildings with exterior woodwork with insulation fittings
low buildings with controlled ventilations and exterior woodwork with special insulation measures

supplementary infiltrations of cold air are added). This coefficient is calculated with the formula:

$$G = \frac{A}{R'_M \cdot V} + 0.34 \cdot n \quad (\mathrm{W/(m^3 \cdot K)}) \tag{36}$$

where: A—area of the building's envelope (m^2); R'_M—corrected average thermal resistance of the building's envelope ($m^2 \cdot K/W$); n—is the number of hourly air exchanges (h^{-1}).

The corrected thermal resistance is calculated with the formula:

$$R'_M = \frac{A}{\sum \frac{A_j \cdot \tau_j}{R'_j}} \quad (\mathrm{m^2 \cdot K/W}) \tag{37}$$

where: A—envelope area (m^2); A_j—area of construction elements [m^2]; τ_j—correction coefficients for cases when the j surfaces do not come into contact with the exterior air; R'_j—corrected average thermal resistances of the construction elements, for the whole building.

Table 16 Minimum thermal resistances R_{min} ($m^2 \cdot K/W$) of the construction elements, for the whole building

No.	Construction element	R_{min} (m^2 K/W)	
		Buildings designed before 1 January 1998	Buildings designed after 1 January 1998
1	Exterior walls (excluding glass surfaces, including walls adjacent to open joints)	1.2	1.4
2	Exterior woodwork	0.4	0.5
3	Platforms above the last storey, under terraces or bridges	2	3
4	Platforms above non-heated basements and cellars	1.1	1.65
5	Walls adjacent to closed joints	0.9	1.1
6	Platforms which delineate the building at the lower end, for exterior (at the lower end, for exterior–, passages)	3	4.5
7	Ground plates (above STQ)	3	4.5
8	Plates at the lower end of basements or heated underground floors (below STQ)	4.2	4.8
9	Exterior walls, below STQ, in basements and heated underground floors	2	2.4

Similarly is calculated the annual cold consumption, Q_C. The difference resides only in that the formula for calculation becomes the following:

$$Q_C = 0.024 \frac{G \cdot V}{\eta_{cooting}} CDD \quad \text{(kWh/year)} \tag{38}$$

where: $\eta_{cooling}$ is the efficiency of the cooling device.The Tables 15, 16 and 17 present the data necessary for applying this method [8].

5.2 Analytically Estimation of the Heat Consumption for Domestic Hot Water (DHW)

Preparing the hot water in household purposes and not only is the second consumption of thermal energy as amplitude, after the consumption of heat for heating place. This consumption for thermal energy is more amplified in the residential sector, comparatively to the industrial sector. The last informative newsletter available and used is the report about the energy consumed in EU15 households and some CEE countries (Fig. 16). The average hot water consumption is estimated to be 60 l/day per capita.

Table 17 Norm thermal insulation global coefficients, G (W/m^3 K), for residences

No. of stories	A/V (m^2/m^3)	G (W/m^3 K)	No. of stories	A/V (m^2/m^3)	G (W/m^3 K)
1	0.8	0.77	4	0.25	0.46
	0.85	0.81		0.3	0.5
	0.9	0.85		0.35	0.54
	0.95	0.88		0.4	0.58
	1	0.91		0.45	0.61
	1.05	0.93		0.5	0.64
	>1.1	0.95		>0.55	0.65
2	0.45	0.57	5	0.2	0.43
	0.5	0.61		0.25	0.47
	0.55	0.66		0.3	0.51
	0.6	0.7		0.35	0.55
	0.65	0.72		0.4	0.59
	0.7	0.74		0.45	0.61
	> 0.75	0.75		> 0.50	0.63
3	0.3	0.49	>10	0.15	0.41
	0.35	0.53		0.2	0.45
	0.4	0.57		0.25	0.49
	0.45	0.61		0.3	0.53
	0.5	0.65		0.35	0.56
	0.55	0.67		0.4	0.58
	>0.6	0.68		> 0.45	0.59

A envelope area, *V* heated volume [8]

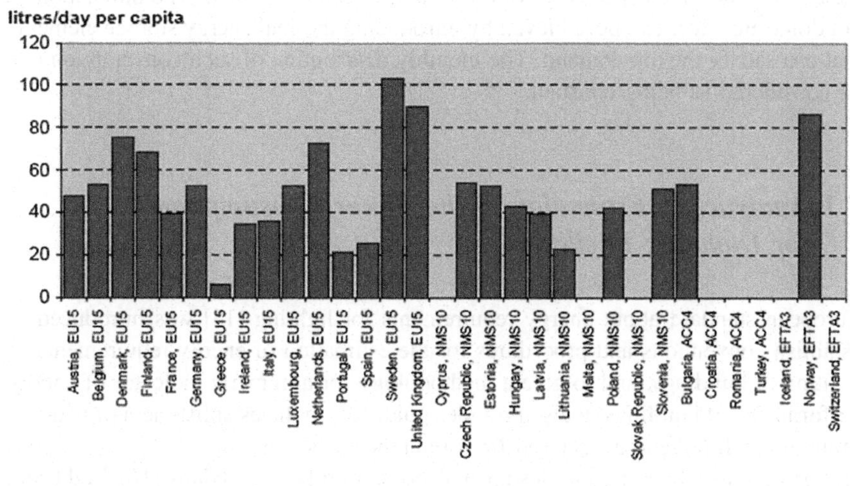

Fig. 16 Hot water consumption for 1995–1996 [1]

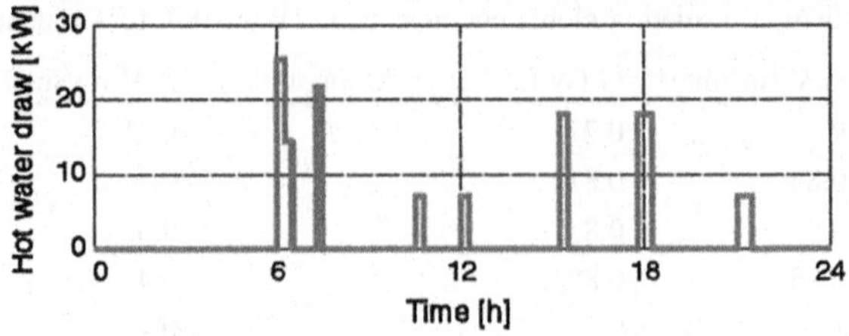

Fig. 17 Daily hot water consumption

Assuming a temperature difference of 50 °C between the hot and cold water sides we can determine, based on the number of persons, the yearly quantity of heat consumed for domestic hot water:

$$Q_{\mathrm{DHW}} = c_w \frac{m \cdot n \cdot N_d}{3,600} \Delta\vartheta \quad (\mathrm{kWh}) \tag{39}$$

where:

Q_{DHW}	energy needed to produce hot water (kWh),
c_w = 4.187 (kJ/kg °C)	specific heat capacity of water,
m	daily quantity of water consumption (kg/capita),
n	number of building occupants,
N_d	number of days,
$\Delta\vartheta$	temperature difference

This hot water household consumption has a daily variation according to Fig. 17.

The heat consumption for domestic hot water has an hourly variation; nevertheless, a daily/monthly average of consumption can be defined. The unification of hot domestic water can be achieved by introducing thermal energy storage elements that can satisfy the top demand. The monthly distribution of such consumption can be considered as being constant.

5.3 Analytically Estimation of the Power Consumption for Domestic Facilities

European standard about energy requirements for lighting [15] was introduced to establish conventions and procedures for the estimation of energy requirements of lighting in buildings, and to give a methodology for a numeric indicator of energy performance of buildings. It also provides guidance on the establishment of national limits for lighting energy derived from reference schemes.

For new installations, the design will be to European standard [16] Light and Lighting—Lighting of work places—Part 1: Indoor work places, what specifies lighting requirements for indoor work places, which meet the needs for visual comfort and performance.

5.3.1 Procedure Based on Global Consumption Estimation

The amount of electricity used for lighting can be based on energy consumption per year (kWh per year).

An estimate of the annual lighting energy required fulfilling the illumination function and purpose in the building (E_L) and annual parasitic energy (E_P) required providing charging energy for emergency lighting and for standby energy for lighting controls in the building shall be established by Eqs. 40 and 41 respectively.

The total estimated energy required for a period in a room or building shall be estimated by the equation:

$$E_t = E_{L,t} + E_{P,t} \quad [\text{kWh}] \tag{40}$$

where:

$E_{L,t}$ is the energy consumed in period t, by the luminaries to fulfill the illumination function and purpose in the building,

$E_{P,t}$ is the parasitic energy consumed in period t, by the charging circuit of emergency lighting and by the standby control system controlling the luminaires.

Energy is the product of installed powers with use time of the light. Because not all the installed power is used it follows that the energy must be adjusted with dependency factors. These factors are:

- constant illuminance factor (Fc) is factor relating to the usage of the total installed power when constant illuminance control is in operation in the room or building;
- occupancy dependency factor (Fo) is factor relating the usage of the total installed lighting power to occupancy period in the room or building;
- daylight dependency factor (F_D) is factor relating the usage of the total installed lighting power to daylight availability in the room or building;
- absence factor (F_A) is factor relating to the period of absence of occupants. This factor has values 0.9 for bathroom, 0.5 for rest room, 0.3 for living room, and 0.2 for kitchen.

An estimate of the lighting energy required to fulfill the illumination function and purpose in the building (E_{Lt}) shall be established using the following equation [17]:

$$E_{L,t} = \frac{\Sigma(P_n \cdot F_c) \cdot [(t_D \cdot F_O \cdot F_D) + (t_N \cdot F_O)]}{1000} \quad (\text{kWh}) \tag{41}$$

where

- $P_n = \sum P_i$ is the total installed lighting power in the room or building power is of all luminaires in the room or building, measured in watts.
- Daylight time usage (t_D) represent operating hours during the daylight time, measured in hours.
- Non-daylight time usage (t_N) represent operating hours during the non-daylight time, measured in hours. Default annual operating hours for t_N is 250 h.

An estimate of the parasitic energy ($E_{P,t}$) required to provide charging energy for emergency lighting and for standby energy for lighting controls in the building shall be established using the equation:

$$E_{P,t} = \frac{\sum \{P_{pc} \cdot [(t_y - t_D - t_N)] + P_{\mathrm{em}} \cdot t_e\}}{1000} \quad (\mathrm{kWh}) \qquad (42)$$

where

- P_{pc} total installed parasitic power of the controls in the room or building is the input power of all control systems in luminaires in the room or zone, measured in watts. This power is a sum of electrical power from the mains supply consumed by the charging circuit of emergency lighting luminaires and the standby power for automatic controls in the luminaire when lamps are not operating ($P_{pc} = \sum P_{p,i}$).
- Standard year time (t_y) is time taken for one standard year to pass, taken as 8 760 h.
- P_{em} total installed charging power of the emergency lighting luminaires in the room or building is input charging power of all emergency lighting luminaires in the room or building, measured in watts. This power is a sum of input power (P_{ei}) to the charging circuit of emergency luminaires.
- Annual operating time (to) is annual number of operating hours of the lamps and luminaires with the lamps on $t_o = t_D + t_N$ (h). This number is determined depending on the building use. Default annual operating hours for t_o is 2,500 h.
- t_e emergency lighting charge time represent operating hours during which the emergency lighting batteries are being charged.

Table 18 shows the standard values for the periods until the next recharging (loading level < nominal value) and the possible times to provide electricity for various sizes of the (acid) batteries based on average consumption.

European standard about energy requirements for lighting: EN 15193-2006 gives a methodology for a numeric indicator of energy performance of buildings. It is Lighting Energy Numeric Indicator, (LENI) described by the equation:

$$\mathrm{LENI} = E/A\left(\mathrm{kWh}/\left(\mathrm{m}^2\mathrm{year}\right)\right) \qquad (43)$$

Table 18 Standard values for battery sizing [18]

Battery voltage	Battery capacity (A h)	Average consumption (kW)	Estimated time for recharging (h)	Duration of the energy delivering for a fully charged battery (h)
48 V	600	0.5	9.5	35
		1	5.5	19
		2	3	10
		4	1.5	4
		6	1	2
48 V	800	0.5	13	48
		1	7	26
		2	4	14
		4	2	6
		6	1.2	3.5
48 V	1000	0.5	16	57
		1	9	29
		2	5	16
		4	2.5	8
		6	1.5	4.5

where:
E is the total annual energy used for lighting, (kWh/year);
A is the total floor area of the building, (m^2).

Determination of dependency factors

- *constant illuminance factor* F_C. The constant illuminance factor is the ratio of the average input power over a given time to the initial installed input power to the luminaire. Normally, the time is taken to be the period of one complete maintenance cycle. Therefore:

$$Fc = (1 + \text{MF})/2 \tag{44}$$

where
MF is the maintenance factor for the scheme.

All lighting installations, from the instant they are installed, start to decay and reduce their output. In the design of the lighting scheme the decay rate is estimated and applied in the calculations known as the Maintenance Factor (MF). The maintenance factor is the ratio between maintained illuminance and initial illuminance. Its value is dependent value by time in use (hours) having 0.9 value for 3,000 h of the time use and 0.8 value for 6,000 h of time use. In installations where a dimmable lighting system is provided, it is possible to automatically control and reduce the initial luminaire output to just provide the

Table 19 F_{OC} values [17]

F_{oc}	Area
Systems without automatic presence or absence detection	
1	Manual on/off Switch
0.95	Manual on/off Switch + additional automatic sweeping extinction signal 0.95
Systems with automatic presence and/or absence detection	
0.95	Auto on /dimmed
0.9	Auto on /auto off
0.9	Manual on /dimmed
0.8	Manual on /auto off

required maintained illuminance. Such schemes are known as "controlled constant illuminance systems". As the light output decays with time, the controls raise input power to the luminaire to compensate.

- *occupancy dependency factor* (*Fo*) can be:
  - $Fo = 1$ if:

    the lighting is switched on 'centrally', i.e., in more than one room at once (e.g., a single automatic system—for instance with timer or manual switch for an entire building, or for an entire floor, or for all corridors, etc.). This applies whatever the type of 'off-switch' (automatic or manual, central or per room, etc.).
    the area illuminated by a group of luminaires that are (manually or automatically) switched together, is larger than 30 m^2.

  - $Fo < 1$ if:
    - in meeting rooms1 (whatever the area covered by 1 switch and/or by 1 detector), as long as they are not switched on 'centrally', i.e., together with luminaires in other rooms;
    - in other rooms, if the area illuminated by a luminaire or by a group of luminaires that are (manually or automatically) switched together, is not larger than 30 m^2, and if the luminaires are all in the same room. In addition, in the case of systems with automatic presence and/or absence detection the area covered by the detector should closely correspond to the area illuminated by the luminaires that are controlled by that detector.

In both cases, also the conditions with respect to timing and dimming level outlined below should be fulfilled. In these instances, F_O should be determined as follows:

$$F_O = F_{OC} + 0.2 - F_A \tag{45}$$

The default value of F_{OC} is fixed as a function of the lighting control system, as given in Table 19.

Table 20 Coefficients for determining the daylight supply factor [17]

A maintained illuminance Em (lux)	Daylight penetration	*a*	*b*
300	Weak	1.2425	−0.0117
	Medium	1.3097	−0.0106
	Strong	1.2904	−0.0088
500	Weak	0.9432	−0.0094
	Medium	1.2425	−0.0117
	Strong	1.3220	−0.0110

Table 21 F_{DC} as a function of daylight penetration [17]

Control of artificial lighting system	F_{DC}		
	Weak	Medium	Strong
Manual	0.2	0.3	0.4
Automatic, daylight dependent	0.75	0.77	0.85

- *daylight dependency factor* (F_D) is shown by equation:

$$F_D = 1-(F_{DS} \cdot F_{DC} \cdot C_{DS}) \tag{46}$$

For each month must be determined monthly daylight supply factor, denoted F_{DS}, and the impact, denoted F_{DC}.

The daylight supply factor F_{DS} can be approximated as a function of latitude γ_{site}, for latitudes ranging from 38° to 60° North, by the relation:

$$F_{DS} = a + b \cdot \gamma_{site} \tag{47}$$

where a and b are shown in Table 20.

The daylight supply factor F_{DS} is valid for a daily operation hour period of 800 h to 1,700 h. For longer daily day time operating periods the values should be multiplied by a correction factor of 0.7. For longer non-daylight periods during the operating time the following applies F_{DS} = 0, i.e. F_D = 1.

F_{DC}, describes the efficiency of how a control system or control strategy exploits the given saving potential, i.e. the daylight supply in the considered space, described by (F_{DS}. F_{DC}) does not consider the power consumption of the control gear itself. Table 21 provides the correction factor of the daylight supply.

C_{DS} is the monthly redistribution factor as a function of latitude and given values for example 46°N latitude shown in Table 22.

Table 22 Monthly redistribution factor C_{DS} as function of daylight penetration for 46°N

	January	February	March	April	May	June	July	August	September	October	November	December
Weak	0.49	0.74	1.09	1.26	1.35	1.41	1.38	1.31	1.09	0.87	0.56	0.42
Medium	0.59	0.84	1.11	1.21	1.25	1.27	1.26	1.25	1.11	0.94	0.66	0.51
Strong	0.70	0.92	1.10	1.14	1.17	1.16	1.17	1.17	1.10	0.98	0.76	0.63

5.3.2 Procedure Based on Specific Consumption Estimation

Generally, visual comfort has a number of aspects: avoiding glare, a balanced luminance distribution on the working plane and within the entire visual field, and color balance of the indoor luminous environment. Too low or too high levels of illuminance have a negative impact on visual comfort, improper light output distribution or unwanted light direction may lead to glare, and light color also affects visual comfort due to its psychological effects on people.

The electrical power requirement can be calculated from the $p_{j,lx}$ (W/m^{2}lx) values as follows:

$$p_j = p_{j,lx} \cdot E_m \cdot k_A \cdot k_L \cdot k_R \quad \mathrm{W/m^2} \tag{48}$$

where:

p_j is the electrical power requirement;

$p_{j,lx}$ is the lamp power required measured in W/(m^{2}lx). The value of lamp power required by lamp type is shown in Table 23.

$\bar{E}m$ is the maintained illuminance (Tables 24 and 25);

k_A is the reduction factor to account for the proportion of the task area.

Table 23 Lamp power required [19]

Lamp type	$p_{j,lx}$ [W/m^{2}lx]
Incandescent	
Open enameled reflector	0.150
General diffusing	0.175
Mercury	
Industrial reflector	0.065
Fluorescent, white	
Open trough	0.040
Enclosed diffusing	0.050
Louvred, recessed	0.055
Fluorescent deluxe warm white	
Enclosed diffusing	0.080
Louvred, recessed	0.090

Table 24 Residential illuminance [19]

Area	Illuminance (lux) E_m	Minimum color rendering (R_a)
Lounge	100–300	80
Kitchens	150–300	80
Bathrooms	150	80
Toilets	100	80

Table 25 Illuminance per activity [19]

Illuminance (lux) E_m	Activity	Area
100	Casual seeing	Corridors, changing rooms, stores
150	Some perception of detail	Loading bays, switch rooms, plant rooms
200	Continuously occupied	Foyers, entrance halls, dining rooms
300	Visual tasks moderately easy	Libraries, sports halls, lecture theaters
500	Visual tasks moderately difficult	General offices, kitchens, laboratories, retail shops

k_L is the correction factor taking into account the type of lamp (for lamps other than tubular fluorescent lamps from Table 26);

k_R is the correction factor taking into account the type of space, given in the Table 27.

The efficacy of a light source is the ratio of the light output to the power consumed. It is given in lumens per Watt. Luminaire efficacy is the light output of the entire luminaire (light fitting) divided by the total power consumed by the lamps and ballasts. It is equal to the lamp efficacy multiplied by the light output ratio of the luminaire and is measured in lumens per watt (lm/W). The higher the efficacy value is, the more the energy-efficient lamps or lighting systems will be. For example, the efficacy of an incandescent light bulb of 60 W is 12 lm/W and of a compact fluorescent lamp (CFL) of 11 W is 55 lm/W. For a 36 W fluorescent tube it is 91 lm/W. The role of lighting systems is to provide adequate visual conditions for human activities to be carried out efficiently and comfortably.

Table 26 Correction factor k_L [18]

Type of lamp	Factor k_L			
	Ballast			
	–	EB[a]	LLB[b]	CB[c]
Incandescent	6	–	–	–
Tungsten halogen	5	–	–	–
Tubular fluorescent	–	1.0	1.14	1.24
Compact fluorescent, with external EB	–	1.2	1.4	1.5
Compact fluorescent, with integrated EB	–	1.6	–	–
Metal halide high-intensity discharge	–	0.86	–	1
High-pressure sodium vapor	–	–	–	0.8
High-pressure mercury vapor	–	–	–	1.7

[a] *EB* Electronic ballast; [b] *LLB* Low loss ballast; [c] *CB* Conventional ballast

Table 27 Correction factor k_R to account for the effect of the space geometry in relation to the room index k [18]

Illumination type	Correction factor k_R											
	Room index k											
	0.6	0.7	0.8	0.9	1	1.25	1.5	2	2.5	3	4	5
Direct	1.08	0.97	0.89	0.82	0.77	0.68	0.63	0.58	0.55	0.53	0.51	0.48
Direct/indirect	1.3	1.17	1.06	0.97	0.90	0.79	0.72	0.64	0.58	0.56	0.53	0.53
Indirect	1.46	1.25	1.08	0.95	0.85	0.69	0.60	0.52	0.47	0.44	0.42	0.39

Note Intermediate values can be obtained by interpolation of the room index values

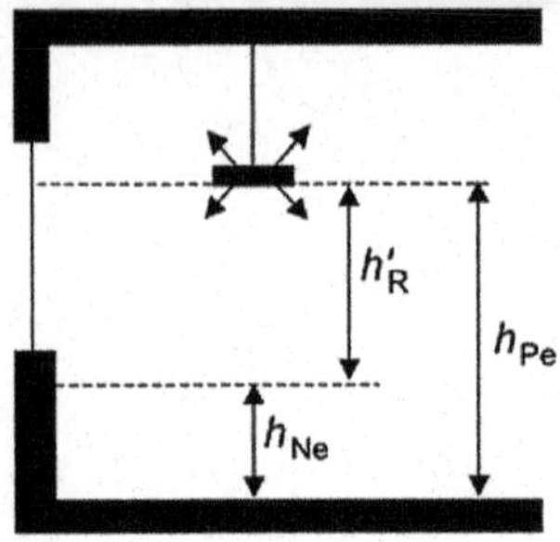

Fig. 18 Determination of height h'_R (schematic diagram)

Room index is:

$$k = \frac{a_R \times b_R}{h'_R \times (a_R + b_R)} \tag{49}$$

If the calculation results in a room index of less than 0.6, a value of $k = 0.6$ shall be used for simplified calculations,

where:

- a_R is the depth of the space;
- b_R is the width of the space;
- h'_R is the difference between the height of the lighting level and the work plane (see Fig. 18).

5.3.3 Simplified Procedure

The specific electrical power can be calculated with equation:

$$p_j = \frac{k_A \cdot E_m}{\text{MF} \cdot \eta_s \cdot \eta_{\text{LB}} \cdot \eta_R} \quad (\text{W/m}^2) \tag{50}$$

where:

- p_j is the electrical power in relation to the floor area for artificial lighting;
- k_A is the reduction factor to take into account the task area;
- MF is the maintenance factor which takes into account system aging processes up to the time of the next maintenance (MF $\in$ (0.5–0.8)) standard value MF = 0.67 shall be assumed. Value 0.80 choose whether if lamps with only a slight decrease in luminous flux and a low failure rate and if the lamps used have a pronounced decrease in luminous flux and a high failure rate, the maintenance factor shall be reduced to a value up to 0.50;
- η_s is the luminous efficacy of the light source and the associated operating devices used;
- η_{LB} is the light output ratio (luminaire efficiency) of the type of luminaire used;
- η_R is the utilization factor given in the Table 28.

Table 28 Utilization factor [18]

Illumination type	ϕ	Utilization factor η_R									
		Room index k									
		0.6	0.8	1	1.25	1.5	2	2.5	3	4	5
Direct	≥ 0.7	0.48	0.59	0.67	0.76	0.82	0.89	0.94	0.98	1.02	1.05
Direct/indirect	$0.1 \leq \phi < 0.7$	0.23	0.30	0.36	0.43	0.48	0.56	0.62	0.67	0.73	0.77
Indirect	<0.1	0.17	0.23	0.29	0.36	0.41	0.48	0.53	0.57	0.62	0.65

Note Intermediate values can be obtained by interpolation of the room index values
Relative luminous flux ϕ into the lower half-space of the luminaire

In the above equation, the characteristic parameters of the individual lamps and luminaires are usually given in manufacturers' datasheets. Utilization factors as a function of the illumination type, luminaire, space reflectance, and room index can be found in technical publications. For simplified estimates, the utilization factors η_R for three types of lighting ("direct", "direct/indirect" and "indirect") given in Table 28 can be used. The characteristic parameter in this case is the relative luminous flux ϕ into the lower half-space of a luminaire. The values are referred to the standard values of the light reflectance combination of the boundary surfaces of the space, i.e., 0.2 for floors, 0.5 for the walls, and 0.7 for the ceiling. The values can be used for estimates of other reflectances to facilitate the calculation procedures. If various types of luminaires are used in one and the same evaluation area, the electrical evaluation powers shall be superimposed in relation to their numbers.

5.3.4 Fast Procedure

Power density can be calculated for office, storage, and industrial spaces so divide the illuminance appropriate to the activity area by 100, and then multiply by 3.75 W/m^2 per 100 lux. For other spaces so divide the illuminance appropriate to the activity area by 100, then multiply by 5.2 W/m^2 per 100 lux.

Lighting power density limits are only one issue influencing the lighting energy use. The other important issues are the control of time of use and the use of daylight. The metric, which includes all these elements and represents the lighting system's performance, is the annual lighting energy intensity, expressed in annual lighting energy consumption per unit area [kWh/m^2 year]. This metric promotes the use of efficient light sources and effective control systems by considering occupancy and the use of daylight.

The electrical energy consumption of the residential consumer is dependent on the endowment of electrical devices in the house and can be calculated by multiplicity the specific power with use surface of the building and time of use. Residential electrical energy consumption depends on the residence's number of appliances. If the instantaneous consumed power load curve is known (Fig. 19), the daily, weekly, or monthly consumption of electricity can be determined.

By introducing elements of storing electric energy, top electric energy demands can be ensured. Integrating power demand curve at daily level or integrating power demand curve at monthly level is possible to determine daily or monthly the

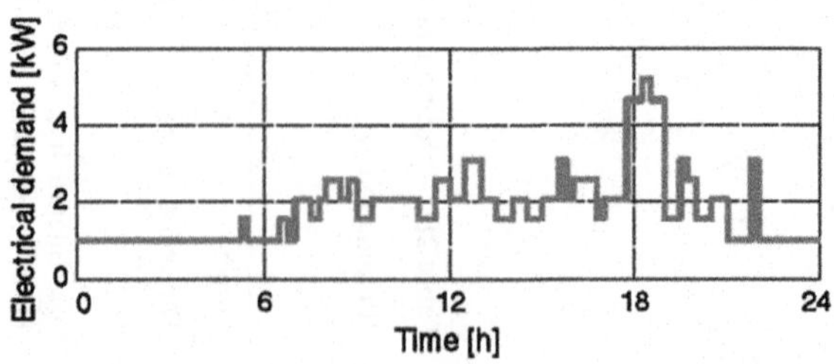

Fig. 19 Typically, power curve of consumer

electricity consumption of the residence. The monthly average consumption of electric energy can be considered constant for a residence that does not use air conditioning inside the residential space during the summer period. The standard monthly consumption of a Romanian residence is (100–300) kWh and can be considered constant for a residence.

5.4 Synthetical Estimation Based on the Residence Energy Certificate

5.4.1 Energy Certificate of Building

Chapter 5 of EN 15217 states that "the energy performance of a building" is represented by an overall indicator energy performance (EP) determined according to EN 15603. The overall indicator is related to the total floor area A_{TFA} in order to facilitate the comparison of the energy performance between buildings. The type of dimensions used to calculate A_{TFA} is not standardized yet because there are many different conventions in the member states. Therefore, internal dimensions, external dimensions, and overall internal dimensions may still be used. According to EN 15217 the type of dimensions shall be specified in the certification procedure. The type of dimensions has a high impact on the indicator EP. The indicator obtained using internal dimensions could be 20 % higher than the one obtained using the external dimension of the same house. The type of dimensions used in the calculation has an impact not only on the indicator EP, but also on the values calculated for the heat transfer, the hot water demand, lighting, etc. This should be remembered when choosing the type of dimensions and also when setting up the calculation methods to be used when including these building services.

Different indicators can be used on a certificate. The certificate shall contain an easy-to-understand global indicator of the energy consumption of the certified building. Different forms of energy can be delivered to a building: e.g., gas, electricity, wood. The indicator will be a weighted sum of these delivered energies. Depending on the weight chosen, the indicator can represent either primary energy, CO_2 emissions, and total energy cost, or a weighted sum of the net delivered energy weighted by any other parameter defined by national energy policy.

In selection of the relevant indicators, the following points should be taken into account:

- For new buildings, a measured energy indicator is not available, so a calculated rating based on design data is the only practical means of assigning an indicator.
- A measured energy indicator will no longer be valid after a change of building occupant or a change in the pattern of use of the building.
- In existing public buildings where there is no change in ownership, the measured energy indicator can be a measure of the quality of the facility management and can be used to motivate building operators and users.

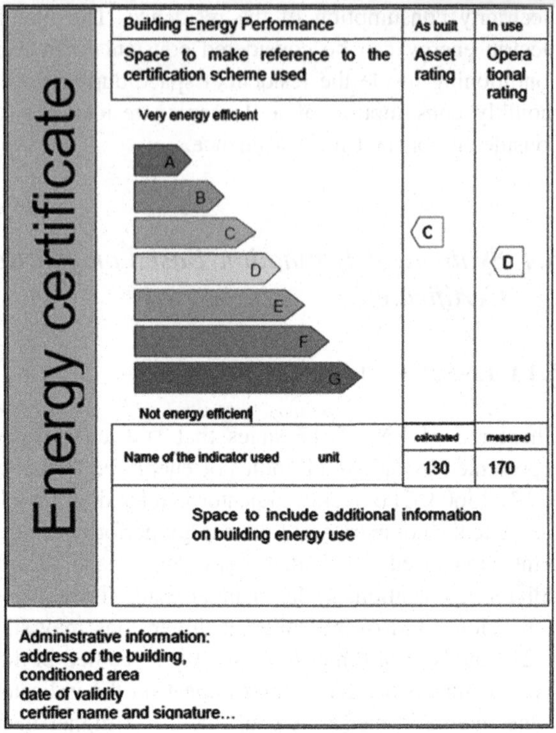

Fig. 20 Example of an energy certificate [20]

- A standard calculated energy indicator requires the collection of data on the building (insulation, heating system, etc.), which will be useful for giving advice on the improvement of its energy performance.
- For managers of buildings, a measured energy indicator can often be easily obtained from the data that is already available in their information systems (energy bills, areas, etc.).
- Measured energy indicators and standard calculated energy indicators are not necessarily based on the same uses of energy.

In addition to the requirements of the EN 15217 indicates that the energy certificate may contain energy classes (Fig. 20). The performance scale shall range from *A* (buildings of highest energy performance) to *G* (buildings of lowest energy performance).

Reference values. According to the requirements, standard specifies that reference values shall be defined for classes of buildings having different functions (e.g., single family houses, apartment blocks, office buildings, educational buildings, hospitals, hotels and restaurants, sports facilities, wholesale and retail trade service buildings, other types). The function refers to the different building services that are required (heating, domestic hot water, air-conditioning), different specifications for the internal climate and different occupant densities and occupancy schedules (buildings used 5, 6, or 7 days a week).The current legal status (e.g., required EP,

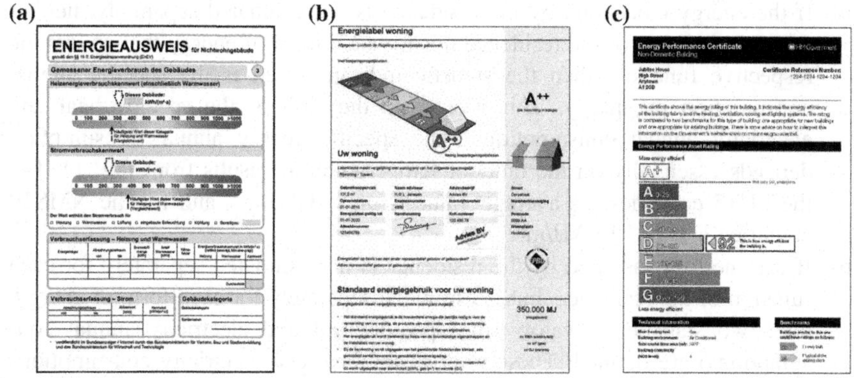

Fig. 21 Examples of different ways to display the energy performance rating and classification that are used in the Member States. **a** Energy certificate based on measured energy use displayed in public buildings in Germany, **b** energy certificate based on calculated energy use for non-residential buildings in The Netherlands **c** energy certificate based on measured energy use displayed in public buildings in the UK

minimum EP for new buildings) and the building stock must be used as references. The reference values shall be documented, not on the individual certificate but in a report that can be easily consulted (e.g., one that is available on a website), taking into account the following aspects: type of reference value; building function; uses considered; assumptions regarding internal and external climate; procedure to be used to select the correct reference value.

Certificate shape. The format of the certificate is very important to enable an easy understanding by non-specialists. The CEN standards offer three examples of certificate layout, which can be used as a basis by Member States (Fig. 21):

- the first example includes a calculated rating and energy classes,
- the second includes a calculated and a measured rating,
- the third includes a continuous scale instead of energy classes.

5.4.2 Estimation of the Specific Consumption

In this respect, *the specific heat, cold, and power annual consumption for building* is obtained by dividing the annual consumption to the *total floor area of the residence* A_{TFA}.

Depending on how global the estimation is, one may distinguish between several cases, which differ in that the evaluation is carried out separately or not for each function of the residence, and in that the evaluation is carried out separately or not for each form of energy.

(a) If the energy consumed by the residence is not evaluated separately, neither for each function of the residence nor for each form of energy that covers the respective function, then the statistic indicator used is the specific annual consumption of energy. This represents the sum of the specific heat and electricity annual consumptions. The specific energy annual consumption depends essentially on the outdoor temperature, as results from Fig. 11. For the EU15 countries, its average value is of 589 MJ/m^2, and for the NMS10 countries it is of 713 MJ/m^2.

(b) If the energy consumed by the residence is not estimated separately for each function of the residence, but is estimated separately for each form of energy, then two statistic indicators (the specific heat and electricity annual consumptions) are defined. These refer to the heat and electricity consumptions corresponding to all the functions of the residence, such as space heating, hot water preparation, ventilation, cooling, illumination, appliances for cooking, etc.

 The values of these specific annual consumptions are statistical determined at national level.

(c) If the energy consumed by the residence is estimated separately for every group of residence functions, but is not estimated separately for every form of energy, then four statistic indicators are the specific annual consumptions corresponding to the four groups of functions (heating, cooling, hot water, and illumination), regardless of the form of energy consumed.

On the other hand, according to the Romanian methodology for calculating the energetic performances of buildings, indicative Mc 001/5—2009, buildings are classified in keeping with the energetic class that their specific demands belong to. In this case, the values of these statistic indicators are determined in keeping with the building's energetic class (Table 29).

For example, one residence in **A** energetic class (see Table 29) have per year the specific heat consumed for space heating q = 70 (kWh/m^2); the specific heat consumed for hot water q_{dhw} = 15 (kWh/m^2); the specific heat consumed for space cooling q_{C} = 20 (kWh/m^2) and the electrical corresponding to specific useful electricity in the residence e = 40 (kWh/m^2).

Table 29 Annual specific consumption, corresponding to the different groups of functions, depending on the energetic classes of buildings

Energetic class	A	B	C	D	E	F	G
Heating annual specific consumption (kWh/m^2 year)	70	117	173	245	343	500	>500
Cooling annual specific consumption (kWh/m^2 year)	20	50	87	134	198	300	>300
Hot water annual specific consumption (kWh/m^2 year)	15	35	59	90	132	200	>200
Illumination annual specific consumption (kWh/m^2 year)	40	49	59	73	91	120	>120

In the model for the specific energy of the residence is necessary to present their evolution in time for each month. The calculus starts from the specific annual consumptions of heat and cold, and aims at finding the specific monthly consumption. The determination of the specific monthly consumption is done with statistical indicators heating degree-day, HDD, and cooling degree-day, CDD. Degree Day method determines monthly the heat and cold consumption, per unit of total floor area of the building with equation:

$$q_{\text{monthly}} = \frac{q \cdot \text{HDD}_{\text{monthly}}}{\text{HDD}_{\text{year}}} \tag{51}$$

Respectively

$$q_{C\cdot\text{monthly}} = \frac{q_C \cdot \text{CDD}_{\text{monthly}}}{\text{CDD}_{\text{year}}} \tag{52}$$

The specific monthly consumption of electricity and domestic hot water (DHW) is considered the same (constant) for each month, namely:

$$e_{\text{monthly}} = \frac{e}{N_{\text{month}}} \tag{53}$$

$$q_{\text{DHW monthly}} = \frac{q_{\text{DHW}}}{N_{\text{month}}} \tag{54}$$

where the number of months is $N_{month} = 12$.

Table 30 The monthly specific consumption (illustrative exercise)

Month	Heating degrees days °C-d	Cooling degrees days °C-d	Heat q (kWh/m^2)	Cold q_c (kWh/m^2)	DHW q_{DHW} (kWh/m^2)	Electricity *e* (kWh/m^2)
January	611	0	14.69	0.00	1.25	3.33
February	521	0	12.52	0.00	1.25	3.33
March	403	0	9.69	0.00	1.25	3.33
April	207	33	4.98	0.42	1.25	3.33
May	40	208	0.96	2.68	1.25	3.33
June	0	300	0.00	3.86	1.25	3.33
July	0	363	0.00	4.67	1.25	3.33
August	0	363	0.00	4.67	1.25	3.33
September	6	234	0.14	3.01	1.25	3.33
October	195	53	4.69	0.68	1.25	3.33
November	390	0	9.38	0.00	1.25	3.33
December	539	0	12.96	0.00	1.25	3.33
Annually	*2,912*	*1,554*	*70.00*	*20.00*	*15.00*	*40.00*

Let us consider a residence in the energy class A (with values according to Table 29), located in Galaţi area, with climate data given in the Table 2. Results of the specific distribution are given in Table 30. Thus, the cold, heat, and power models will be obtained.

5.5 Consumption Aggregating

The residence functional needs are met by incorporating into the system of several consumers as home appliance, chiller with mechanical or thermal compression (MCC or TCC), radiator, heater, and domestic facility. These components of the system will be powered either by electricity (voltage 220 V) or by heat (temperature of thermal agent for 60–90 °C). The other parameter values of energy or the other forms of energy were excluded when it was determined the search space that defines the conceptual framework.

Consumptions occasioned by their operation will be aggregated, resulting in heat and power consumption at the system level. To cover these consumptions will be incorporated in the system structure only two storages, power storage and heat storage.

As examples, below are presented two cases of consumption aggregating, namely case of a system with mechanical compression chiller (MCC) and case of a system with thermal compression chiller.

5.5.1 Consumption Aggregating in Case of Mechanical Compression Chiller

The structure of a mCCHP system with mechanical compression chiller is presented in Fig. 22.

In this case, electric energy is used to condition the residential space. The process of conversion is characterized by the coefficient of performance COP, defined as ratio between the quantity of cold produced by the chiller, q_c, and the quantity of electric energy consumed. The usual value of COP may be considered as being 3. Based on this value, the power consumption of the chillers in the system is determined. The power consumption of the system, e_{sys}, is determined as the sum of the power consumption of the residence, e, and the power consumption of the chillers. As regards the heat consumption of the system, q_{sys}, it is equal to the heat consumption of the residence, which is determined by summing up the heat for heating, q, and the heat for hot water, q_{DHW}.

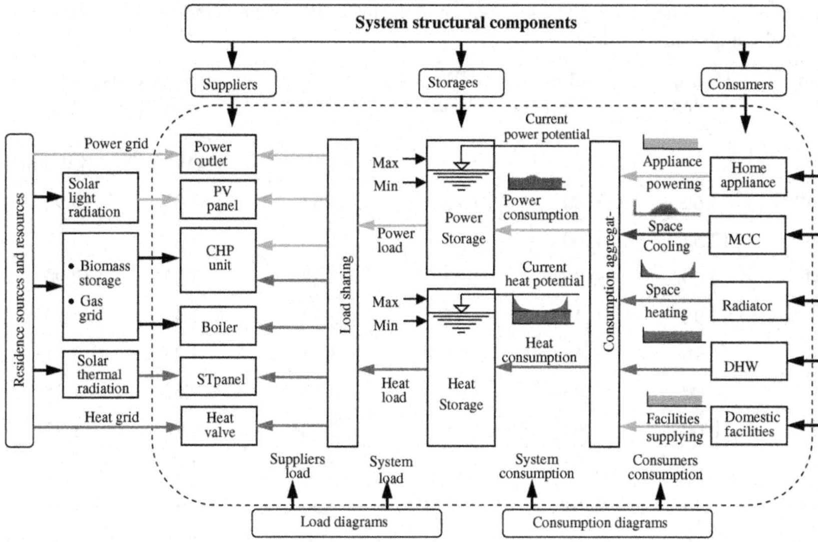

Fig. 22 Structural model of the CCHP system with MCC

The relations for determining the energy consumption of the system in the case of cooling with mechanical compression chillers are as follows:

$$q_{\text{sys}} = q + q_{\text{DHW}} \tag{55}$$

$$e_{\text{sys}} = e + \frac{q_c}{\text{COP}} \tag{56}$$

For example, the monthly values of the heat and power-specific consumption for a mCCHP system, located in Galati area, are recorded in the Table 31 (when COP = 3). These values are obtained by applying the relations 55 and 56 to the data given in the Table 30.

The diagrams of the mCCHP system power and heat loads in the case of cooling with mechanical compression chillers are presented in Fig. 23.

What may be observed is that the power consumption and heat consumption of the mCCHP system are not uniform during the whole year. In the case of systems with a mechanical compression chiller, the power consumption is low during the cold season and high during the hot season. The heat consumption of the mCCHP system is low during the hot season and very high during the cold season.

Table 31 Monthly energy specific consumption of the mCCHP system in the case of cooling with mechanical compression chillers

Month	Heat q (kWh/m^2)	Cold q_c (kWh/m^2)	DHW q_{DHW} (kWh/m^2)	Electricity e (kWh/m^2)	Energy consumption of the system	
					e_{sys} (kWh/m^2)	q_{sys} (kWh/m^2)
January	14.69	0.00	1.25	3.33	3.33	15.94
February	12.52	0.00	1.25	3.33	3.33	13.77
March	9.69	0.00	1.25	3.33	3.33	10.94
April	4.98	0.42	1.25	3.33	3.47	6.23
May	0.96	2.68	1.25	3.33	4.23	2.21
June	0.00	3.86	1.25	3.33	4.62	1.25
July	0.00	4.67	1.25	3.33	4.89	1.25
August	0.00	4.67	1.25	3.33	4.89	1.25
September	0.14	3.01	1.25	3.33	4.34	1.39
October	4.69	0.68	1.25	3.33	3.56	5.94
November	9.38	0.00	1.25	3.33	3.33	10.63
December	12.96	0.00	1.25	3.33	3.33	14.21
Total	*70.00*	*20.00*	*15.00*	*40.00*	*46.67*	*85.00*

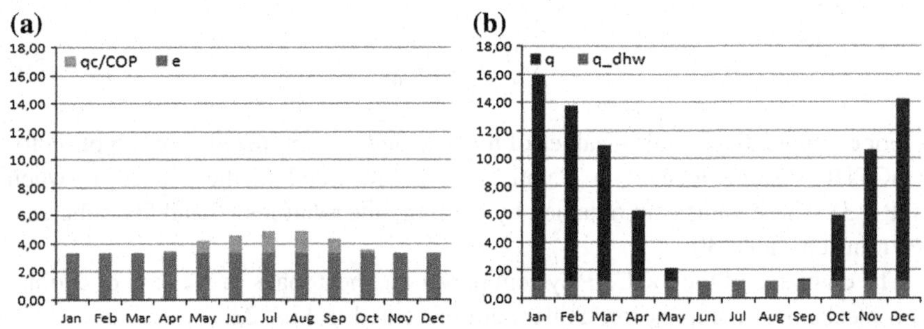

Fig. 23 Diagrams of the system's power and heat consumption in the case of cooling with mechanical compression chillers; **a** diagrams of the system's power consumption **b** diagrams of the system's heat consumption

5.5.2 Consumption Aggregating in Case of Thermally Compression Chiller

The structure of a mCCHP system with a thermally compression chiller is presented in Fig. 24. The heat load of the mCCHP system must cover the heat consumption of the residence, as well as the heat consumption for cooling. The heat consumption

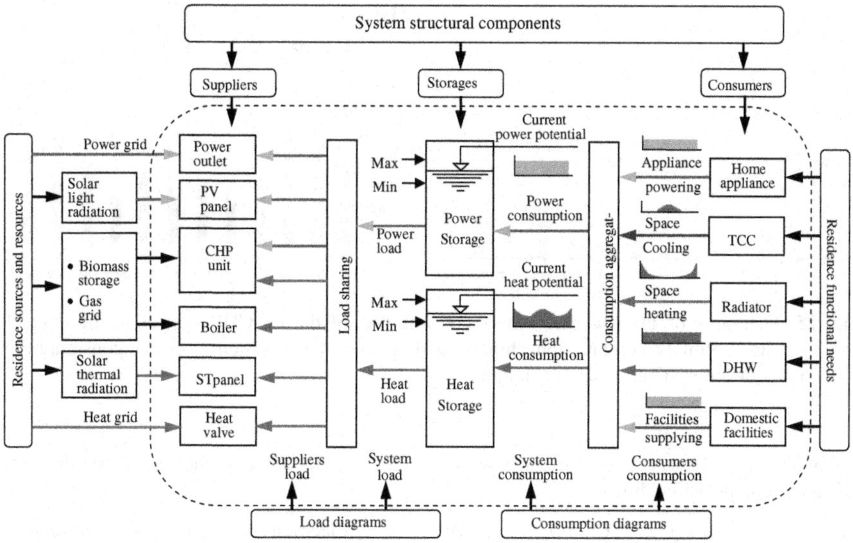

Fig. 24 Structural model of the mCCHP system with thermally compression chiller

for cooling depends on the coefficient of performance, COP_a, of the thermally compression chiller. In general, the COP_a of these cooling devices is between 0.6 and 0.8. The monthly power consumption of the residence is considered constant, the entire year. The relations for determining the energy consumption of the system in the case of cooling with thermal compression chillers are as follows:

$$q_{\text{sys}} = q + q_{\text{DHW}} + \frac{q_c}{\text{COP}_a} \tag{57}$$

$$e_{\text{sys}} = e \tag{58}$$

The diagrams of the power and heat consumption of the mCCHP system in the case of cooling with thermal compression chillers, located in Galati area, are presented in Fig. 25. These values are obtained by applying the relations 57 and 58 to the data given in the Table 29, when $COP_a = 0.8$.

The heat consumption of the mCCHP system for certain specific climate zones can be balanced between cool and warm season.

The monthly values of the heat and power loads for a mCCHP system, located in Galati area, are recorded in the Table 32, when $COP_a = 0.8$.

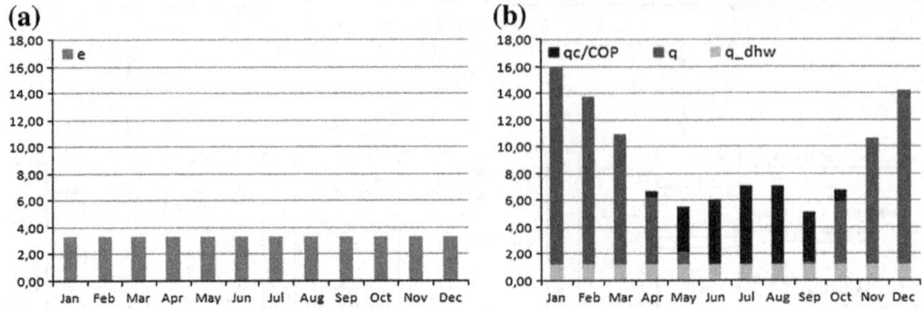

Fig. 25 Diagrams of the heat and power consumption of the mCCHP system in the case of cooling with thermally compression chillers; **a** diagrams of the system's power consumption; **b** diagrams of the system's heat consumption

Table 32 Monthly energy specific consumption of the mCCHP system in the case of cooling with thermally compression chillers

Month	Heat q (kWh/m^2)	Cold q_c (kWh/m^2)	DHW q_{DHW} (kWh/m^2)	Electricity e (kWh/m^2)	Energy consumption of the system	
					e_{sys} (kWh/m^2)	q_{sys} (kWh/m^2)
January	14.69	0.00	1.25	3.33	3.33	15.94
February	12.52	0.00	1.25	3.33	3.33	13.77
March	9.69	0.00	1.25	3.33	3.33	10.94
April	4.98	0.42	1.25	3.33	3.33	6.76
May	0.96	2.68	1.25	3.33	3.33	5.56
June	0.00	3.86	1.25	3.33	3.33	6.08
July	0.00	4.67	1.25	3.33	3.33	7.09
August	0.00	4.67	1.25	3.33	3.33	7.09
September	0.14	3.01	1.25	3.33	3.33	5.16
October	4.69	0.68	1.25	3.33	3.33	6.79
November	9.38	0.00	1.25	3.33	3.33	10.63
December	12.96	0.00	1.25	3.33	3.33	14.21
Total	*70.00*	*20.00*	*15.00*	*40.00*	*40.00*	*110.00*

6 Load Estimation

6.1 Load Versus Consumption

For each supplier, the output is the total energy produced, while *the load* represents a first part only of the output, namely, that which is used to satisfy the residence demands. A second part is that used to cover the supplier's own energetic consumptions (such as the consumption of the recirculation pumps or the consumption of the heat exchanger ventilators). Also, a third part might exist, namely that which is dissipated on the outside, as it cannot be used. To exemplify, notice that the output of the CHP unit consists in electric and thermal output, produced in a fixed ratio γ. If, at a certain point, the ratio between the electric demand and the thermal demand is not equal with γ, then the CHP unit produces energy in excess under one of these forms. This excess could be stored or dissipated on the outside, last case representing a loss. On the other hand, for each consumer, the input is the total energy used to satisfy the residence needs, including the consumer's own energetic needs, and is called *consumption*.

Between consumption and load there is a causal relationship according to which the load must cover consumption. If every time this causal relationship is satisfied, then the load diagram would be identical to the consumption diagram.

It is difficult (and sometimes impossible) to control the system so that this causal relationship to be constantly respected. Therefore, between suppliers and consumers is interposed storage and the causal relationship is substituted with two others, namely:

(a) permanent, the storage must cover the consumption, and
(b) periodically, the suppliers should fill storage, so that the amount of energy contained in it to not exceed two limits considered acceptable.

Therefore, although the system controller continuously adjusts the load so that these conditions are observed, however temporarily, the values of consumption and load come to be different. This is because that *in the short periods of time* between consumption and load there is not a direct and immediate relation. As a result, the load diagram is different from the consumption diagram.

More specifically, the system controller controls the load according to its control law. This law establishes the momentary load depending on the amount of energy that is available at that moment in storage. In turn, the control law is established from the basic condition that must be followed, namely *every time the cumulative difference between consumption and load does not exceed the capacity* ΔE *of storage* (see Fig. 26a).

In order to get a more practical formulation of the basic condition let us analyze the evolution of consumption and load during one day (see Fig. 26b).

It can be seen that this condition is met if:

- maximum value of any of the hatched area is ΔE;
- two successive areas must have signs opposing values.

At the same time we should note that the PQRS surface area is the daily consumption and is equal to the P_1Q_1RS surface area representing the daily load and also with the P_2Q_2RS surface area that represents both, the daily load and the daily consumption. Also P_2Q_2 line represents both average load and average consumption. Note that the ratio of the maximum load and the average load could be considered as a measure of the level of smoothing of the consumption diagram when it is transformed into the load diagram.

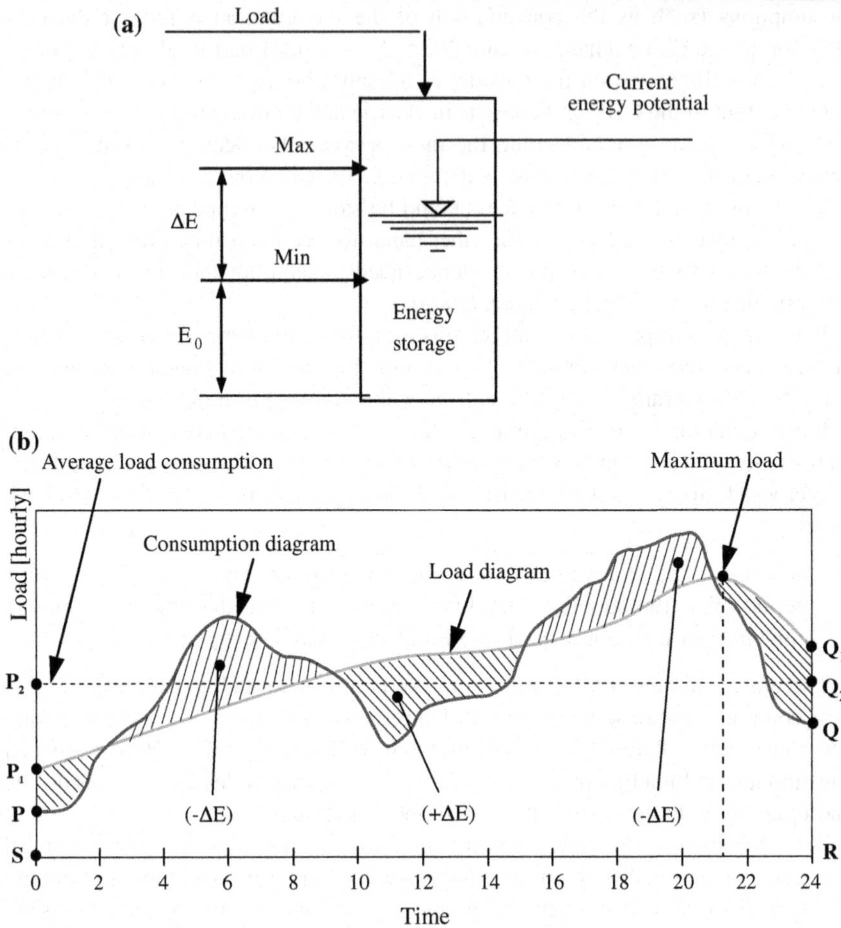

Fig. 26 Load evolution during one day; **a** storage capacity, ΔE; **b** load diagram versus consumption diagram

In short, due to the fact that the storage covers consumption, the load diagram is a form more or less smoothed of the consumption diagram. Mainly, the smoothing depends on both the control law and storage capacity ΔE.

Above, was discussed the load diagram for one day. In structural design, it is accepted the assumption according to which all days of a month have the same consumption diagram and the same load diagram. Therefore, in the particular case where ΔE is higher than the daily consumption (very common case in practice), the shape of load diagram might even be independent of the consumption diagram. The only remaining condition is that the surface area P_1Q_1RS must be equal to the daily consumption. For example, the load diagram might be even line P_2Q_2. If, on the contrary, ΔE is very small, then the load diagram becomes identical with the consumption diagram. In conclusion, it follows that for the supplier to be able to cover the consumption throughout the month, just as their installed capacity can support the daily load.

6.2 System Load Estimation

The system load is defined as the load necessary to fill the storage. It is estimated in two situations, namely during the system operation and during the system design.

During *system operation*, it is estimated *the momentary load command*, so that the suppliers to cover the demand occurs at the moment. If ΔE is less than the daily consumption, then the load is even the P_1Q_1 line in Fig. 26b. In this case, the estimation is based on the consumption diagram resulting from the system monitoring. If ΔE is higher than the daily consumption, then estimation of the momentary load command is performed on other bases, which will be presented in Chap. Functional Design of the mCCHP-RES System.

During *system design,* it is estimated *the monthly load*, namely that corresponding to each month of the year. No matter how big is ΔE the estimation is carried out on the base of the monthly consumption, which was obtained by applying the estimation methods presented in Sect. 5.

6.3 Load Sharing

During system design, the ultimate goals of the load estimation are: (a) *sizing the suppliers*, so that they can support the estimated load, and (b) *assess the performance indicators* for each structural model in order to eliminate models less efficient. To achieve these goals the system load should be split between suppliers, to thus find the daily load *for each supplier*. As a graphical example, let us consider the following cases: (a) structural model with mechanical compression chillers and (b) thermally compression chillers shown in Figs. 27 and 28. This problem is called *load sharing* and has the following features.

- The solution of the problem depends on *the system structure*. Therefore, for each given structural model the problem resolves separately.
- Solving the problem does not lead to a unique solution (except for very simple cases). In order to select a solution, it should adopt a certain *load sharing strategy*, previously defined. Therefore, the solution depends on the strategy adopted.
- Level of performance indicators depends not only on the given structural model and the adopted load sharing strategy, but also on *the consumption diagram*, which varies from month to month. Thus, the performance indicators change from month to month of the year. Therefore, an optimal solution of the problem is valid only for the month for which was determined.

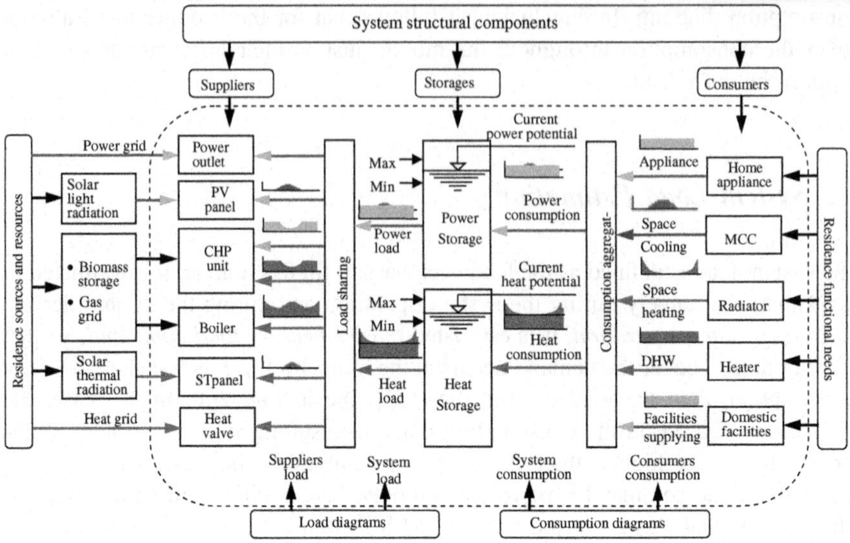

Fig. 27 Load sharing of the mCCHP system with mechanical compression chiller

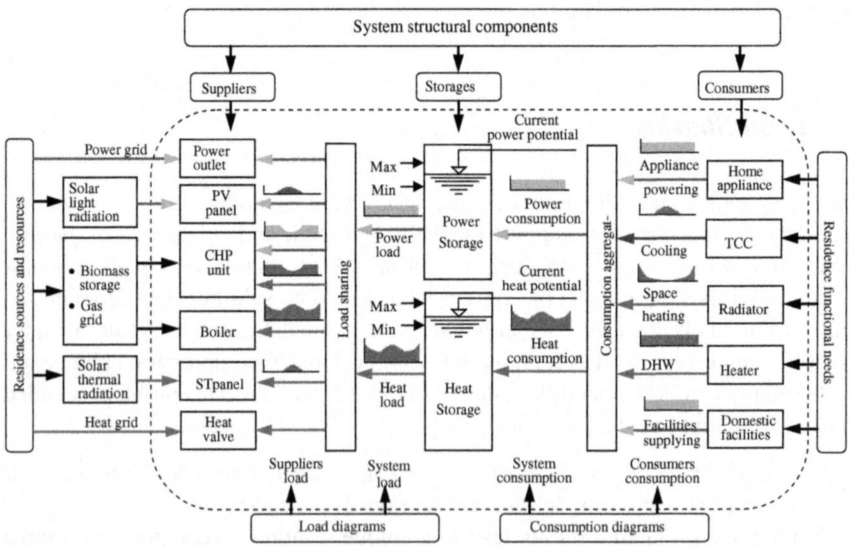

Fig. 28 Load sharing of the mCCHP system with thermally compression chiller

It follows that, in structural design, the performance indicators can be improved by finding an appropriate association between structural system model, consumption diagram, and load sharing strategy.

Many load sharing strategies can be imagined. To exemplify, two typical strategies are presented, namely:

- Power driven load sharing (when excess power does not appear), and
- Heat driven load sharing (when excess heat is avoided).

Further, the Sect. 7 shows their application to the evaluation of the potential structural models (models that were developed in Sect. 4). The aim is to identify and eliminate those models that appear as less efficient.

A. Power driven load sharing for an off-grid system is based on the following basic assumptions:

- First, the power load will be shared on the power suppliers so that any excess power did not occur.
- Then the heat load will be shared by the heat suppliers so as not to appear any heat deficit.
- The risk of a possible excess of heat load will be mitigated by modulating one of the suppliers, according to complementary assumptions.

As a theoretical exercise, in Fig. 29 it shows the application of the power driven load sharing in a hypothetical case.

Here:

- Lines PQ and MN show the evolution of the monthly power load and heat load for entire system, during a calendar year.
- Surfaces PQRS and MNRS are the annually power load and heat load of the system, respectively.
- (PV panels load)$_{\text{full}}$ and (ST panels)$_{\text{full}}$ are the surfaces UaVU and UbVU respectively, and represent the annually load of the PV panels and ST panels when they are working at their full installed capacity.
- (SE power load) is the surface PQRVaUS, represents the annually power load of the Stirling engine, and is given by relation:

$$(\text{SE power load}) = (\text{Power load PQRS}) - (\text{PV panels load})_{\text{full}}$$

- (SE *heat load*) is the surface IJKLRS, represents the annually heat load of the Stirling engine, and is given by relation:

$$(\mathit{SE\ heat\ load}) = (\mathit{SE\ power\ load})/\gamma$$

- (*Boiler load*) is the pair of surfaces JIM and KLN, represents the annual heat load of the boiler, and is given by relation:

$$(\mathit{Boiler\ load}) = (\mathit{Heat\ load\ MNRS}) - (\mathit{SE\ heat\ load}),$$

  only positive values.
- (*Heat excess*) is the surface JcKdJ and represents the heat remaining.

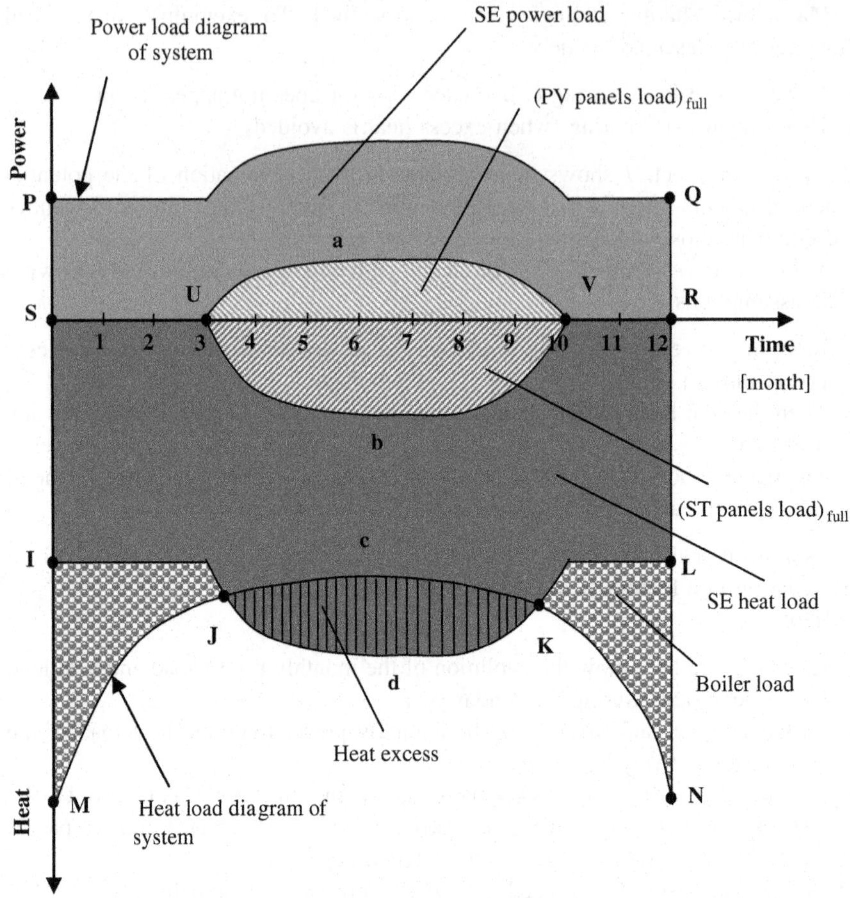

Fig. 29 Power driven strategy of load sharing

If this heat cannot be valorized otherwise, then the excess heat risk could be mitigated as follows:

(a) by modulation of the ST panel load when the $(ST\ panels)_{full}$ will be replaced with $(ST\ panels)_{modulated}$, given by relation

$$(ST\ panels)_{modulated} = (ST\ panels)_{full} - (Heat\ excess),$$

(b) by a corresponding decrease of the ST panels area, A_{ST}.

B. Heat driven load sharing for an off-grid system is based on the following basic assumptions:

- First, the heat load will be shared on the heat suppliers so that any excess heat does not occur.

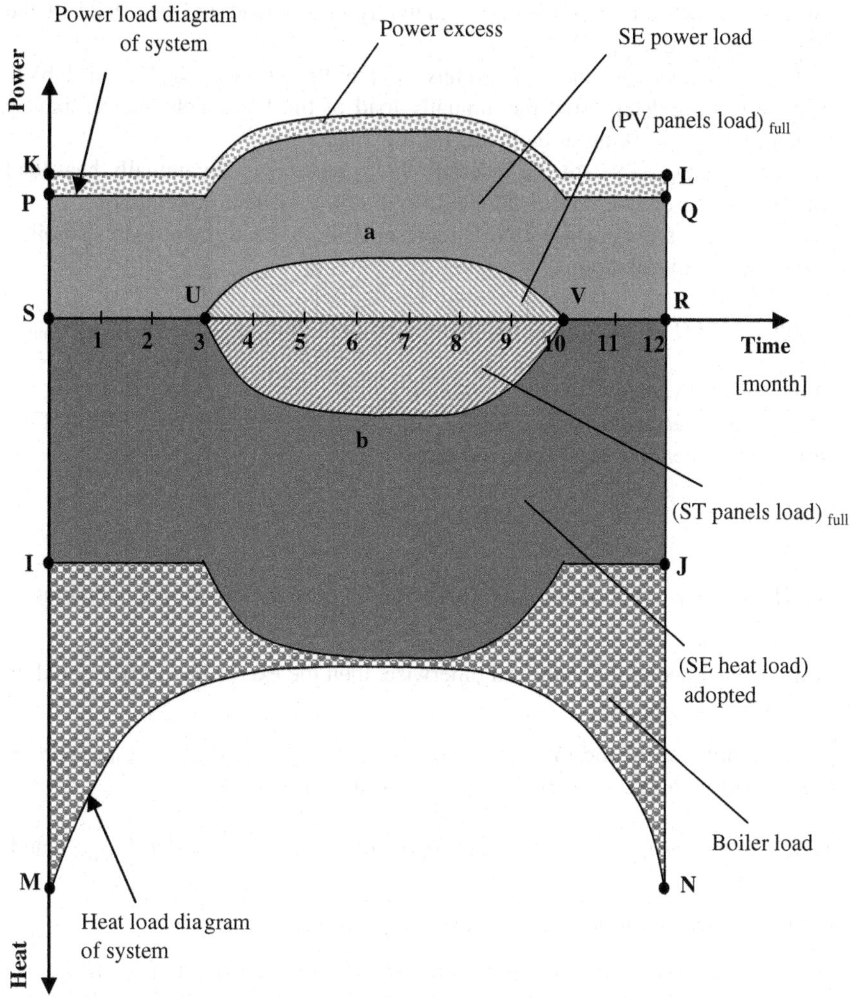

Fig. 30 Heat driven strategy of load sharing

- Then the power load will be shared by the power suppliers so as not to appear any power deficit.
- The risk of a possible excess of power load will be mitigated by modulating one of the suppliers, according to complementary assumptions.

As a theoretical exercise, in Fig. 30 it shows the application of the heat driven load sharing in a hypothetical case. Here:

- Lines PQ and MN show the evolution of the monthly power load and heat load for entire system, during a calendar year.

- Surfaces PQRS and MNRS are the annually power load and heat load of the system, respectively.
- (*PV panels load*)$_{full}$ and (*ST panels*)$_{full}$ are the surfaces UaVU si UbVU respectively, and represent the annually load of the PV panels and ST panels when they are working at their full installed capacity.
- (*SE heat load*)$_{adopted}$ is the surface IJRVbUS, represents the annually heat load of the Stirling engine, and is adopted on the base of other assumptions.
- (*Boiler load*) is the surface IJNM, represents the annually heat load of boiler, and is given by relation:

$$(Boiler\ load) = (Heat\ load\ MNRS) - (SE\ heat\ load)_{adopted} - (ST\ panels)_{full},$$

  only positive values.
- (*SE power load*) is the surface KLRS, represents the annually power load of the Stirling engine, and is given by relation:

$$(SE\ power\ load) = \gamma (SE\ heat\ load)_{adopted}$$

  – (*Power excess*) is the surface PQLK and represents the power remaining.

If this power cannot be valorized otherwise, then the excess power risk could be mitigated as follows:

(a) by modulation of the (*SE heat load*) when the (*SE heat load*)$_{adopted}$ will be replaced with (*SE heat load*)$_{modulated}$, given by relation:

$$(SE\ heat\ load)_{modulated} = \left[(Power\ load\ PQRS) - (PV\ panels\ load)_{full}\right]/\gamma, \text{ and}$$

(b) by a corresponding decrease of the PV panels area, A_{PV}.

To determine the electrical energy produced by the cogeneration unit, e_{cg}, we will use the power balance equation for the stationary regime at batteries level and result:

$$e_{cg} = e_{sys} - e_{PV} \tag{59}$$

If result $e_{cg} < 0$, then the cogeneration unit does not operate, the required energy is provided by PV panels and the excess is stored in the battery. Further, because e_{cg} and q_{cg} are connected by the power to heat ratio, γ, the specific heat produced by the cogeneration unit is given by the relation:

Table 33 Load sharing of the mCCHP system with mechanical compression chiller

Month	System load		PV and ST panels load		CHP unit load		Boiler load
	e_{sys} (kWh/m^2)	q_{sys} (kWh/m^2)	e_{PV} (kWh/m^2)	q_{TP} (kWh/m^2)	e_{cg} (kWh/m^2)	q_{cg} (kWh/m^2)	q_{ad} (kWh/m^2)
January	3.33	15.94	1.24	4.14	2.09	6.27	5.52
February	3.33	13.77	1.76	5.88	1.57	4.71	3.19
March	3.33	10.94	2.91	9.69	0.42	1.27	0.00
April	3.47	6.23	3.47	6.23	0.00	0.00	0.00
May	4.23	2.21	4.23	2.21	0.00	0.00	0.00
June	4.62	1.25	4.62	1.25	0.00	0.00	0.00
July	4.89	1.25	4.89	1.25	0.00	0.00	0.00
August	4.89	1.25	4.86	1.16	0.03	0.09	0.00
September	4.34	1.39	3.59	0	0.75	1.39	0.00
October	3.56	5.94	2.51	7.7	1.05	3.15	0.00
November	3.33	10.63	1.34	4.47	1.99	5.97	0.18
December	3.33	14.21	1.01	3.35	2.33	6.98	3.87
Total	*46.67*	*85.00*	*39.31*	*130.98*	*10.23*	*30.68*	*12.77*

$$q_{cg} = \frac{e_{cg}}{\gamma} \tag{60}$$

To determine the heat produced by the boiler, q_{ad}, we will use the heat balance equation for the stationary regime at heat storage level and result:

$$q_{ad} = q_{sys} - q_{cg} - q_{ST} \tag{61}$$

The additional boiler does not operate if $q_{ad} \leq 0$. The required energy is provided by thermal solar panels, q_{ST}, and cogeneration unit, while the excess is stored in the heat storage.

Using data from Table 31 for monthly energy specific consumption of the mCCHP system in the case of cooling with mechanical compression chillers and from Table 6 for monthly specific power and heat production of the panels and building we can determine the load sharing of the CHP and backup boiler. The results of the load sharing of the mCCHP system with mechanical compression chiller are shown in Table 33.

Similar, using data from Table 32 and from Table 6 results the load sharing of the mCCHP system with thermally compression chiller, which are shown in Table 34.

Table 34 Load sharing of the mCCHP system with thermally compression chiller

Month	System load		PV and ST panels load		CHP unit load		Boiler load
	e_{sys} (kWh/m^2)	q_{sys} (kWh/m^2)	e_{PV} (kWh/m^2)	q_{TP} (kWh/m^2)	e_{cg} (kWh/m^2)	q_{cg} (kWh/m^2)	q_{ad} (kWh/m^2)
January	3.33	15.94	1.24	4.14	2.09	6.27	5.52
February	3.33	13.77	1.76	5.88	1.57	4.71	3.19
March	3.33	10.94	2.91	9.69	0.42	1.27	0.00
April	3.33	6.76	3.33	6.76	0.00	0.00	0.00
May	3.33	5.56	3.33	5.56	0.00	0.00	0.00
June	3.33	6.08	3.33	6.08	0.00	0.00	0.00
July	3.33	7.09	3.33	7.09	0.00	0.00	0.00
August	3.33	7.09	3.33	7.09	0.00	0.00	0.00
September	3.33	5.16	3.33	5.16	0.00	0.00	0.00
October	3.33	6.79	2.51	8.37	0.82	2.47	0.00
November	3.33	10.63	1.34	4.47	1.99	5.97	0.18
December	3.33	14.21	1.01	3.35	2.33	6.98	3.87
Total	*40.00*	*110.00*	*39.31*	*130.98*	*9.23*	*27.68*	*12.77*

7 Evaluation and Improving of the Structural Models Performance

7.1 Indicators for Performance Evaluation at System Level

During system design, the ultimate goals of the load estimation were (a) sizing the suppliers, so that they can support the estimated load, and (b) assess the performance indicators for each structural model in order to eliminate the models less efficient. The performance is evaluated using the following two indicators:

- The first indicator is *Percentage of energy saving* or *Primary Energy Save* (PES) that refers basically to the percentage of fuel saved from the energy production of the CCHP system compared to the same energy produced by the reference system. Conventional, the reference system is the one, which produces heat and power with efficiency reference values η_{Href} and η_{Eref}, respectively. PES is calculated with relation [21]:

$$\mathrm{PES} = \left(1 - \frac{q_F}{\frac{e_{sys}}{\eta_{Eref}} + \frac{q_{sys}}{\eta_{Href}}}\right) \times 100\,\% \tag{62}$$

where:

$\eta_{Href} = 0.8$ is the efficiency reference value for heat production;
$\eta_{Eref} = 0.33$ is the efficiency reference value for power production;
e_{sys} is the annual specific electricity production from CCHP system;
q_{sys} is the annual specific heat production from CCHP system;
q_F is the specific fuel consumption of CCHP system.

PES does not point if this production of the CCHP system is useful or not. The energy could be exceedingly produced and dissipated in the environment (especially the thermal energy).

- The second indicator namely *Energy efficiency* (EFF) sets the correlation between the primary and useful energy and shows how much of the primary energy is found to be useful energy and is given by the relation:

$$\mathrm{EFF} = \frac{e_{sys} + q_{sys}}{q_F + e_{PV} + q_{ST}} \times 100\,\% \tag{63}$$

where:

e_{PV} is the annual specific electricity production from PV panels;
q_{ST} is the annual specific heat production from ST panels.

7.2 Performance Evaluation and Improving in the Case of Structural Models with Mechanical Compression Chiller

To calculate the two indicators PES and EFF, we need to determine the specific fuel consumption of CCHP system. Specific fuel consumption is the sum of fuel input for cogeneration unit and additional boiler. Its determination involves knowledge of cogeneration unit and boiler energy production, and cogeneration unit and boiler efficiency.

To exemplify, let us consider the illustrative exercise from the paragraphs above. Using data corresponding to the illustrative exercise, namely from Table 31 for monthly energy specific consumption of the mCCHP system in the case of cooling with mechanical compression chillers, and from Table 6 for monthly specific power and heat production of the panels, we can determine the energy production of the CHP and backup boiler. The results obtained for the mCCHP system with mechanical compression chiller are shown in Table 35, then PES = 73 % and EFF = 55 % for the whole CCHP system. Even though the PES is high, the efficiency is low indicating an excessive energy production.

Table 35 Energy production of the mCCHP system with mechanical compression chiller

Month	Energy specific consumption of the mCCHP system		Specific power and heat production of the PV and ST panels		Specific power and heat production of the CHP unit		Specific heat production of the boiler
	e_{sys} (kWh/m^2)	q_{sys} (kWh/m^2)	e_{PV} (kWh/m^2)	q_{TP} (kWh/m^2)	e_{cg} (kWh/m^2)	q_{cg} (kWh/m^2)	q_{ad} (kWh/m^2)
January	3.33	15.94	1.24	4.16	2.09	6.26	5.52
February	3.33	13.77	1.76	5.90	1.57	4.71	3.16
March	3.33	10.94	2.91	9.73	0.42	1.27	0.00
April	3.47	6.23	4.04	13.53	0.00	0.00	0.00
May	4.23	2.21	5.02	16.81	0.00	0.00	0.00
June	4.62	1.25	5.47	18.32	0.00	0.00	0.00
July	4.89	1.25	5.54	18.56	0.00	0.00	0.00
August	4.89	1.25	4.86	16.27	0.03	0.10	0.00
September	4.34	1.39	3.59	12.02	0.75	2.26	0.00
October	3.56	5.94	2.51	8.40	1.05	3.15	0.00
November	3.33	10.63	1.34	4.49	1.99	5.97	0.17
December	3.33	14.21	1.01	3.37	2.32	6.97	3.87
Total	*46.65*	*85.01*	*39.26*	*131.57*	*10.24*	*30.71*	*12.71*

A more detailed analysis made at monthly level shows the yearly evolution of the excessive electrical and thermal energy produced. Thus, based on the data recorded in Table 35 the electricity and heat excess production are shown in Figs. 31 and 32.

Excess energy is the difference between energy production (denoted with "load") and useful energy (denoted with" sys").

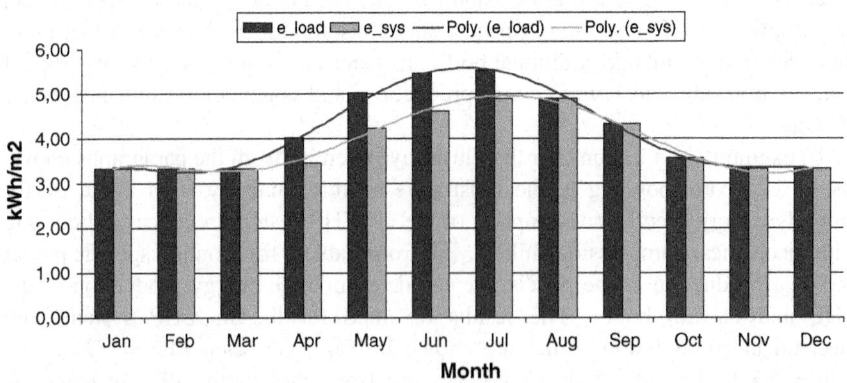

Fig. 31 Excess electricity production

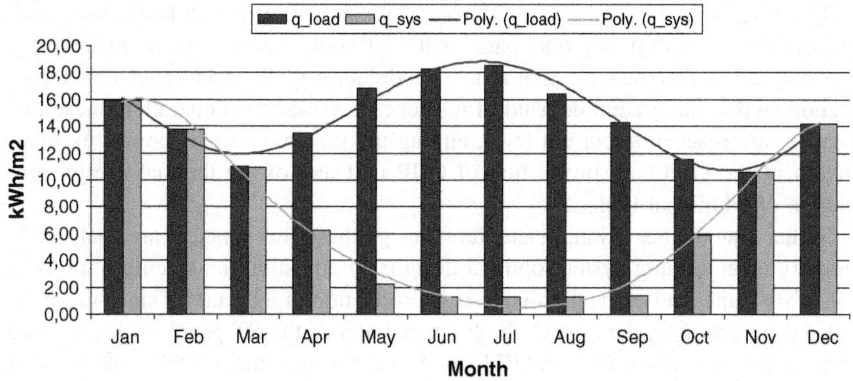

Fig. 32 Excess heat production

In addition, these figures show the monthly useful and produced specific energies as well as their trend during the entire year.

Improving the performance indicators of the system is achieved by analyzing the specific energy production system components and through actions to reduce losses but especially by diminishing the excess energy produced.

Analysis of the monthly power and heat consumption, and monthly energy production of PV and ST panels (Table 35), highlights big excess heat produced by thermal panels in summer season and small excess production of photovoltaic panels.

Linking the renewable energy production with the energy consumption of residence could be done by changing the surface area of the thermal and photovoltaic panels. To find the concrete actions, which could lead to performance improving, two cases may be considered, namely, case of the CCHP system type off-grid and case of the CCHP system type on-grid. Finally, the results of these two cases shall be comparative analyzed in order to make a decision.

7.2.1 Structural Model Type off-Grid with Mechanical Compression Chiller

The off-grid system supplies the thermal and electrical energy for own residential consumption only. During operation, the whole CCHP system control is based on balancing the electrical energy production and consumption. Control of this balance consists in appropriate modulation of the electrical energy obtained from CHP unit, so that, together with the current PV panel production, to cover the electrical energy current demands.

But, by producing electricity, the CHP unit produces thermal energy in mean time. The amount of this energy cannot be modulated because it is proportional with the amount of electricity produced.

This is why the thermal energy balance is reached by using an additional boiler which, together with the thermal panel and CHP unit, supplies the thermal energy. The balance is obtained by appropriate modulation of the additional boiler production so that the current demand of residence is covered. In operation, these two balances are reached at current level, aiming the system optimal control, by modulation of the electricity production of CHP unit and of the thermal energy production of additional boiler.

Rather than the operation, in structural design these two balances are analyzed at monthly level, aiming system optimal design, by adjusting the electricity production of PV panel and of the thermal energy production of ST panel. This adjusting is realized by changing the area of the PV panel and of the ST panel so that the two performance indicators, PES and EFF, evaluated for the entire CCHP system and at monthly level, acquire the best values, while the utilization factor satisfies the condition of being less than 40 % (condition imposes by architectural rules).

To find the optimal area of PV and ST panels, for the month with the biggest production of the PV, should eliminate the excess electricity production, by considering it as equal to the residence demand, $e_{PV} = e_{sys}$.

Let us consider the illustrative exercise. In this case the utilization factor of PV panels (see Tables 6 and 35) could be recalculated using the relation:

$$k_{PV}^{r} = \frac{e}{e_{PV}^{r}} \tag{64}$$

Similarly, the the utilization factor of ST panels become:

$$k_{ST}^{r} = \frac{q}{q_{ST}^{r}} \tag{65}$$

The calculation results are $k_{PV}^{r} = 0.2$ and $k_{ST}^{r} = 0.012$. Corresponding to the ground floor area of 100 m^2, the PV and ST panels areas are A_{PV} = 20 m^2 and A_{ST} = 1.2 m^2. The ground floor area was considered equal with the total floor area of the building (building with a floor). The results allow the graphic representation of the electricity balance presented in Fig. 33 and of the thermal balance presented in Fig. 34.

The monthly heat balance has a small excess. The obtained performance indicators point a decrease of PES = 61 % and a significant increase of the global efficiency of the CCHP system, EFF = 86 %. The monthly distribution of PES and EFF are shown in Fig. 35.

7.2.2 Structural Model Type on-Grid with Mechanical Compression Chiller

Unlike the case of the off-grid model, when the imposed condition is the balance between required and produced power and heat, in case of the on-grid system the

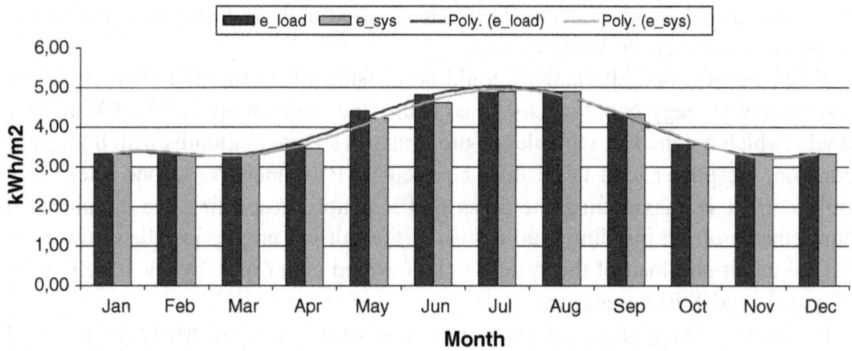

Fig. 33 Electricity balances

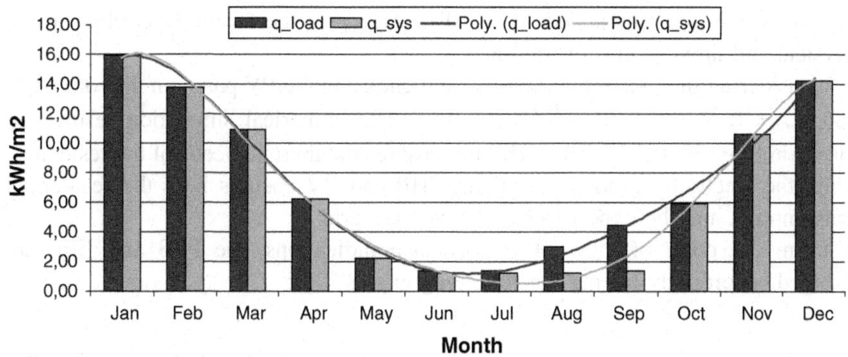

Fig. 34 Heat balances

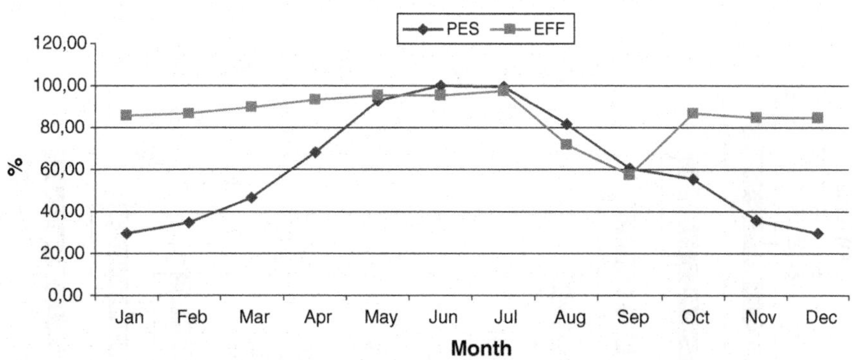

Fig. 35 PES and EFF monthly distribution

imposed condition is the balance between required and produced heat only while the power can be produced in excess.

To exemplify, two alternatives could be considered. In the first alternative, the load sharing strategy could consist in adjusting the nominal area of the PV and ST panels (which are the key variables of the strategy) so that in months with high solar radiation, the power load to be fully covered by PV panels. A second alternative could consist in maximizing the areas of PV panels (respecting the architectural requirements of the building) and adjusting the Stirling engine installed power, so that the maximum load of the system to be covered either only by PV panels or by PV panels and Stirling engine.

In the last alternative, the prosumers can export the electricity to the grid (thermal energy cannot be exported) and the governing strategy for the CHP unit changes so that to supply at any moment the thermal energy needed. Consequently, the additional boiler and thermal solar panels can be discarded. Since the electricity can be exported to the grid, the PV panels area could be $A^r_{\mathrm{PV}} = 0.4\, A_{\mathrm{GFA}}$, for the roof, to which it can be added $\mathrm{A}^{\mathrm{f}}_{\mathrm{PV}} = 0.15\ \mathrm{A}_{\mathrm{GFA}}$, for the façade (according to the physical and architectural restrictions).

Considering the illustrative exercise and choosing the PV panel on the roof only, $A_{\mathrm{PVpanel}} = 0.4*100 = 40\ \mathrm{m}^2\ \left(k^r_{\mathrm{PV}} = 0.4\right)$. The numerical simulation results are shown in Figs. 36 and 37. The CHP unit ensures the thermal needs of the residence, while the electricity production of the CHP and PV ensures both the residential consumption and the export to the low-voltage grid.

From the point of view of performance indications, the PES and EFF are changed accordingly having the following expressions:

$$\mathrm{PES} = \left(1 - \frac{q_F}{\frac{e_{\mathrm{sys}}+e_{grid}}{\eta_{\mathrm{Eref}}} + \frac{q_{\mathrm{sys}}}{\eta_{\mathrm{Href}}}}\right) \times 100\,\% \tag{66}$$

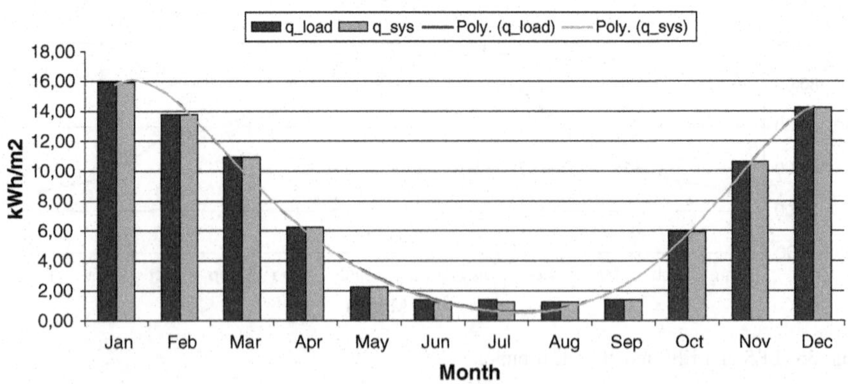

Fig. 36 On-grid thermal balance

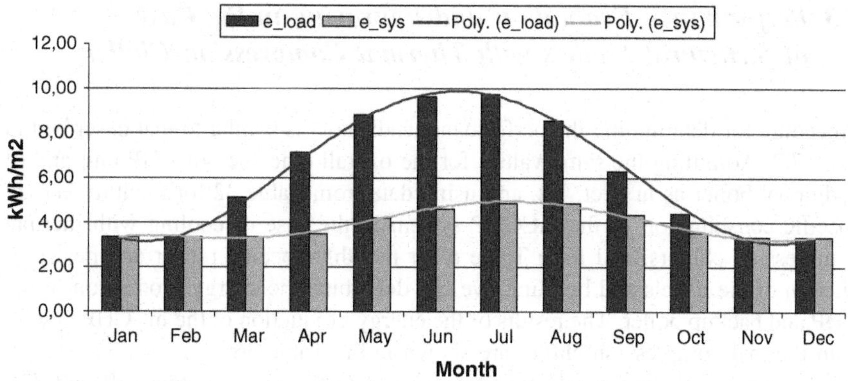

Fig. 37 On grid electricity balance

$$EFF_{CCHP} = \frac{e_{sys} + e_{grid} + q_{sys}}{q_F + e_{PV} + q_{ST}} \times 100\,\% \tag{67}$$

where:

e_{grid} electricity produced by the CCHP system and exported in grid.

The obtained performance indicators became PES = 69.6 % and EFF = 94.6 %. The monthly distribution of PES and EFF are shown in Fig. 38.

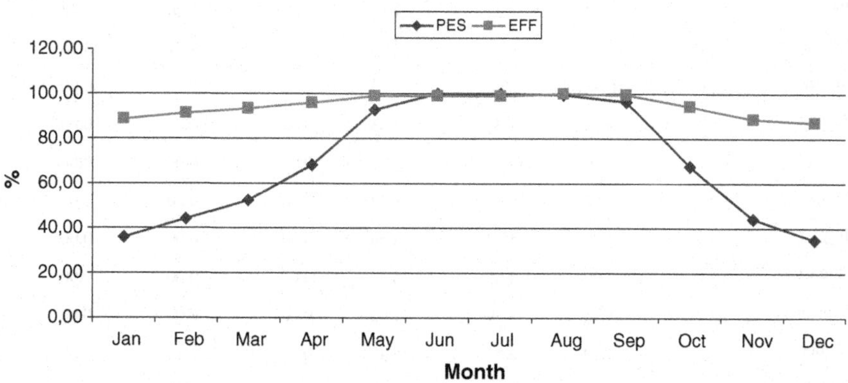

Fig. 38 PES and EFF for on grid system with MCC

7.3 Performance Evaluation and Improving in the Case of Structural Models with Thermal Compression Chiller

Procedure for determining the performance indicators is similar to that described in Sect. 7.2. Admitting the same values for the overall efficiency of CHP unit and of additional boiler as in Sect. 7.2, and using data from Table 32 for monthly energy specific consumption of the mCCHP system in the case of cooling with thermal compression chillers, and from Table 6 for monthly specific power and heat production of the panels and building, we can determine the energy production of the CHP and backup boiler. The results of the energy production of the mCCHP system with thermal compression chiller are shown in the Table 36.

The power and heat production represented in the graphic (Figs. 39 and 40), depending on the number of functioning hours of the system in a year, allows the analysis of excessive electrical and thermal energy produced. The figures show the useful specific energies produced by the system and excessive energy.

The performance indicators of the system determinate for this case, indicate PES = 81 % and EFF = 66 % of the whole CCHP system.

Table 36 Energy production of the mCCHP system with thermal compression chiller

Month	Energy specific consumption of the mCCHP system		Specific power and heat production of the PV and ST panels		Specific power and heat production of the CHP unit		Specific heat production of the boiler
	e_{sys} (kWh/m^2)	q_{sys} (kWh/m^2)	e_{PV} (kWh/m^2)	q_{TP} (kWh/m^2)	e_{cg} (kWh/m^2)	q_{cg} (kWh/m^2)	q_{ad} (kWh/m^2)
January	3.33	15.94	1.24	4.16	2.09	6.26	5.52
February	3.33	13.77	1.76	5.90	1.57	4.71	3.16
March	3.33	10.94	2.91	9.73	0.42	1.27	0.00
April	3.33	6.76	4.04	13.53	0.00	0.00	0.00
May	3.33	5.56	5.02	16.81	0.00	0.00	0.00
June	3.33	6.08	5.47	18.32	0.00	0.00	0.00
July	3.33	7.09	5.54	18.56	0.00	0.00	0.00
August	3.33	7.09	4.86	16.27	0.00	0.00	0.00
September	3.33	5.16	3.59	12.02	0.00	0.00	0.00
October	3.33	6.79	2.51	8.40	0.82	2.46	0.00
November	3.33	10.63	1.34	4.49	1.99	5.97	0.17
December	3.33	14.21	1.01	3.37	2.32	6.97	3.87
Total	40	110	39.26	131.57	9.22	27.65	12.71

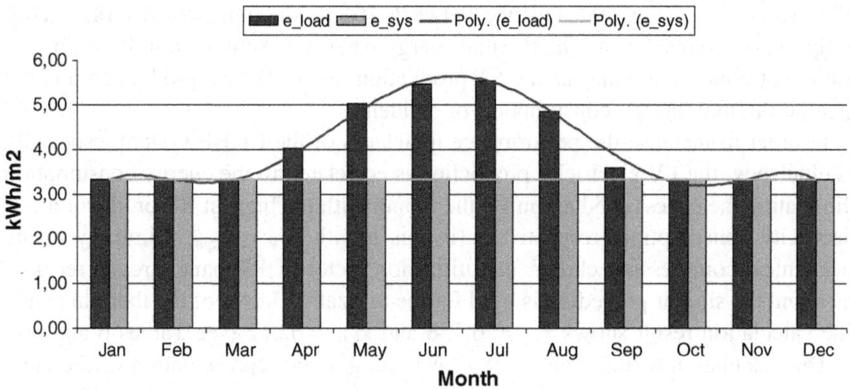

Fig. 39 Excess electricity production (TCC)

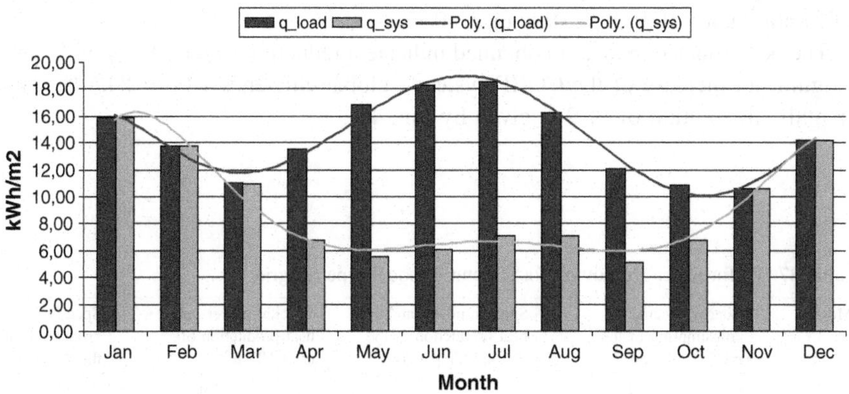

Fig. 40 Excess heat production (TCC)

Increasing the performance indicators of the system is achieved by analyzing the specific energy production of the system components and through actions to reduce losses, but especially to reduce the excess energy produced.

To find the concrete actions that could lead to performance improving, two cases will be considered, namely case of the structural model type off-grid and case of the structural model type on-grid.

7.3.1 Structural Model Type off-Grid with Thermal Compression Chiller

The off-grid system ensures the energy only for own residential consumption in the two main forms, electrical and thermal. The CHP unit is governed by the power drive strategy ensuring the balance between PV production and electricity consumption of the residence. The thermal energy balance is kept in equilibrium by the additional boiler that balances the ST production, the CHP unit production and the specific thermal energy consumption of residence.

In order to increase the performance indicators of the CCHP system, especially its efficiency, the PV electricity production is correlated to the energy consumption eliminating the excess production for the month with the biggest PV production and electricity consumption, respectively (in this month $e_{PV} = e_{sys}$). Similarly to the mechanical compression chiller, the utilization factor of PV panel area is recalculated and the similar procedure is used for the utilization factor of the thermal panel. The calculation result shows $k_{PV}^{r} = 0.136$ and $k_{ST}^{r} = 0.062$ (see Tables 6 and 36).

The calculation result (Table 37) allows the graphic representation of the electricity balance shown in Fig. 41 and the thermal balance shown in Fig. 42.

The monthly heat balance has a small excess. After the calculation of the heat and electricity from solar energy, it is necessary to verify the conditions set about architectural sustainability and utilization factors. For this case the results show a utilization factor of 19.8 % (less than 40 %).

The performance indicators obtained indicate a reduction of the PES = 64 % and a significant increase of the CCHP system's global efficiency EFF = 84.6 %. The monthly distribution of PES is given by Fig. 43.

Table 37 Performance improving for structural model type off-grid

Month	Energy specific consumption of the mCCHP system		Specific power and heat production of the PV and ST panels		Specific power and heat production of the CHP unit		Specific heat production of the boiler
	e_{sys} (kWh/m^2)	q_{sys} (kWh/m^2)	e_{PV} (kWh/m^2)	q_{TP} (kWh/m^2)	e_{cg} (kWh/m^2)	q_{cg} (kWh/m^2)	q_{ad} (kWh/m^2)
January	3.33	15.94	0.74	1.59	2.59	7.76	6.59
February	3.33	13.77	1.06	2.26	2.27	6.82	4.69
March	3.33	10.94	1.74	3.72	1.59	4.77	2.45
April	3.33	6.76	2.42	5.17	0.91	2.74	0.00
May	3.33	5.56	3.01	6.42	0.32	0.97	0.00
June	3.33	6.08	3.27	7.00	0.06	0.17	0.00
July	3.33	7.09	3.32	7.09	0.01	0.03	0.00
August	3.33	7.09	2.91	6.22	0.42	1.26	0.00
September	3.33	5.16	2.15	4.59	1.18	3.54	0.00
October	3.33	6.79	1.50	3.21	1.83	5.48	0.00
November	3.33	10.63	0.80	1.72	2.53	7.58	1.33
December	3.33	14.21	1.50	1.29	1.83	5.50	7.42
Total	40	110	24.42	50.29	15.54	46.63	22.49

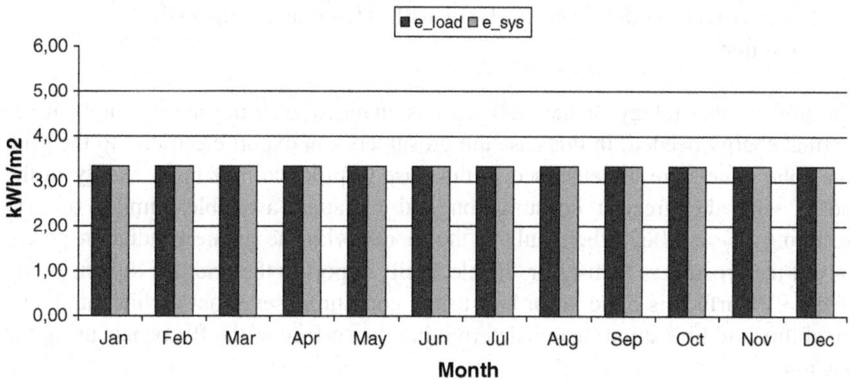

Fig. 41 Optimal electricity balances

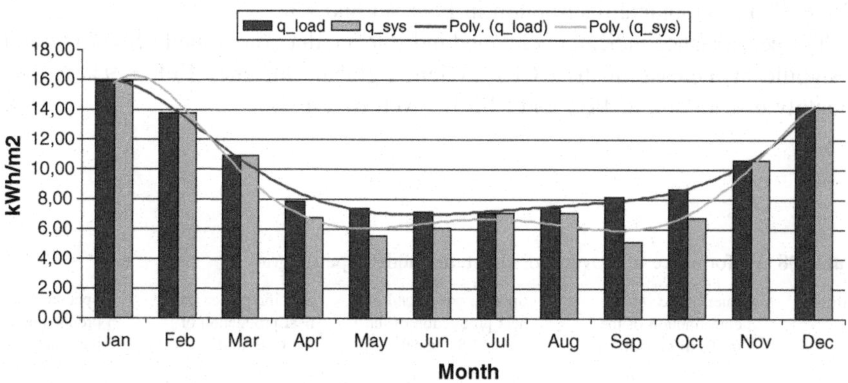

Fig. 42 Optimal heat balances

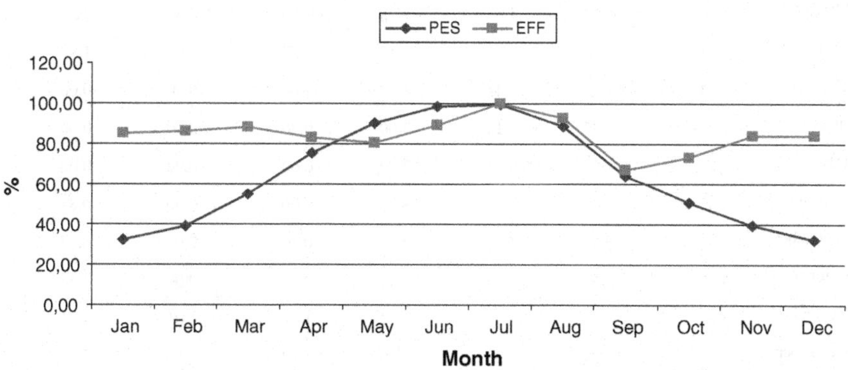

Fig. 43 Monthly distribution of PES and EFF

7.3.2 Structural Model Type on-Grid with Thermal Compression Chiller

The governing strategy of the CHP unit is changed ensuring at any moment the thermal energy needed. In this case the prosumers can export electricity to the grid. The solar panels are dimensioned in this case to produce maximum energy in the month with the greatest consumption and radiation available, similar off-grid (resulting $k^r_{\mathrm{ST}} = 0.062$ where July is the month with the greatest radiation potential). If the prosumers focuses on the electricity export to the grid, the dimensioning of the PV surface is done according to the conditions set about architectural sustainability and utilization factors determining the surface of the PV panels using the relation:

$$k^r_{PV} = 0.4 - k^r_{ST} \tag{68}$$

The results of numerical simulation for the illustrative exercise are shown in Table 38 and graphical distribution in Figs. 44 and 45.

The performance indicators obtained indicate a reduction of the PES = 75 % and a significant increase of the CCHP system's global efficiency EFF = 94 %. The monthly distribution of PES and EFF is given by Fig. 46.

Table 38 Performance improving for structural model type on-grid

Month	Energy specific consumption of the mCCHP system		Specific power and heat production of the PV and ST panels		Specific power and heat production of the CHP unit		Specific heat production of the boiler
	e_{sys} (kWh/m^2)	q_{sys} (kWh/m^2)	e_{PV} (kWh/m^2)	q_{TP} (kWh/m^2)	e_{cg} (kWh/m^2)	q_{cg} (kWh/m^2)	q_{ad} (kWh/m^2)
January	3.33	15.94	1.85	1.59	1.48	4.44	9.91
February	3.33	13.77	2.62	2.26	0.71	2.12	9.39
March	3.33	10.94	4.33	3.72	0.00	0.00	7.22
April	3.33	6.76	6.01	5.17	0.00	0.00	1.59
May	3.33	5.56	7.47	6.42	0.00	0.00	0.00
June	3.33	6.08	8.14	7.00	0.00	0.00	0.00
July	3.33	7.09	8.25	7.09	0.00	0.00	0.00
August	3.33	7.09	7.23	6.22	0.00	0.00	0.87
September	3.33	5.16	5.34	4.59	0.00	0.00	0.57
October	3.33	6.79	3.73	3.21	0.00	0.00	3.58
November	3.33	10.63	1.99	1.72	1.34	4.01	4.91
December	3.33	14.21	1.50	1.29	1.83	5.50	7.42
Total	40	110	58.46	50.29	5.36	16.07	45.45

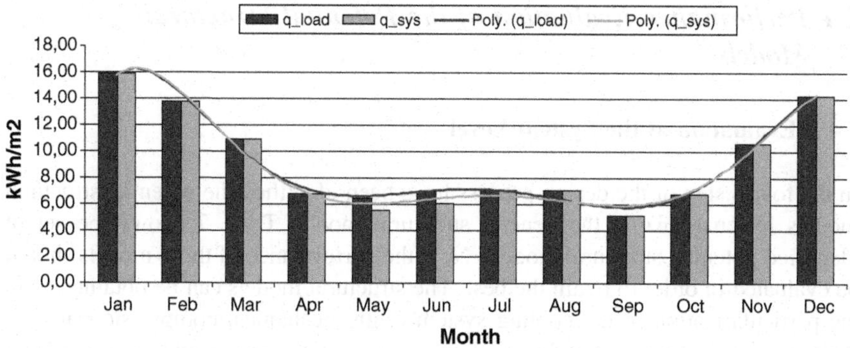

Fig. 44 On-grid heat balances (TCC)

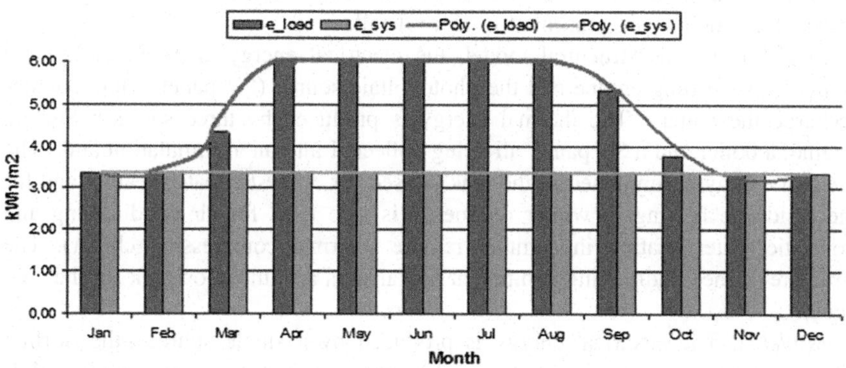

Fig. 45 On-grid electricity balances (TCC)

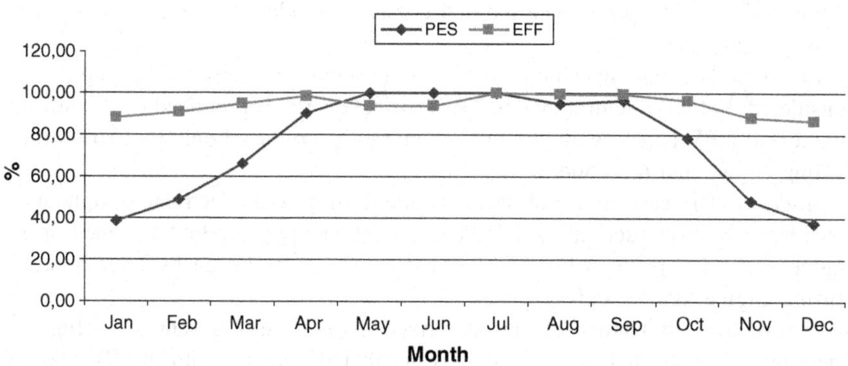

Fig. 46 PES and EFF for on grid with TCC

7.4 *Performance Evaluation of the Potential Structural Models*

7.4.1 Evaluation at the System Level

In the fourth step of the design process have been identified the potential structural models by customizing the general structural model. Table 7 shows the set of identified potential structural models. Now the performance of these models should be evaluated in order to retain the best. The structural models can be obtained from the particularization of the cooling system (with mechanical compression or with thermal compression) and of the auxiliary sources (with thermo solar panels or with photovoltaic panels). With all the structural models, the electric load control is achieved by adjusting the fuel consumption of the Stirling engine, and the thermal load control—by controlling the fuel consumption of the additional boiler. The next set of potential structural models refers to the off-grid case only. In what follows, the retained models are presented in more detail.

Model 1. In this structural model, the electrical energy is produced by two sources: the Stirling engine and the photovoltaic source (PV panel). Both sources recharge the battery. The thermal energy is produced by three sources: Stirling engine, a boiler and a ST panel, all being collected into an accumulation tank. The thermal energy accumulated in the tank is used for domestic water heating and for the residence heating in winter regime. It is also used for air conditioning and domestic water heating in summer regime (thermal compression chiller). The structure of the conditioning equipment contains an accumulation tank for the cold agent.

Model 2. The electrical energy is produced by a single source—the Stirling engine. The thermal energy is produced by three sources, as well as in model 1, namely Stirling engine, boiler and ST panel.

Model 3. The difference as compared to the previous model consists in the fact that, the ST panel is missing.

Model 4. In this structural model, the electrical energy is produced by the Stirling engine and by a PV panel. The thermal energy is produced only by two sources: Stirling engine and boiler.

Model 5. The case in which the air conditioning uses electrical equipment is considered. In this structural model, the electrical energy is produced by the Stirling engine and a PV panel, while the thermal energy is produced only by two sources: Stirling engine and the boiler.

Model 6. The case in which the air conditioning uses electrical equipment is considered. In this structural model, the electrical energy is produced by the Stirling engine and a PV panel, while the thermal energy is produced by three sources: Stirling engine, boiler, and ST panel.

Performance indicators of the described models are presented in Fig. 47 Therefore, we note that the best models are that which use both the PV and ST panels. At the same structure of energy sources, the model with thermal

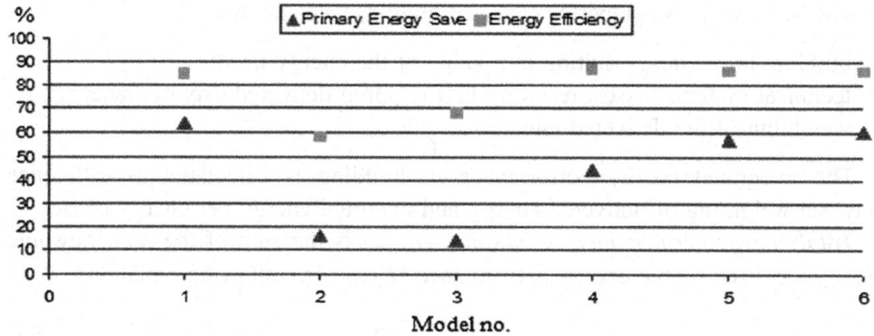

Fig. 47 Performance indicators of the models

compression chiller is more efficient than that with mechanical compression chiller. The model 5 and 6 differ in number of heat sources, but these models have almost identical performance indicators, reason that, in Chap. Functional Design of the mCCHP-RES System will be analyzed only the first 5 models.

7.4.2 Evaluation at the Couple Building-System Level

The energy performance of a building [22] shall be expressed in a transparent manner and shall include an *energy performance indicator* and a *numeric indicator of primary energy use* based on primary energy factors per energy carrier, which may be based on national or regional annual weighted averages or a specific value for on-site production.

The energy performance indicator is energy class of the building, which by Energy Performance Certificates (EPCs) give information only on building properties, resulting in energy needs (used).

For the mCCHP system, the energy efficiency indicator indicate conversion efficiency on the system and the PES indicator indicate the percentage of fuel saved from the energy production of the CCHP system compared to the same energy produced by the reference system, without indicating the share of renewable energy.

The new numeric indicators of primary energy use refer to specific value for on-site production and the share of renewable energy.

A new generation of EN standards [23] for implementing the Energy Performance of Building Directive introduced a clear distinction, between renewable and nonrenewable primary energy. In this sense was defined for assessment of energy performance of building, the two new indicators namely:

- integrated energy performance of building, $E_{p,\mathrm{tot}}$;
- share of renewable or Renewable Energy Ratio (RER).

The integrated energy performance of building depends on:

- building properties, resulting in a value of the energy need;
- technical systems losses, resulting in a required delivered/exported energy;
- weighting of the delivered energy.

The integrated energy performance of building is calculated as difference between weighting of delivered energy and exported energy per energy carrier.

Total energy performance of building $(E_{p,\mathrm{tot}})$ is the sum of the nonrenewable $(E_{p,\mathrm{nren}})$ and renewable energy performance $(E_{p,\mathrm{ren}})$ by following equation:

$$E_{p,\mathrm{tot}} = E_{p,\mathrm{nren}} + E_{p,\mathrm{ren}} \tag{69}$$

By renewable energy use for buildings, the new standard defines the ratio between renewable and total integrated energy performances, namely share of renewable or Renewable Energy Ratio (RER) by equation:

$$RER = \frac{E_{p,\mathrm{ren}}}{E_{p,\mathrm{tot}}} .100\,\% \tag{70}$$

In this way was introduced a clear distinction, between renewable and non-renewable primary energy and their connection with the total primary energy factor, to support renewable energy share evaluation. This is possible by introduce the final weighting factor namely primary energy conversion factors to define the integrated energy performance of buildings, so:

$$E_p = \sum E_{\mathrm{del},i} f_{p,\mathrm{del},i} - \sum E_{\mathrm{exp},i} f_{p,\mathrm{exp},i} \tag{71}$$

where:

- E_p—The primary energy demand,
- $E_{\mathrm{del},i}$—final energy demand of energy carrier i,
- $f_{p,\mathrm{del},i}$—primary energy factor for demand energy carrier i,
- $E_{\mathrm{exp},i}$—exported final energy of energy carrier i,
- $f_{p,\mathrm{exp},i}$—primary energy factor for export energy carrier i,
- i—the current number of the carrier.

For electricity delivered/exported, the primary energy factor is shown in Table 39.

The primary energy weighting is done depends of the fuel type and have the values given in the Table 40.

In the structural model 1 the electrical energy is produced by two sources: the Stirling engine and the photovoltaic source (PV panel) while the thermal energy is produced by three sources: Stirling engine, a boiler and a ST panel.

Table 39 Primary energy factors for electricity

Primary energy factor (fp delivered)			Primary energy factor (fp exported)	
On site	2.3		2.5	Immediate use
Nearby	2		2.2	Temporary exported
Distant	1.6		1.8	Exported (never used in building)

Table 40 Primary energy factors [24]

Type of the energy supplier	Primary energy factor		
	Renewable fp.ren.	Non-renewable-fp.nren	Total fp.tot
District heat/nearby	0	1.3	1.3
Grid electricity	0.2	2.3	2.5
Solar thermal	1	0	1
Photovoltaics	1	0	1
Bio gas	0.2	0.9	1.1
Natural gas	0	1.1	1.1

The results of evaluation with the new numeric indicators of primary energy use at the couple building-system level if the building is off-grid or on-grid (for electricity connection) are shown in Tables 41 and 42.

Table 41 Case 1—Structural model 1 type off-grid

	Energy used (kWh/m^2 year)	Delivered energy (kWh/m^2 year)	fp. ren	Ep.ren (kWh/m^2 year)	fp. nren	Ep.nren (kWh/m^2 year)	fp. tot	Ep.tot (kWh/m^2 year)
Electrical	40.00							
PV		24.42	1.00	24.42	0.00	0.00	1.00	24.42
CHP unit		15.54	0.00	0.00	1.10	17.09	1.10	17.09
Thermal	110.00							
ST		50.29	1.00	50.29	0.00	0.00	1.00	50.29
CHP		46.63	0.00	0.00	1.10	51.29	1.10	51.29
Back-up		22.49	0.00	0.00	1.10	24.74	1.10	24.74
Exported energy	0.00	On site	0.20	0.00	2.30	0.00	2.50	0.00
Energy performance indicator of building			**Ep. ren**	74.71	**Ep. nren**	93.13	**Ep. tot**	**167.84**
Renewable energy ratio			**RER** [%]					**44.51**

Table 42 Case 2—Structural model 1 type on-grid

	Energy used (kWh/ m^2 year)	Delivered energy (kWh/ m^2 year)	fp. ren	Ep.ren (kWh/ m^2 year)	fp. nren	Ep.nren (kWh/ m^2 year)	fp. tot	Ep.tot (kWh/ m^2 year)
Electrical	40.00							
PV		58.46	1.00	58.46	0.00	0.00	1.00	58.46
CHP unit		5.36	0.00	0.00	1.10	5.90	1.10	5.90
Thermal	110.00							
ST		50.29	1.00	50.29	0.00	0.00	1.00	50.29
CHP		16.07	0.00	0.00	1.10	17.68	1.10	17.68
Back-up		45.45	0.00	0.00	1.10	50.00	1.10	50.00
Exported energy	23.82	On site	0.20	4.76	2.30	54.79	2.50	59.55
Energy performance indicator of building			*Ep. ren*	103.99	**Ep. nren**	18.78	**Ep. tot**	**122.77**
Renewable energy ratio			**RER**	%	–	–	–	**84.70**

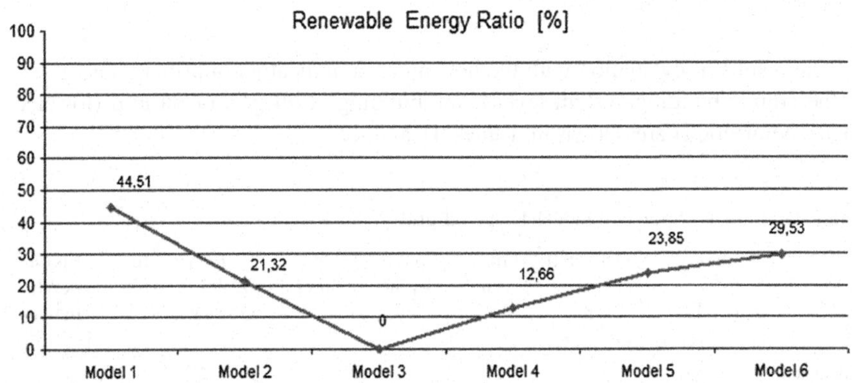

Fig. 48 Renewable energy ratio of the structural models

Evaluation at the couple building-system level for the set of potential structural models (from model 1 to model 6) if the mCCHP system work off-grid. showed as in terms of renewable energy ratio the model 1 is the best. The share of renewable energy for these models is shown in Fig. 48.

References

1. Werner S The european heat market—Ecoheatcool Work Package 1-Intelligent Energy Europe Programme © Ecoheatcool and Euroheat & Power 2005–2006. http://www.euroheat.org
2. Constantinescu N, Dalin P, Nilsson J, Rubenhag A Ecoheatcool Work Package 2-Intelligent Energy Europe Programme © Ecoheatcool and Euroheat & Power 2005–2006. http://www.ecoheatcool.org
3. RETScreen climate database http://www.retscreen.net/ang/d_data_w.php
4. Joint Research Centre—Institute for Energy and transport—PVGIS © European Communities http://re.jrc.ec.europa.eu/pvgis/solres/solres.htm
5. Joint Research Centre—Institute for Energy and transport—PVGIS © European Communities http://re.jrc.ec.europa.eu/pvgis/solres/solreseurope.htm
6. Laustsen J Energy efficiency requirements in building codes, energy efficiency policies for new buildings © OECD/IEA, March 2008
7. Asan H (2006) Numerical computation of time lags and decrement factors for different building materials. Build Environ 41:615–620 (Karadeniz Technical University, Trabzon, Turkey)
8. Norm on thermo calculation of building construction elements. C 107-2005 M.T.C.T.Ordinul no. 2055/29.11.2005
9. High performance buildings and envelopes. Tallinn University of Technology, Estonia, June 2010
10. Raluca Pleşu et. al -Infrared thermography applications for building investigation. Universitatea Tehnică, Gheorghe Asachi din Iaşi Tomul LVIII (LXII), Fasc. 1, 2012 Secţia Construcţii. Arhitectură
11. Albatici R, Tonelli AM (2010) Infrared thermovision technique for the assessment of thermal transmittance value of opaque building elements on site. Energy Build 42:2177–2183
12. DallO G (2013) Infrared screening of residential buildings for energy audit. Energies 6:3859–3878. doi: 10.3390/en6083859
13. Barreira E, de Freitas PV (2007) Evaluation of building materials using infrared thermography. Constr Build Mater 21:218–224
14. Thermal performance of buildings—qualitative detection of thermal irregularities in building envelopes—infrared method; EN 13187:1998—appendix D
15. EN 15193-2006 Energy requirements for lighting
16. EN 12464-1, Light and lighting—lighting of work places—Part 1: indoor work places what specifies lighting requirements for indoor work places, which meet the needs for visual comfort and performance
17. Methodology for assessing the energy performance of buildings. Romanian Ministry of Commerce, Industry and Tourism 2009
18. Karlsruhe Fachinformationszentrum (ed) (2005) Blockheizkraftwerke: Eine Leitfaden für den Anwender. TÜV GmbH, Köln
19. CIBSE (Chartered Institute of Building Services Engineers). Indoor work places and CIBSE code for lighting part 2
20. EN 15217 (2007) Energy performance of buildings—methods for expressing energy performance and for energy certification of buildings
21. Directive 2012/27/EU of the European parliament and of the council of 25 October 2012 on energy efficiency, amending directives 2009/125/EC and 2010/30/EU and repealing directives 2004/8/EC and 2006/32/EC (OJ L 315, 14 November 2012, p 1)
22. Directive 2010/31/EU. Directive on the energy performance of buildings OJ L 153, pp 13–35

23. EN 15603 Energy performance of buildings. Overall energy use and definition of energy ratings
24. High energy performance buildings: design and evaluation methodologies new features in EN 15603, overall energy use and definition of energy ratings Laurent Socal (IT)—Bruxelles 25th of June 2013

Functional Design of the mCCHP-RES System

Nicolae Badea, Alexandru Epureanu, Emil Ceanga, Marian Barbu and Sergiu Caraman

Abstract In order to be shown in detail, the design process algorithm (see Fig. 2 in Chapter "Structural Design of the mCCHP-RES System") was divided into two stages, namely structural design and functional design. In Chapter "Structural Design of the mCCHP-RES System" was presented the first stage while in this chapter is presented the second. Then, in Chapter "Experimental Case Study" it is shown a concrete example. Here, the functional design was divided into five steps and to each step has been devoted a paragraph. Regarding the contents of steps, this is as following: In the first step (described in Sect. 2.1) the acceptable structural models resulted from the structural design stage, are completed with all the complementary components (such as recirculation pumps, expansion tanks, heat exchangers etc.), this way becoming *the functional models of the system*. After that all the components, main or complementary, are dimensioned taking into account both the energetic consumptions of the residence and the internal energetic consumptions of the system. The next two steps (described in Sects. 3 and 4) consist in establishing *the operation and control strategy* of the system, based on which then the *system dynamics* is analyzed numerically. Simulation and analysis of the system operation highlights the system performance in critical moments (such as the coldest/hottest day of the year, for example), in case of each functional model. The step described in Sect. 5 consists in *designing the monitoring and control subsystem*, which includes choosing the field equipments, design of the connection

N. Badea (✉) · A. Epureanu · E. Ceanga · M. Barbu · S. Caraman
"Dunarea de Jos" University of Galati, Galati, Romania
e-mail: nicolae.badea@ugal.ro

A. Epureanu
e-mail: alexandru.epureanu@ugal.ro

E. Ceanga
e-mail: emil.ceanga@ugal.ro

M. Barbu
e-mail: marian.barbu@ugal.ro

S. Caraman
e-mail: sergiu.caraman@ugal.ro

N. Badea (ed.), *Design for Micro-Combined Cooling, Heating and Power Systems*,
Green Energy and Technology, DOI 10.1007/978-1-4471-6254-4_6

devices of these equipments to the data acquisition system, design of the monitoring software of the whole system, as well as designing the numerical controllers of the thermal and electric control loops. The last step (described in Sect. 8) consists in *designing the system interfaces*. Finally, the system designer has at his disposal several functional models, each with advantages and disadvantages, and, on this basis, can make the best decision.

1 Introduction

Nicolae Badea and Alexandru Epureanu

Conceptually, the functional design is the second stage of the design process (see Fig. 2 in Chap. "Structural Design of the mCCHP-RES System") and consists in: (a) conversion of the system structural model (that is the result of first stage) in functional model of the system, (b) adding of an incorporated control subsystem, and (c) drafting the manufacturing documentation of system (that is the result of the second stage). During the functional design, the *completion*, *customization*, and *analysis* of the system model takes place so that the description of the system gets to be sufficiently detailed that it becomes possible to elaborate the manufacturing documentation. In other words, the starting point of the functional design is the structural models resulting in the structural design stage, and the output is a complete description of mCCHP system, based on which the customer can realize it.

Unlike the closed-ended approach, where the functional design has as input only one structural model, in case of open-ended approach (the case presented here), at input there are several structural models all of which being considered acceptable. Therefore, in this case, all structural models will be transformed into functional models, following that the customer will select one of them. For the selected model, the manufacturing documentation of the mCCHP system will be prepared.

Practically, the functional design means undertaking the following actions:

- *Supplementing* the structural model with complementary components required for a system to work, thus this becoming the functional scheme of mCCHP system.
- *Sizing* of both the basic components and the complementary, thereby understanding *the setting of the technical characteristics of components* (such as the installed power of a supplier or consumer, the electric capacity of battery, the diameter of pipes through which circulate the thermal fluid, or the diameter of electrical wires, for example), on the one hand, and *determining of the operating parameters of components* (such as the operating time of a supplier, the pressure, flow, and temperature of the fluid flowing through a pipe, the amount of energy supplied daily by the PV/ST panel, for example), on the other hand. This way the functional scheme becomes the functional model of mCCHP system.
- *Analysis* of the system dynamics (by numerical simulation of the functional model operation, for example).

- *Design* of the control subsystem, including all components of this subsystem, both hard (the set of sensors for monitoring, for example) and soft ones (the numerical controller, for example), followed by incorporation of the entire control subsystem in the functional model of mCCHP system.

Algorithmically, the functional design consists of the following steps:

1. Functional modeling of the system,
2. Elaboration of the operation and control strategies,
3. Analysis of the system dynamics,
4. Design of the control subsystem, and
5. Design of the system interfaces.

In the following, the functional design algorithm is presented in detail. Each paragraph is dedicated to one of the steps above.

2 System Functional Modeling

Nicolae Badea

In establishing the set of structural models and evaluation of their performances (see Chap. "Structural Design of the mCCHP-RES System"), the following simplifications have been taken into account:

- Only the main components of the system have been included in the structure, especially those which generate the thermal or electric energy that the system supplies the residence with.
- Only the relation between the residence consumption and the system's components load has been considered, leaving out the system components' own energetic consumptions.
- Both the consumers' consumption and the suppliers load were evaluated at the level of each month of the year, and there were no established neither the technical characteristics (installed power, for example) nor the operation time of each component (either consumer or supplier).

For the structural design of the system, these simplifications are acceptable; in functional design, however, all the structural components of the system and all the situations they are to work within need to be taken into account. That is why, once selected, the structural models are considered as the starting point for the first step in functional design, which *is building the functional models*. This step includes two main actions:

- Completing the structural models with all the complementary components, so that it might support the concrete functioning mode of the system, this way each of them becoming a *functional scheme* of the system;
- *Dimensioning all the components* (be the main or complementary), of all functional schemes, this way each of them becoming a *functional model* of the system.

2.1 Building the Functional Schemes

The energetic fluxes in the mCCHP system imply generation, transport, and storage of energy. Each of these processes presupposes the manipulation of certain quantities of energy. The energy "carriers" may be fuels, thermal agents (usually fluids like water or air), or electrons, depending on the type of energy—thermal or electric.

The circulation of these "carriers" (which is the concrete form by means of which the energetic fluxes are achieved) presupposes the existence of a number of complementary components, which facilitate this circulation. The transport of pellets, the water recirculation pumps, the ventilators, the fuel-power-heat- and cold storages and the heat exchangers may be considered as examples of complementary components.

As available on the market, the main components of the system incorporate part of the complementary components and that is why they are already included in the structural model of the system. The rest of the complementary components need to be added. Thus, the structural model of the system becomes the functional scheme, which may then be used in the dimensioning of all the components, this way obtaining the functional model of the system.

To exemplify, let us consider the structural models presented in Sect. 7.4 in Chap. "Structural Design of the mCCHP-RES System". With part of them, the chiller is a mechanical compression one, and with the other part, it is thermally activated.

For the structural models with thermally activated chiller was built the global functional scheme presented in Fig. 1.

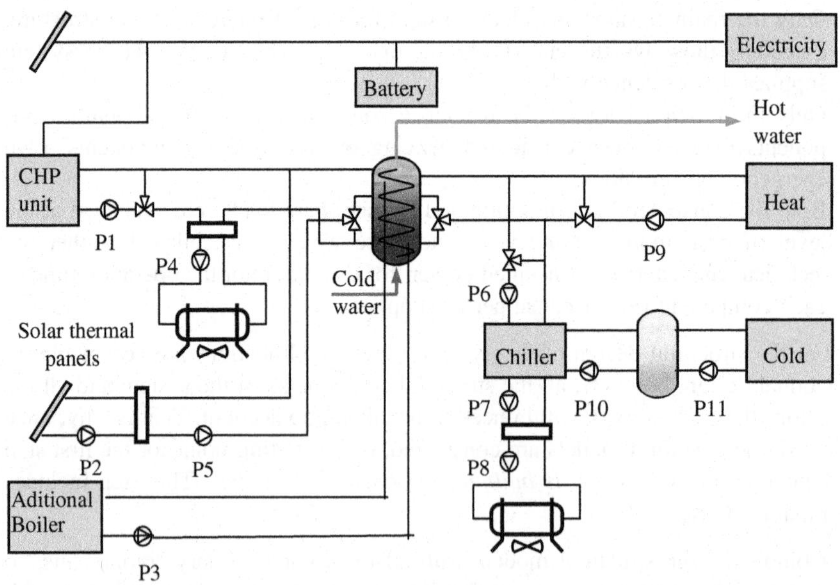

Production	Heat exchanger	Heat storage	Cooling	Cold storage	Consumptions

Fig. 1 Global functional scheme for a mCCHP system with a thermally compression chiller

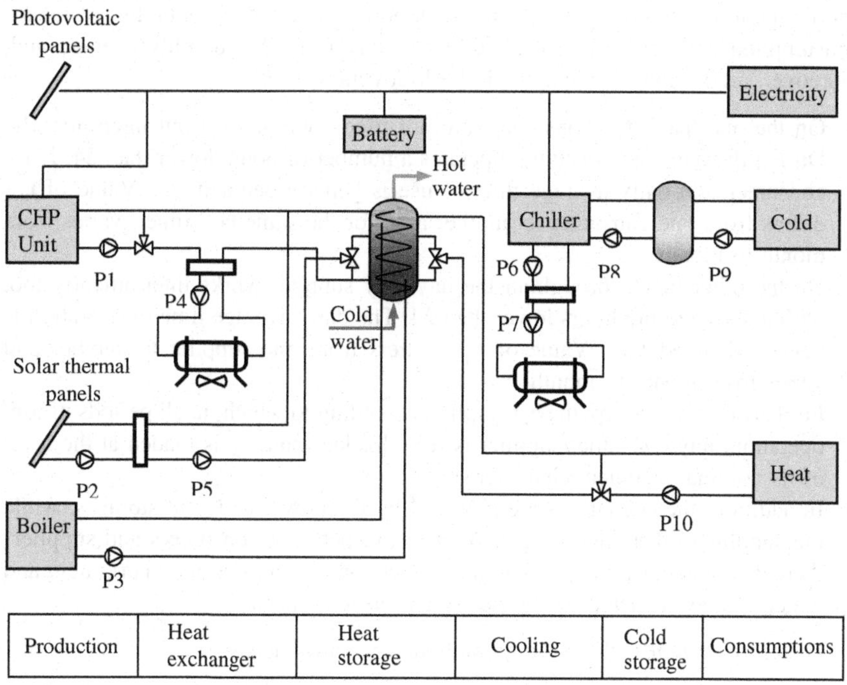

Fig. 2 Global functional scheme of a mCCHP system with a mechanical compression chiller

For the structural models with mechanically activated chiller was built the global functional scheme presented in Fig. 2.

The functional schemes of the system will be obtained by customizing the two global functional schemes.

2.2 *Sizing of the System Components*

For the functional schemes of the system to become functional models, all the components, both the main and the complementary ones, should be dimensioned and chosen from those available on the market. Sizing of the system components starts from the following observations.

In structural design, both the consumers' consumption and the suppliers load were evaluated at the level of each month only. Evaluation result is that the consumers "monthly consumption" was found, i.e., the amount of energy that each consumer consumes every month of the year. Also, it was found the suppliers monthly load, i.e., the amount of energy that each supplier must give to each month of the year so that, along with others, always cover this variable consumption.

In functional design one take into consideration that although at the level of each month, the daily residence consumption and system load are equal, though in shorter periods of time (e.g., at different hours of the day) both the residence consumption and the system load differs greatly from the monthly average and, moreover, are no longer equal, from the following causes.

- On the one hand, the consumers are not used continuously but intermittently. During the day, any consumer operates a number of hours lower than 24. For a consumer, the daily average time of use is generic denoted t_{use} .Value of t_{use} differs from one consumer to another and, for the same consumer, varies from month to month.
- On the other hand, throughout the day, any supplier works intermittently too, with total operating hours lower than 24. The daily average time of operation is generic denoted t_{oper}. Value of t_{oper} differs from one supplier to another and varies from month to month.
- Furthermore, we accept the assumption according to which, in all periods of use/ operation, any consumer/supplier is fully loaded (namely is loaded at the level of its nominal technical characteristics).
- In addition, the consumers are powered by the power and heat storages, while the suppliers fill these storages. As a result, between consumers and suppliers there does not exist a permanent and direct connection, but one intermittent and which is intermediated by the power and heat storages.

Therefore, *actually,* the system components sizing means:

- establishing of *the nominal values* for *certain defining technical characteristics* of each system component (whether it is main or complementary and whether it is of consumer, supplier, or storage type), namely those characteristics, called *characteristics required,* whose values are imposed by the system functioning, and based on which the component is chosen from the market and embedded into the system,
- for each component, choosing those products which *meets the required characteristics* of the component, and
- evaluating the daily *operation time*, t_{oper}, of each supplier, and the daily *use time*, t_{use}, of each consumer,

taking into account:

- *the monthly quantities of energy* resulting from any kind of evaluation (whether it refers to consumption or load, and whether it refers to the main components that have been evaluated in Chap. "Structural Design of the mCCHP-RES System" or to the complementary components that will be evaluated in this chapter), and
- the nominal technical characteristics of *the components available in the market.*

What follows presents the dimensioning of the main and complementary components.

2.2.1 Electric Subsystem Components

Power generating need have two essential functions: first, the generating unit is required to be equipped with a "kilowatt function" to generate sufficient power when necessary, and second, must possess a "frequency control function," fine-tuning the output so as to follow minute-by-minute and second-by-second the fluctuations in demand, using the extra power from the "kilowatt function" if necessary. Renewable energy facilities such as solar sources do not possess both a kW function and a frequency control function, because they have DC current. Electrical energy storage is expected to be able to compensate for such difficulties with a kW function and a frequency control function. Stationary batteries are utilized to support renewable energy output with their quick response capability. To supply residence is needed AC power at the industrial frequency (50 Hz). The switch from DC to AC is realized using mono-phase or three-phase inverter depending on the installed power. The stand-alone micro CCHP supplies the building (Fig. 3) with electricity by means of a battery and three- or one-phase inverters to build up a supply network. If the batteries are charged, the power output of the inverters has to guarantee electrical power supply of the building. When the CHP unit is in operation, its power output must cover the residential consumption and must charge the batteries (if the PV panel is not in operation). When the CHP unit is not in operation, the total supply period is determined by the battery capacity and depends on the total consumption. Additionally, a battery management system is recommended to guarantee the special operation procedures of the batteries resulting in improved lifetimes. In case that the produced electrical energy (especially photovoltaic energy) cannot be used by consumers or stored in the batteries,

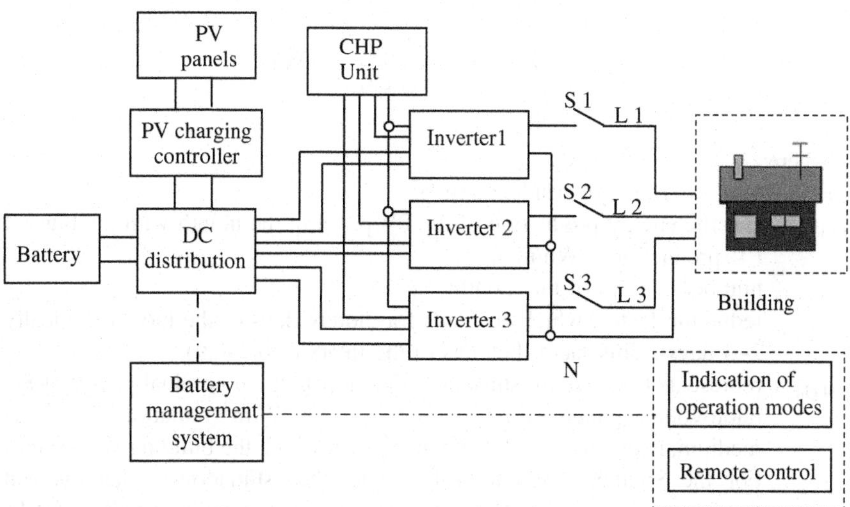

Fig. 3 Electric subsystem scheme

the energy will be used by an electrical heating element in order to produce hot water in the back-up boiler.

A. *Photovoltaic panel*

Photovoltaic panel are comprised of solar modules connected in series, thus making up a branch, while the several branches are connected in parallel. Series connection of modules leads to the output voltage setting and the parallel connection of branches leads to the current setting.

The starting data for sizing the PV panel include the specific PV production in the month with the biggest PV production and the medium sunshine time at the nominal power of the PV panel, corresponding to the same month.

The characteristics required are the following:

- The maximum power, P_{SMn}, and voltage, V_{SMn}, of the solar modules;
- The power, $P_{PV_{panel}}$, voltage, V_{CC}, and topology of the PV panel.

In sizing the photovoltaic panel, the following steps are required:

(a) Determining the amount of energy produced by the photovoltaic panel, $E_{PV_{max}}$, in the month in which the photovoltaic production is the biggest:

$$E_{PV_{max}} = A_{TFA} \cdot e_{PV_{max}} \quad (\text{kWh/month}) \tag{1}$$

(b) Determining the produced energy in a day of this month, $E_{PV_{day}}$:

$$E_{PV_{day}} = \frac{E_{PV\max}}{k \cdot N_{days}} \quad (\text{kWh/d}) \tag{2}$$

(c) Determining the power of the photovoltaic panel $P_{PV_{panel}}$:

$$P_{PV_{panels}} = \frac{E_{PV\text{day}}}{t_{\text{sunlight}}} \quad (\text{kW}) \tag{3}$$

where:

A_{TFA} total floor area of residence (m^2);

$e_{PV\max}$ specific energy production of the PV panel in the month with the biggest PV production (kWh/m^2);

N_{days} number of days of the month;

k reduction factor, which accounts for cloudy days of the month (typically 2–3 days). This factor belongs to the interval (0.7–0.9).

$t_{sunlight}$ represents the medium sunshine time, in [h], at the nominal power of the photovoltaic panels corresponding to 1,000 W/m^2 radiation (4–8 h in medium, depending on the climate area in which the building is situated). For the Southeast region of Romania, the estimation is that the sun furnishes in 1 year 1.5 MWh/m^2, which means a medium value of 4.11

kWh/m^2/day. We can assimilate that one day is equivalent to 4.11 h of solar radiation at 1 kW/m^2. It follows that:

$$t_{\text{sunlight}} = \frac{H}{1{,}000} \quad (\text{h}) \tag{4}$$

where:

H = daily solar energy expressed in Wh/m^2;
1,000 = reference instant solar radiation expressed in W/m^2.

Then, from the available products in the market are chosen a solar module with P_{SMn} nominal value of the maximum power and V_{SMn} nominal value of the voltage.

Electric performance of a solar photovoltaic module is represented by the current–voltage (I–V) curve, which represents the current versus voltage for a given solar irradiation.

This curve depends on two main factors: the solar irradiation (illumination) received by the solar cells and the temperature. For a given solar module, the generated maximum current is directly proportional to the solar irradiation, and the terminal voltage is slightly reduced along with an increase in temperature. The "load" determines the operating point of the panel. The optimal load corresponds to the maximum power charged by a photovoltaic panel. For example, in the product data sheet a solar module with area 1.275 m^2 has the nominal features given in Tables 1 and 2, for Standard Test Conditions (STC) and Normal Operating Cell Temperature (NOCT).

(d) The PV panel voltage, V_{CC}, is chosen according to the recommendation given in Table 3, the inverters input/output capabilities, and the consumers' voltage.
(e) The topology of the photovoltaic panel is established according to the power and voltage of both the panel and its modules.

The number of modules installed, n_{SM}, has to be the even integer, superior to the number resulted from the following relation:

$$n_{SM} = \frac{P_{PV_{\text{panel}}}}{P_{\text{SMn}}} \tag{5}$$

Table 1 Standard test conditions (STC)

Electrical features in STC conditions	Irradiation 1,000 W/m^2; module temperature 25 °C
Optimal operating voltage V_{SMp}	36.6 V
Optimal operating current I_{mp}	5.2 A
Open circuit voltage V_{oc}	45.2 V
Shortcircuit current I_{sc}	5.62 A
Maximum nominal power P_{SMn}	190 W
Efficiency η	14.8 %

Table 2 Normal operating cell temperature (NOCT)

Electrical features in NOCT conditions	Irradiation 800 W/m^2; ambient temperature 20 °C
Optimal operating voltage V_{SMp}	33.1 V
Optimal operating current I_{mpps}	4.19 A
Open circuit voltage V_{ocs}	41.3 V
Shortcircuit current I_{scs}	4.56 A
Maximum nominal power P_{SMn}	

Table 3 Recommended voltages for photovoltaic systems

Power of the photovoltaic system (Wp)	Recommended voltage (Vc.c)
0–500	12
500–2000	24
2000–10000	48
>10000	>48

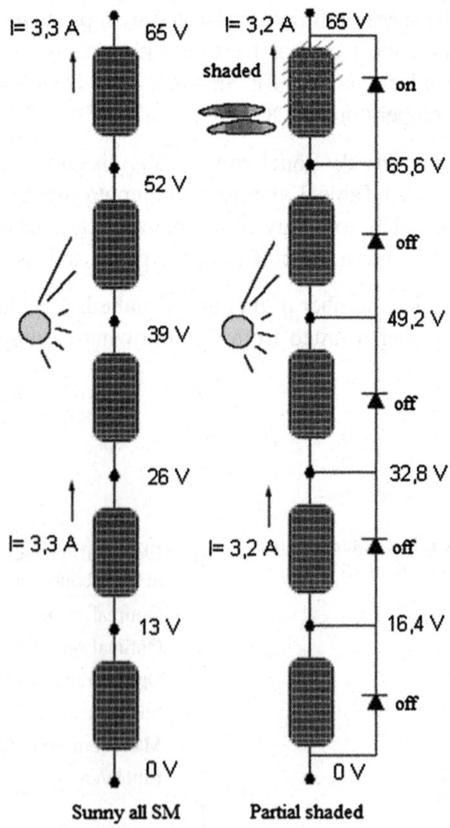

Fig. 4 Series connection with and without bypass diode

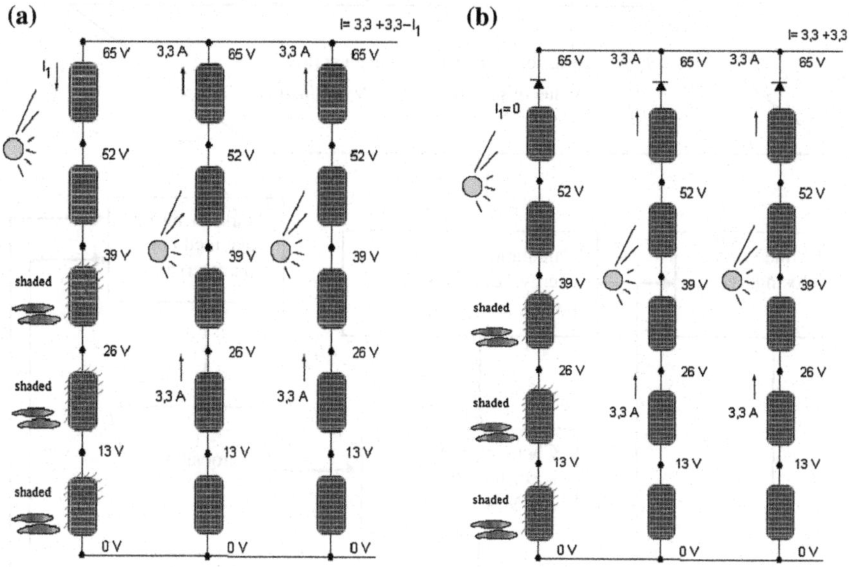

Fig. 5 Parallel connection. **a** without blocking diodes, **b** with blocking diodes

The number of modules connected in series, n_{SMs}, making up a branch, is given by relation:

$$n_{SMs} = V_{CC}/V_{SMn} \tag{6}$$

The number of branches connected in parallel, n_{SMp}, is given by relation:

$$n_{SMp} = n_{SM}/n_{SMs} \tag{7}$$

There is a probability, that not all solar modules will receive the same amount of solar irradiation, due to shadows. There is a risk of overheating and permanent damage when a single SM within a PV panel is affected by shadow. Bypass diodes are used in PV panel to protect the SMs from overheating (Fig. 4).

When connected in parallel several reverse circulation currents can occur on shaded modules series. To avoid the reverse circulation is necessary to use the blocking diodes (Fig. 5).

The algorithm for photovoltaic panel choosing is shown in Fig. 6.

B. *Batteries capacity*

Battery system design and selection involve many decisions and tradeoffs. Choosing the right battery for a PV application depends on many factors. In PV systems, lead–acid batteries are most common due to their wide availability in many sizes, low cost, and well-understood performance characteristics. In a few

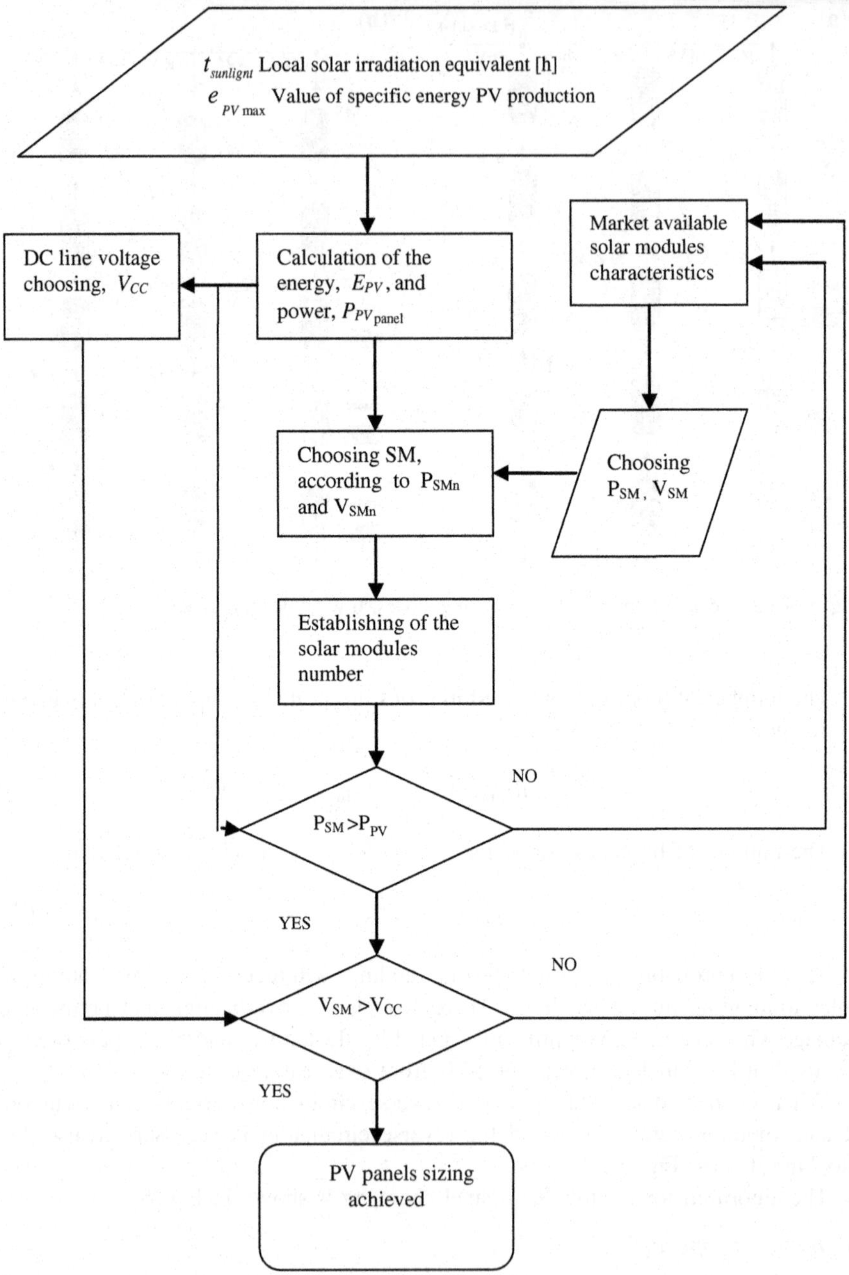

Fig. 6 Algorithm for photovoltaic panels sizing

Table 4 Standard values for battery sizing

Minimum storage capacity (Ah) at 48 V batteries voltage (Ah)	Average consumption (kWh/day)
420	Up to 15
490	Up to 20
600	Up to 30
800	Up to 50
1,000	Up to 70
1,200	Up to 90

critical, low-temperature applications nickel–cadmium cells are used, but their high initial cost limits their use in most PV systems. There is no "perfect battery" and it is the task of the PV system designer to decide which battery type is most appropriate for each application. Battery capacity is the amount of energy a battery contains. Lead–acid batteries are really the only type to consider for home energy storage at the present time. The important characteristics of the battery are capacity C = It expressed in (Ah) and voltage battery $V_{battery}$, where I is the current and t is time.

Capacity is usually rated in ampere-hours at a given voltage. A battery rated at 100 Ah will deliver 1 A of current for 100 h. The capacity of batteries is determined by two factors, namely how much energy is needed and how long must the battery supply this energy. Generally, the batteries' capacities are dimensioned according to the electric peak or to the electric base load. In the first case, the electric peak has to be covered by the CHP unit. About how long must the batteries supply this energy, the answer is that alternative energy systems work best with between 1 and 3 days of storage potential. Standard values for the batteries' sizing are presented in Table 4 [1].

If a battery with higher capacity than needed is selected, the cycles of loading/unloading can be diminished, thus increasing its lifetime. Batteries store power as direct current (DC). Regular household power is AC, the reason for which between battery and household should be interposed an inverter. For dimensioning the batteries, the following steps are required:

(a) Determining the energy stored in the batteries (E_{need}), based on the condition imposed, namely that in the month with the highest consumption of electricity, this energy has to cover the consumption for ($N_{dayneed}$) days (usually 1–-2 days). The relation used is:

$$E_{need} = \frac{A_{TFA} \cdot e_{max}}{N_{days}} \cdot N_{day\,need} \tag{8}$$

where :
A_{TFA} total floor area of residence [m^2];
e_{max} specific power consumption in the month with the highest consumption of electricity (kWh/m^2);
N_{days} number of days in this month.

(b) Choice the voltage for the batteries bank $V_{batteries} = V_{cc}$ and for the battery $V_{battery}$.

Technical characteristics of the batteries are *voltage* and battery *capacity*. The battery's voltage depends on the rate, in relation to the battery's capacity, that energy is either being withdrawn or added to the battery. The voltage of a lead–acid battery gives a read out of how much energy is available from the battery. Figure 7 illustrates the relationship between the battery's state of charge and its voltage. This graph is based on a 48-volt battery at room temperature. Rates of charge and discharge are expressed as ratios of the battery's capacity in relation to time. Rate (of charge or discharge) is equal to the battery's capacity in ampere-hours divided by the time in hours it takes to cycle the battery. If a completely discharged battery is totally filled in a 10 h period, this is called a C/10 rate.

Deep cycle lead–acid batteries [2] which are continually recharged at rates faster than C/10 will have shortened lifetimes. The literature [3] indicates that the best overall charging rate for deep cycle lead–acid batteries is the C/20 rate. The C/20 charge rate assures good efficiency and longevity by reducing plate stress.

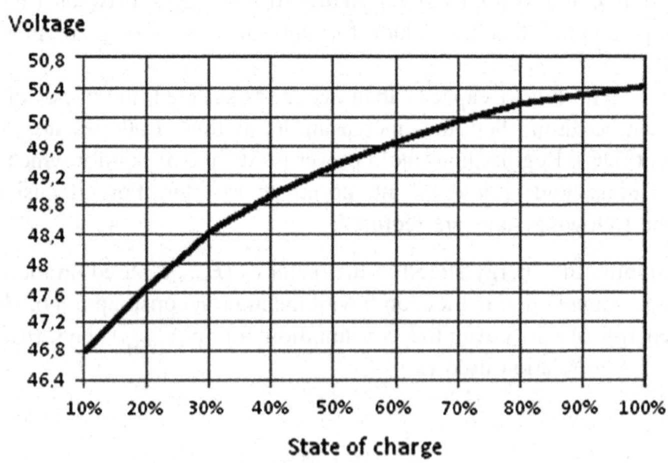

Fig. 7 Rest voltage versus state of charge

In order to obtain the voltage *Vcc*, that is the DC voltage on the lines, will be connected n_1 batteries in series that have:

$$n_1 = \frac{V_{cc}}{V_{\text{battery}}} \tag{9}$$

So, will be obtained a system of batteries which will have the input voltage Vcc and capacity is C_{battery} which represents the nominal capacity of a series battery. These batteries are designed to provide electricity over long periods, and can repeatedly charge and discharge up to 80 % of their capacity. Automotive batteries, which are shallow-cycle (and therefore prone to damage if they discharge more than 20 % of their capacity), should not be used. A battery should be completely filled each time it is cycled. They are dependence on climatic conditions for energy input. We have found that all alternative energy systems need some form of backup CHP power. The CHP source can provide energy when the alternative energy source is not operating. The CHP source can also supply the steady energy necessary for complete battery charging. The addition of a CHP source also reduces the needed battery capacity. Solar sources need larger battery capacity to offset their intermittent nature.

(c) Determining the capacity of the batteries system using the relation:

$$C_{\text{batteries}} = \frac{E_{\text{need}}}{V_{cc}} \quad \text{(Ah)} \tag{10}$$

The available capacity of an accumulator is not the rated capacity C20 (corresponding to a discharge during 20 h at 25 °C), but the real available capacity all the time in work. Also, an accumulator cannot be discharged more than a certain level, otherwise it may be damaged. In the absence of problems related to low temperatures and normal use, it is considered as acceptable a level of discharge of ND = 0.7–0.8, depending on the secondary battery model: 0.7 for batteries that support a small number of recharges and 0.8 for batteries that supports a large number of recharges. A charged battery at 70 % is at a discharge level of 30 % (ND = 0.3); a charged battery at 30 % is at a discharge level of 70 % (ND = 0.7). If the battery will be recharged frequently, ND will be reduced in order to extend the battery life. Depending on the minimum temperature that can be supported by the system, the coefficient of reduction of capacity is RT.

Taking into account both phenomena due to temperature and the maximum level of discharge, the rated capacity is calculated:

$$C_{20} = \frac{C_{\text{batteries}}}{ND \cdot RT} \tag{11}$$

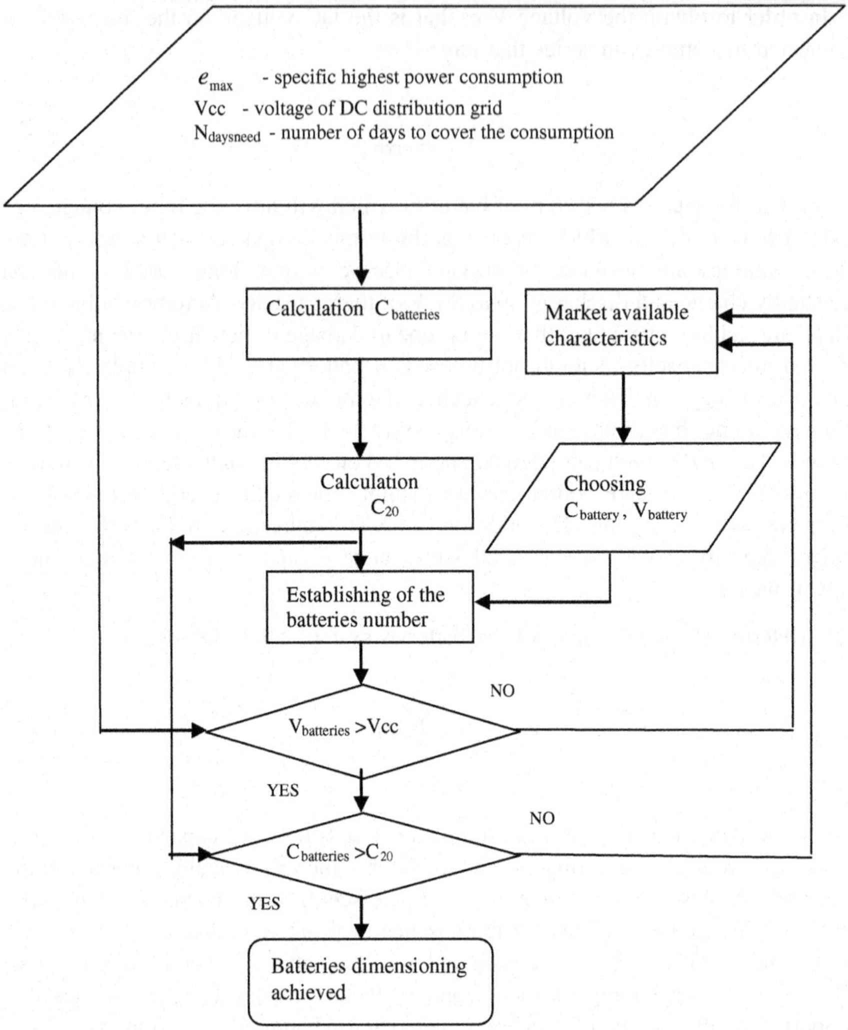

Fig. 8 Algorithm for batteries sizing

where:

C_{20} = rated capacity; ND = maximum discharge allowed level; RT = coefficient of capacity reducing due to temperature.

Because the accumulator system will be placed in an area inside the building, result that in this area there are no low temperature, so the coefficient of capacity decreasing due to temperature will be RT = 1. The maximum discharge level coefficient will be chosen in the middle of the acceptable range (ND = 0,75). For obtaining the C_{20} capacity, it will be connected n_2 battery rows in parallel with $C_{battery}$ capacity of each row.

(d) Determining the numbers of battery rows need using the relation:

$$n_2 > \frac{C_{20}}{C_{\text{battery}}}. \tag{12}$$

(e) Determining the number of batteries need using the relation:

$$n_{\text{batteries}} = n_1 \cdot n_2 \tag{13}$$

It will be chosen batteries for that balance price/lifetime is optimal, depending on available funds dedicated to installation. The algorithm for batteries choosing is shown in Fig. 8.

C. *Solar charge controller*

Electrical energy is a very specific product. The possibility for storing electricity in any significant quantity is very limited so it is consumed at the instant it is generated. When the system is not producing electricity, it needs batteries and a charge controller. Charge controllers for stand-alone system regulate rates of flow of electricity from the source to the battery and the load. The controller keeps the battery fully charged without overcharging it. The controller allows the current to flow from the generation source into the battery, the load, or both. When the controller senses that the battery is fully (or nearly fully) charged, it reduces or stops the flow of electricity from the source, or diverts it to an auxiliary or "shunt" load (most commonly an electric water heater). Many controllers will also sense when loads have taken too much energy from batteries and will stop the flow until sufficient charge is restored to the batteries. This last feature can greatly extend the battery's lifetime. In case photovoltaic panels are used for off-grid applications, the most important parameter that has to be met is the value of the voltage for maximum power. Modern charging controller for photovoltaic panels is capable to perform maximum power point tracking. Total efficiency of PV panel is highly dependent on the maximum power point tracking performance, which determines the maximum amount of produced electrical energy. In PV panel without maximum power point tracking, the energy losses can be up to 30 % higher than in the PV panel with maximum power point tracking. Different methods are used for maximum power point tracking. The charging controller, transfers the electric energy from the photovoltaic panels to the batteries array, following the optimal operating point of the photovoltaic panels. The solar charge controller is defined by a specific set of technical characteristics as is shown in Table 5. Their values obtained from the photovoltaic panels design are compared with the values given in the product data sheet. On this basis, the solar controller is chosen from those available in the market.

Table 5 Technical characteristics of the solar charge controller

Technical characteristics	Values from product data sheet
Nominal battery voltage	48 Vdc
Maximum PV array voltage operating	140 Vdc
Maximum PV array open circuit voltage	150 Vdc
Array short circuit current	60 Adc

According to the battery voltage Vcc, its selection involves choosing:

- input voltage in solar charge controller (Vsc) higher than the maximum voltage of the PV panel so, Vsc > Vcc;
- input power in solar charge controller (Psc) higher than the maximum power of the PV panel so, Psc > P_{PVP};
- Maximum charging current to the batteries.

D. *Cogeneration unit*

The electric power of the CHP unit is obtained from the relation:

$$P_{CHP} = \frac{A_{TFA} \cdot e_{\max}}{N_{\text{days}} \cdot t_{\text{oper}}} \quad [\text{kW}] \tag{14}$$

where:

A_{TFA} total floor area of residence (m^2);

$e_{CHP\max}$ value of specific energy in the month with the biggest load of the CHP unit (kWh/m^2);

N_{days} number of days of in the month;

t_{oper} daily time of operation [h] (6–8 h on average in correlation with estimated time for battery recharging Table 17 in Chap. "Structural Design of the mCCHP-RES System").

After that is chosen from the market, a CHP unit with nominal power P_{CHP_n} satisfies the relation:

$$P_{CHP_n} \geq P_{CHP} \tag{15}$$

E. *Inverters*

To connect the photovoltaic system to the electric grid or domestic appliances, we will need an additional equipment called inverter to condition the electricity and safely transmit the electricity to the load that will use it. Most electrical appliances and equipments in the home run on alternating current (AC) electricity. Virtually all the available renewable energy technologies produce direct current (DC) electricity. To run standard AC appliances, the DC electricity must first be converted to AC electricity using inverters and related power conditioning equipments [4].

There are three basic elements to power conditioning:

- Conversion of the constant DC power to oscillating AC power at 50 Hz,
- Voltage consistency, i.e., the extent to which the output voltage fluctuates,
- Quality of the AC sine curve whether the shape of the AC wave is jagged or smooth.

The DC voltage available will affect the rating and efficiency of an inverter due to the higher currents being drawn from low voltage power sources. Typical input ranges are detailed in the Table 6.

For example, the technical specification of the inverter is:

(a) Rated Output Capacity: 3 kVA;
(b) Rated Battery Voltage: 48 ± 10 % V_{DC};
(c) Rated Output Voltage: 220–230 V_{AC};
(d) Total Harmonic Distortion: <4 %;
(e) Output Frequency: 50 ± 0.05 Hz;
(f) Dynamic Response: 5 %;
(g) Overload Ability: 120 % 1 min, 150 % 10 s.

Bidirectional inverters are used in PV systems and provide, besides converting DC energy stored in batteries to AC, power and control voltage and current of the battery charge. They are recommended for mono-phase and three-phase systems rated at more than 2 kW.

The most advanced inverters are equipped with electrical network synchronization algorithm equipment and can be used both to generate the three-phase system as well as to function as independent systems. Choosing the right inverter is based on the battery voltage and the maximum power of consumers, their type, and their working regimes.

These equipments are called "balance-of-system." Typical balance-of-system equipment for an off-grid system includes batteries, charge controller, power conditioning equipment (inverter), safety equipment, and meters and instrumentation.

Table 6 Requirements for input voltage inverter

Nominal battery voltage (V)	Input voltage ranges (V)
12	10.2–14.4
24	20–32
48	40–60
110	88–132
220	176–264

A grid-connected system that is connected to the electric grid requires balance-of-system equipment that allows to safely transmit electricity to the loads and to comply with power provider's grid-connection requirements. In this case, the system will need power conditioning equipment (inverter), safety equipment, and power meters and instrumentation. In CCHP systems on grid type, the prosumer provides electricity via a public distribution system. According to standard [5], the supplier is the party who provides electricity via a public distribution system, and the user or customer is the purchaser of electricity from a supplier. The user is entitled to receive a suitable quality of power (PQ) from the supplier. Measurement and evaluation of the quality of the supplied power has to be made at the instant of its consumption. The measurement of PQ is complex, since the supplier and user, whose sensitive electrical equipment is also a source of disturbances, have different perspectives. In practice, the level of PQ is a compromise between user and supplier. The main document dealing with requirements concerning the supplier's side is the standard EN 50160, which characterizes voltage parameters of electrical energy in public distribution systems. The correct operation of electrical equipment requires a supply voltage that is as close as possible to the rated voltage (Fig. 9). For the majority of equipments, voltage changes in the range (0.9–1.1)U_n, do not cause any negative consequences, especially for common heating devices. For other electrical equipments, the relationship between supply voltage and their power or efficiency may be significant. Even relatively small deviations from the rated value can cause suboptimal operation of equipment, e.g., operation at reduced efficiency, or higher power consumption with additional losses and shorter service life. For an equipment with higher sensitivity to the supply voltage proper protection should be installed.

Correct equipment operation requires the level of electromagnetic influence on equipment to be maintained below certain limits. Equipment is influenced by

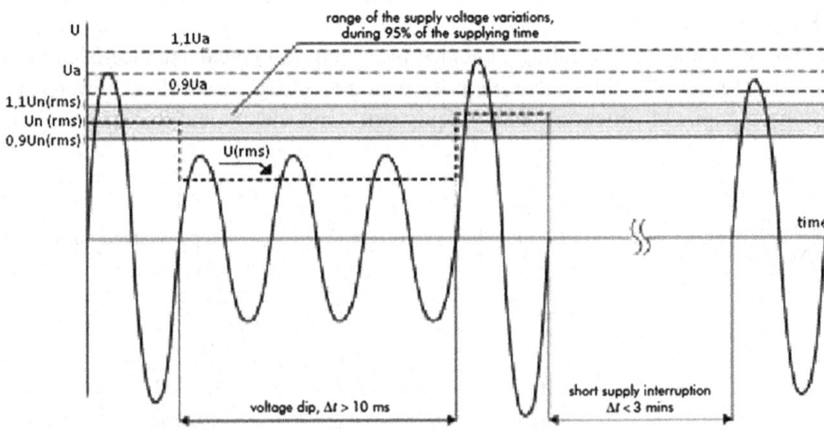

Fig. 9 Illustration of a voltage dip and a short supply interruption

Table 7 Requirements for supply and utility low voltage

Supply voltage characteristics according to EN 50160 THD	Requirements for utility low voltage EN 61000 series THD	
	EN 61000-2-2	EN 61000-3-2
5 % 3rd, 6 % 5th, 5 % 7th, 1, 5 % 9th, 3.5 % 11th, 3 % 13th, 0, 5 % 15th, 2 % 17th, 1, 5 % 19th, 0,5 % 21th, 1.5 % for >23th	6 % 5th, 5 % 7th, 3.5 % 11th, 3 % 13th, THDu < 8 %	5 % 3rd, 6 % 5th, 5 % 7th, 1.5 % 9th, 3.5 % 11th, 3 % 13th, 0.3 % 15th, 2 % 17th

disturbances on the supply and by other equipment in the installation, as well as itself influencing the supply.

These problems are summarized in the EN 61000 series of EMC standards, in which the limits of conducted disturbances are characterized. Standard IEC 038 [6] distinguishes two different voltages in electrical networks and installations:

- *supply voltage* which is the line-to-line or line-to-neutral voltage at the point of common coupling (PCC), i.e., main supplying point of installation;
- *utility voltage* which is the line-to-line or line-to-neutral voltage at the plug or terminal of the electrical device.

On the user's side, it is the quality of power available to the user's equipment that is important. The EMC standards concern the utility voltage, according to IEC 038, while EN 50160 deals with the supply voltage. The differences between these voltages are due to voltage drops in the installation and disturbances (see Table 7) originating from the network and from other equipments supplied from the installation. Because of this, in many standards of the EN 61000 series, the equipment current is an important parameter, while this current is not relevant according to EN 50160.

In conclusion, the electrical part of the CCHP system for on-grid operation must satisfy the EN 50160 standard, having the main voltage parameters and their permissible deviation ranges at the customer's point of common coupling in public low voltage (LV), under normal operating conditions. The electrical equipments, especially the inverter, must satisfy EN 61000 series of EMC standards, having the main voltage parameters and their permissible deviation ranges.

In short, for sizing an inverter the following conditions must be satisfied:

- Rated Output Capacity of inverter $P_{output} > P_{CHP}$;
- Input Voltage of inverter $V_{input} = V_{CC}$ (equal with rated batteries voltage);
- Rated Output Voltage V_{output}: 220–230 V_{AC};
- Total Harmonic Distortion THD : <4 %;
- Output Frequency: 50 ± 0.05 Hz.

2.2.2 Thermal Subsystem Components

As shown in [7], basically a heating system that uses solar radiation has the components shown in Fig. 10, where **A**-Solar collector, **B**-Pump station, **C**-Hot water tank, **D**-Backup heat source (and CHP unit), **E**-Hot water distribution system, **F**-Solar controller, and **G**-Expansion tank.

All components work together to heat water as follows.

First the sunlight hits the solar collector and heats the thermal transfer liquid passing through it. A pump circulates the heated thermal transfer liquid from the solar collector to the water tank coil, where its heat transfers to water within the tank. The pump returns the cooled thermal transfer liquid to the solar collector for reheating.

Second, when the CHP unit works, the thermal energy is stored in the tank. When needed, a backup heat source, such as a boiler, on-demand water heater, or electrical immersion element, boosts the tank water to the desired temperature before it is distributed to its end use. The differential temperature controller monitors and regulates the circulation rate of the thermal transfer liquid, based on weather and hot water demand. The expansion tank regulates system pressure as the thermal transfer liquid expands and contracts.

A. *Solar thermal panel*

The main technical characteristics of the solar thermal panel are:

- Collector area,
- Performance data:

Fig. 10 Solar heating system [7]

- Efficiency η_{ST} [%],
- Nominal flow rate [l/h],
- Maximum operation pressure [bar].

The required characteristic which is imposed by the system functioning, and based on which the component is chosen from the market and is embedded into the system, is the absorber area (A_{ST}) of the ST panel. To size the solar thermal panel, the absorber area should be determined by the following steps:

(a) Determining the amount of energy produced by the ST panel, $Q_{ST_{max}}$, in the month in which the ST panel production is the biggest:

$$Q_{ST_{max}} = A_{TFA} \cdot q_{ST_{max}} \quad (\text{KWh/month}) \tag{16}$$

(b) Determining the produced energy in a day of this month, $Q_{ST_{day}}$:

$$Q_{ST\,day} = \frac{Q_{ST_{max}}}{k \cdot N_{days}} \qquad [\text{kWh/d}] \tag{17}$$

(c) The heat flow $\dot{Q}_{day}$ can be calculated with the following relation. Their obtaining value will be used for choosing the circulation pump.

$$\dot{Q}_{day} = \frac{Q_{ST day}}{t_{sunlight}} \quad [\text{kW}] \tag{18}$$

where:

A_{TFA} total floor area of residence [m^2];

$q_{ST_{max}}$ value of specific heat production in the month with the biggest production of the ST panel [kWh/m^2];

N_{days} number of days of the month;

k reduction factor which account for cloudy days of the month (typically 2–3 days). This factor belongs to the interval (0.7–0.9).

$t_{sunlight}$ represents the medium sunshine time at the nominal thermal power of the solar thermal panels corresponding to a 1,000 W/m^2 radiation (4–6 h in medium depending on the climatic in which the building is situated).

(d) Determining the absorber area (A_{ST}), with the following formula:

$$A_{ST} = \frac{Q_{ST_{day}}}{H} \quad [\text{m}^2] \tag{19}$$

where:
H is the solar daily irradiation (kWh/m^2).

From the product data sheet choose the solar thermal collector with nominal absorber area A_{ST_n} and optical efficiency $\eta_{ST\,o}$. By knowing these data of the solar collector, it is possible to determine the number of the solar thermal collectors which is given by the equation:

$$n_{ST} > \frac{A_{ST}}{\eta_{STo} \cdot A_{ST_n}} \tag{20}$$

All solar thermal collectors can operate at a variety of flow rates in [l/h·m^2]. The flow rate will affect the pressure drop of the collectors and also change the temperature increase across the panel. Generally a low flow rate (typically 30 l/h·m^2) is desired because it minimizes the pressure drop through the system resulting in smaller pipes and less energy to run the pump. The flow rate selected needs to take the heat exchanger into account to ensure sufficient heat transfer from the solar system to the tank. The temperature increase across the collectors is driven by the level of irradiation, the angle of the sun, and the outdoor temperature. Once the energy output of the collector is known, the increase in temperature can be computed with equation:

$$\Delta\vartheta = \frac{\dot{Q}_{\text{day}}}{\rho \cdot c \cdot \dot{V}} \tag{21}$$

This increase in temperature across the solar collector is used in setting of the solar controllers while the pump speed is varied to maintain this value.

B. *Backup boiler (additional boiler)*

The main technical characteristics of the back-up boiler are:

- Thermal power [kW],
- Efficiency η_{ST} [%],
- Nominal volume flow of the hydraulic circuit [l/h],
- Max. operation pressure [bar].

The required characteristic which is imposed by the system functioning, and based on which the backup boiler is chosen from the market and embedded into the system, is the *maximum thermal power* that must be supplied.

For sizing the backup boiler, the maximum thermal power should be determined in the following steps:

(a) Determining the amount of energy produced by the backup boiler $Q_{ad.\max}$, in the month in which its load is the biggest:

$$Q_{ad.\max} = A_{TFA} \cdot q_{ad.\max} \quad [\text{KWh/month}] \tag{22}$$

(b) Determining the produced energy in a day of this month, $Q_{ad.\,\text{day}}$:

$$Q_{ad.\,\text{day}} = \frac{Q_{ad.\max}}{N_{\text{days}}} \quad [\text{kWh/d}] \tag{23}$$

(c) The thermal power $\dot{Q}_{ad}$ can be calculated with the following relation:

$$\dot{Q}_{ad} = \frac{Q_{ad.\text{day}}}{t_{\text{oper}}} \quad [\text{kW}] \tag{24}$$

where:

A_{TFA} total floor area of residence (m^2);
$q_{ad\cdot\max}$ value of the specific heat production in the month with the biggest production of the additional boiler (kWh/m^2);
N_{days} number of days of the month; and
t_{oper} represents the daily operation time of the additional boiler at the nominal thermal power (6–8 h on average).

After that choose from the market a boiler with nominal power $\dot{Q}_{ad.n}$ that satisfies the relation:

$$\dot{Q}_{ad.n} \geq \dot{Q}_{ad} \tag{25}$$

C. *Chiller*

A cooling system based on the absorption/adsorption technology consists of three hydraulic circuits: driving heat, cold distribution, and heat rejection (Fig. 11).

The technical characteristics for absorption/adsorption chillers are presented in Table 8 [8].

The required characteristic, which is imposed by the system functioning, and based on which the chiller is chosen from the market and embedded into the system is the maximum cooling power. For chiller sizing, the maximum cooling power should be determined in the following steps:

(a) Determining the amount of cold produced by the chiller, $Q_{\text{cooling}\cdot max}$, in the month in which the chiller load is the biggest:

$$Q_{\text{cooling}\cdot max} = A_{TFA} \cdot q_{\text{cooling}\cdot\max} \quad (\text{KWh/month}) \tag{26}$$

(b) Determining the cold produced in a day of this month, $Q_{\text{cooling}\cdot day}$:

$$Q_{\text{cooling}.\,\text{day}} = \frac{Q_{\text{cooling.max}}}{k \cdot N_{\text{days}}} \quad (\text{kWh/day}) \tag{27}$$

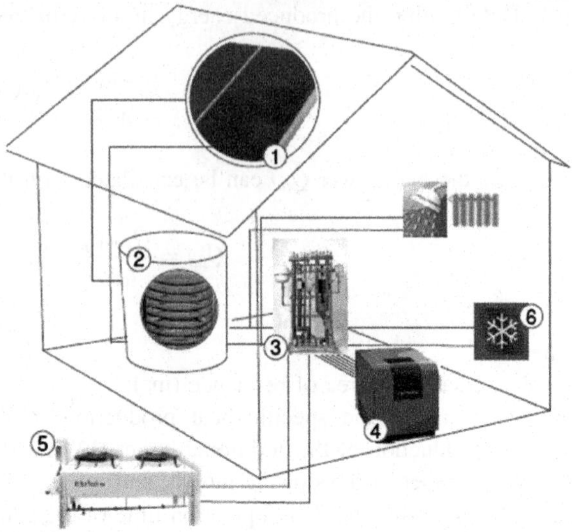

Fig. 11 Cooling system [8] *1* Heat source, *2* storage tank, *3* pump system, *4* thermal chiller, *5* heat rejection, and *6* cold distribution

Table 8 Technical characteristics of a cooling system (example)

Basic performances data	Statistical evaluation
Cooling capacity. max	23 Kw
COP_{therm} max	0.65
Cooling capacity nominal	15 kW
$COP_{therm.}$nominal	0.6
Nominal working conditions	*Temperature in/out* (°C)
Driving heat circuit	72/65
Heat rejection circuit	27/32
Chilled water circuit	18/15
Temperature range (out)	6–20

(c) The maximum cooling power $\dot{Q}_{\text{cooling}}$ can be calculated with the following relation:

$$\dot{Q}_{\text{cooling}} = \frac{Q_{\text{cooling.day}}}{t_{\text{oper}}} \quad (\text{kW}) \tag{28}$$

where:

A_{TFA} total floor area of residence (m^2);

$q_{\text{cooling.max}}$ value of specific cooling in the month with the biggest load of the chiller (kWh/m^2);

N_{days} number of days of the month;

t_{oper} represents the daily operation time of the chiller at the nominal value of the cooling power (8–10 h on average in the hottest month);

k reduction factor which accounts for cloudy days of the month (typically 2–3 days). This factor belongs to the interval (0.7–0.9).

After that choose from the market a chiller with nominal cooling capacity $\dot{Q}_{cooling.n}$ that satisfies the relation:

$$\dot{Q}_{cooling.n} \geq \dot{Q}_{cooling} \tag{29}$$

A cold water tank with hydraulic separation is used for air conditioning in summer time. A fan coil system assumes the cold distribution of the building.

D. *Heat storage*

The technical characteristics of the heat storage tank are the following:

- Temperature domain,
- Storage capacity,
- Storage thermal agent,
- Size,
- Investment costs,
- Energy resources used, and
- Safety in exploitation.

From these, the first two are those called the required characteristics, on which the tank selection is based.

Thus, the first step in the heat storage sizing is to establish the temperature domain ϑ_{max}, ϑ_{min} from the condition of compatibility between the storage tank, on the one hand, and those suppliers and consumers with which it exchange heat, on the other hand. The maximum cylinder temperature is dictated by the water. The minimum possible cylinder temperature is therefore the decisive variable for determining the required cylinder volume. To size DHW cylinders, the average cold water temperature (e.g., 10 °C) is taken as the minimum temperature.

The second step consists in evaluation of the storage capacity $Q_{max.stored}$ that means the maximum amount of heat stored in tank. Evaluation of the storage capacity presupposes knowing exactly the temporal evolution of the produced and consumed thermal flows, because the difference between the two should be covered by the storage capacity. Evaluation of these two required characteristics need simulation of the mCCHP system's functioning in various scenarios, and determining, for each thermal circuit, the maximum quantity of heat which must be stored by the tank.

For sizing the heat/cold storage tank, is necessary the following:

(a) Determining the energy demand $Q_{\text{demand.max}}$, in the month in which the demand is the biggest:

$$Q_{\text{demand.max}} = A_{TFA} \cdot q_{sys.\max} \quad \text{(KWh/month)} \tag{30}$$

(b) Determining the produced energy in a day of this month, $Q_{\text{demand. day}}$:

$$Q_{\text{demand. day}} = \frac{Q_{\text{demand.max}}}{N_{\text{days}}} \quad \text{(kWh/d)} \tag{31}$$

On this basis, the tank is then dimensioned, that is the volume of the thermal agent (water) in the tank is determined. On the other hand, the maximum quantity of heat which may be stored by the thermal agent is given by the equation:

$$Q_{\text{demand,day}} = Q_{\text{max.stored}} = \rho \cdot V \cdot c_p \cdot \Delta\vartheta \quad \text{(kWh)} \tag{32}$$

where: $Q_{\text{max.stored}}$—maximum stored heat, [kWh]; ρ—density of thermal agent, [kg/m^3]; V—volume of the thermal agent, [m^3]; c_p—specific heat of the thermal agent, [kJ/kg·K]; $\Delta\vartheta = \vartheta_{\max} - \vartheta_{\min}$ is the width of the temperature domain $[\vartheta_{\max}, \vartheta_{\min}]$ for the thermal agent, [°C].

For *simple systems*, with one source and one consumer only, dimensioning is simply based on the application of this relation. For example, in case of heat for heating tank, the volume of tank is given by the relation:

$$V_H = \frac{Q_{H.\max}}{\rho \cdot c_p \cdot \Delta\vartheta_H} \tag{33}$$

where: V_H—volume of the tank for heating, [m^3]; $\Delta\vartheta_H$ is the temperature domain of the thermal agent which fill this tank.

With certain simple systems, there is the possibility of using a simplified relation, which has in view the experience accumulated. For example, in the case of *solar thermal panel*, dimensioning may be achieved according to the relation:

$$V = k \cdot A_{ST} \tag{34}$$

where *k*—dimensioning coefficient, [l/m^2], A_{ST},—absorber area of the ST panel, [m^2].

For most applications, k = 50–100 l/m^2. For applications perfected especially through systems of energy management, k = 25 l/m^2.

For combi type systems, there are more sources of heat, which supply water at different temperatures, and more consumers, which ask for water at different temperatures. An example of such a combi system is that in which the sources are solar thermal panel and boiler while the consumers are the residence heating systems and the domestic hot water systems.

On the other hand, for the same volume of hot water resulting from the calculus, a single storage tank or more may be chosen. The use of one tank only (usually a cylinder—vertical one) has the disadvantage associated with the difficulty in positioning and transportation, but has the advantage of a reduced specific loss of heat. If two or more smaller tanks are used, then the surface that heat is lost through is wider. A better insulation of the tanks might diminish the losses. Another idea is that of applying the solution of "tank inside tank." In this case, a large tank is chosen, with a V_H volume, inside which the tank for domestic hot water, with a V_{DHW} volume, is inserted. The total volume is the sum of the two:

$$V_H = \frac{Q_{H.\max}}{\rho \cdot c_p \cdot \Delta\vartheta_H} \tag{35}$$

$$V_{\mathrm{DHW}} = \frac{Q_{\mathrm{DHW.max}}}{\rho \cdot c_p \cdot \Delta\vartheta_{\mathrm{DHW}}} \tag{36}$$

If the two tanks are co-axial, then the following relation applies to their temperatures:

$$\Delta\vartheta_H = \Delta\vartheta_{\mathrm{DHW}}$$

The total volume is:

$$V_T = \frac{1}{\rho \cdot c_p \cdot \Delta\vartheta_H}\left(Q_{\mathrm{DHW.max}} + Q_{H.\max}\right) \tag{37}$$

E. *The hydraulic circuits*

In an mCCHP system, the thermal energy (similarly with the electrical energy) must be transported from suppliers to heat storage and further to consumers. The energy "carriers" are the thermal fluids (similarly with the electrons, in case of electrical energy) that flow along the hydraulic circuits, especially designed for the transport of heat (similarly with the electrical circuits).

Each hydraulic circuit is designed to connect two main components of the mCCHP system in order to transfer the heat from one to other. The thermal fluid takes the heat from one component (e.g., from the boiler), so that its temperature increases with $\Delta\vartheta$, and transfers it to the other (e.g., heat storage tank), after which the temperature will drop with the $\Delta\vartheta$, also.

Outside the two components, along the circuit are placed fittings, devices for monitoring, control, and protection, and other elements, as well as a pump that moves the fluid.

Sizing a hydraulic circuit means (i) establishing the technical characteristics of the elements that make up the circuit, and (ii) finding the real values of the operating parameters of the circuit.

The following variables represent the operating parameters of a hydraulic circuit when it is intended to carry thermal energy:

$\dot{Q}$ Heat flow (kW);
$\dot{V}$ Volume flow (m^3/h);
$\Delta\vartheta$ Temperature difference (°C);
ρ Density of thermal fluid (kg/m^3);
Ho Pressurization pressure of the thermal fluid (Pa);
c Specific heat capacity of the thermal fluid (kWh/kg °C);
v Flow velocity (m/s).

In circuit sizing, *the required values of operating parameters* are known from the beginning, as there are among the technical characteristics of those components of the system mCCHP that the circuit connects them. Connection is only possible if these values are the same for all connected components.

At the same time, the required values of operating parameters represent the input dataset in the circuit sizing problem.

Sizing of such hydraulic circuit requires the following steps:

(a) *The circuit establishing* This consists of establishing the site of each element that is placed on the circuit, and choosing the appropriate pipelines (including the pipe and fittings).

For example, the pipe is chosen according to the required internal diameter, which can be determined using the equation:

$$DN = 4,6\sqrt{\frac{\dot{V}}{v}} \qquad \text{(mm)} \tag{38}$$

where:

DN Internal diameter (mm),
$\dot{V}$ Volume flow (Ltr/min),
v Flow velocity (m/s).

Note:

1. In order to minimize the pressure loss through the pipe, it is recommend that the flow velocity through the pipe should not exceed 1 m/s. Ideally, flow velocities between 0.4 and 1 m/s should be used, resulting in a pressure drop of between 1 and 2.5 mbar/m pipe length.
2. It is possible to determine the volume flow with the equation :

$$\dot{V} = \frac{\dot{Q}}{\rho \cdot c \cdot \Delta\vartheta} \qquad (\mathrm{m^3/h}) \tag{39}$$

Table 9 Typical domestic pipe sizing [7]

Collector area (m^2)	Flow rate (l/min)	Pipe diameter copper external (mm)	Pressure (mbar)
2	120	15 × 1	8.5
3	180	15 × 1	12.5
4	240	15 × 1	17
5	300	22 × 1	21
6	260	22 × 1	25
8	240	22 × 1	33

Table 9 shows the recommended pipe sizing for typical domestic systems.

(b) *Finding the circuit curve* This curve is defined as the relationship between the delivery head (total pressure required for fluid moving) and the volume flow. Conceptually, any operation regime of the circuit is described by a particular values set of the operating parameters and corresponds to a particular point on the circuit curve. In concrete, at the required values of the operating parameters the required operation point is corresponding. Entire curve is analytically described by the following equation:

$$H_{Ges} = H_{geo} + H_A \quad \text{(Pa)} \tag{40}$$

where:

$$H_A = H_{VL} + H_{VA} \quad \text{(Pa)} \tag{41}$$

$$H_{VL} = R \cdot L \quad \text{(Pa)} \tag{42}$$

$$H_{VA} = \sum \zeta \frac{\rho v^2}{2} \quad \text{(Pa)} \tag{43}$$

H_{Ges} Delivery head [Pa],
H_{geo} Geodetic pressure difference [Pa],
H_A Pressure loss along the hydraulic circuit [Pa],
H_{VL} Pipeline pressure loss [Pa],
H_{VA} Fitting pressure loss [Pa],
R Pipe friction resistance in Pa/m, given in diagrams like that presented in Fig. 12 ,
L Pipe length [m],
ζ Resistance coefficient, which is specific to each element that is placed on the circuit (be it component of system, device for measuring, control, protection, or just a fitting). For every element, this coefficient is a technical feature whose value is given in the product data sheet.

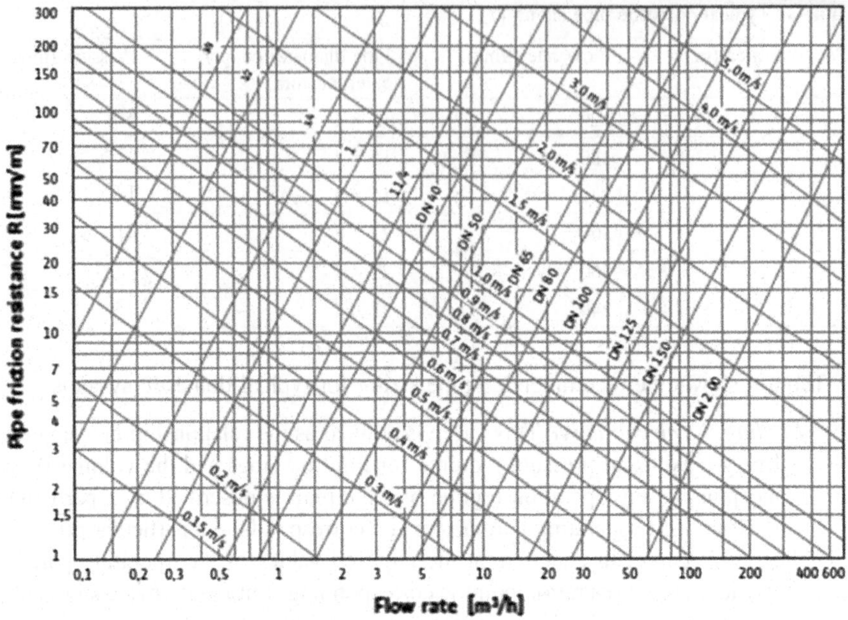

Fig. 12 Pipe friction resistance (*example steel pipe* [8])

While H_{geo} (static) remains constant, independent of the volume flow, H_{VL} and H_{VA} (dynamic) increase quadratically due to the widely varying losses in the pipelines and fittings, as well as due to the increased friction, as seen in Fig. 12. As almost always, the hydraulic circuits for heat transport are closed, the geodetic pressure difference $H_{geo} = 0$. The pressure loss of the hydraulic circuit, H_A in (Pa), corresponds to the differential height, h in [m], according to the relation:

$$h = \frac{H_A}{\rho \cdot g} \quad \text{(m)} \tag{44}$$

where g is the gravity (9.81 m/s^2).

(c) *Pump selection* and setting the actual values of the circuit operating parameters.

Similar to circuit curve, the pump curve is defined also as the relationship between the delivery head (total pressure required for fluid moving) and the volume flow. But this time, the relationship is expressed not analytic, because it is deduced experimentally and is the main feature of the centrifugal pumps used in hydraulic circuits for heat transport.

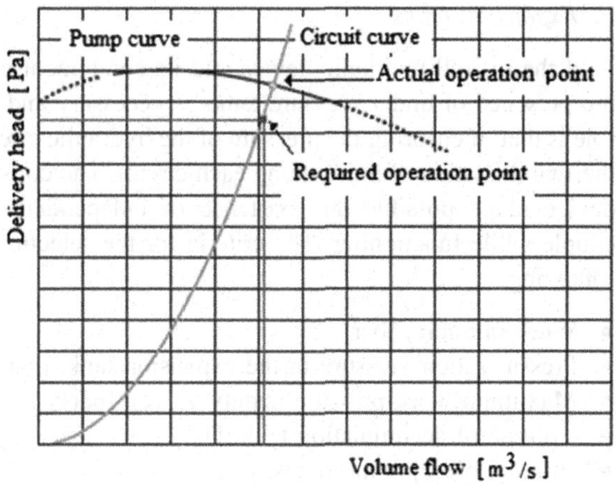

Fig. 13 Circuit and pump curves with the required and actual operation points

The operation parameters of the circulation pumps are:

- Volume flow of the pump $\dot{V}$(m^3/h);
- Delivery head (total pressure required for fluid moving) H_{Ges} (Pa); and
- Power consumption P (W).

Any operation regime of the pump is described by a particular values set of these parameters and corresponds to a particular point on the pump curve.

The volume flow of the pump $\dot{V}$[m^3/h] can be calculated with Eq. (35) and, the hydraulic pump power is calculated with the relation:

$$P_h = \dot{V} \cdot H_{\text{Ges}} / \left(3.6 \times 10^6\right) \quad (\text{kW}) \tag{45}$$

where: $\dot{V}$—volume flow [m^3/h] and H_{Ges}—pressure in the hydraulic circuit [Pa].

If the pump works in a certain circuit, then the pump and circuit have the same operating point, namely the point of intersection of the pump curve with the circuit curve, as shown in Fig. 13. To this point, the actual values of the operating parameters are corresponding.

The shaft power P_s—the required power, transferred from the engine to the shaft of the pump—depends on the efficiency of the pump η and can be calculated as:

$$P_s = P_h / \eta \quad (\text{kW}) \tag{46}$$

The criterion for choosing the pump is its curve. The pump whose curve is close to the required operation point of the hydraulic circuit will be chosen, so that the actual values of the operating parameters are higher than the required values (as seen in Fig. 13). In this way, the capacity of the circuit to transport heat is greater than is strictly needed and the pump may work intermittently. Energy consumption of the pumps should be added to the electricity consumption of building.

F. *Expansion tanks*

If the overall hydraulic device and its component circuits work under pressure, the presence of under pressure tanks is necessary and compulsory. The functional role is that of ensuring the pressure of the hydraulic circuits and the absorbing of the thermal dilatation of fluids along each circuit. The construction particularities of the device make possible the existence of independent circuits and circuits which couple while functioning. The criteria for the selection of expansion tanks are the following:

- Work pressure, [bar];
- Pressurization pressure in the expansion tank, [bar];
- Maximum work pressure (safety valve), [bar];
- Volume of accumulation tank, [1];
- Volume of adjoining pipes, [1];
- Maximum/minimum work temperature, (°C).

The characteristics of expansion tanks are the following:

- Volume of expansion tank, [1];
- Maximum work pressure, (°C);
- Maximum/minimum expanded volume, [1];
- Tank diameter and height, [mm].

Dimensioning of the expansion tanks consists in determining their volume. To this end, two procedures are used:

The analytical procedure, which uses different forms of the mathematical model of the liquid volume dilation. One form is the following:

$$V_{et} = kV_w[(v_1/v_0) - 1]/[1 - (p_0/p_1)] \tag{47}$$

where V_{et} is the volume of the expansion tank, k = safety factor (approximately 2 is common), V_w = water volume in the system, [liter]; v_0 = specific volume of water at initial (cold) temperature, (m^3/kg); v_1 = specific volume of water at operating (hot) temperature, (m^3/kg); p_0 = system initial pressure—cold pressure, (Pa); p_1 = system operating pressure—hot pressure, (Pa).

Another form of the mathematical model based on which the expansion tanks are dimensioned is:

$$V = \frac{e \cdot V_w}{1 - \frac{p_c}{p_h}} \tag{48}$$

where: V_w—volume of water in the circuit, [liter]; e—water dilatation coefficient; p_c—absolute pressure of the water in the system at initial temperature (cold), [bar]; p_h—absolute pressure of the water in the system at maximum work temperature (hot), (bar). The coefficient of water dilation at atmospheric pressure is given in the Table 10.

Table 10 Variation of water dilatation coefficient with work temperature

Working temperature (°C)	Expansion coefficient of water (%)
85	0.0324
90	0.0359
95	0.0396
100	0.0434

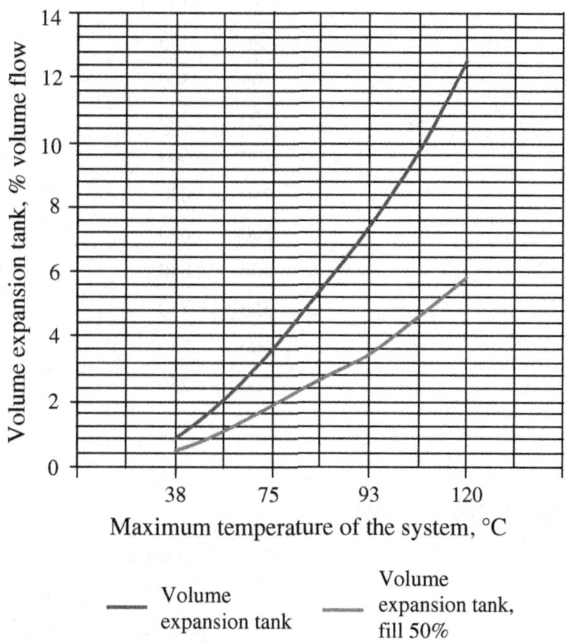

Fig. 14 Diagram for dimensioning expansion tanks

The graphic procedure is faster if the total volume of liquid, the nature of the liquid, the work pressure, and the min–max temperature interval in which it works are known. This procedure is based on previous experience and consists in using the graph given in Fig. 14.

The volume of water in the circuit is calculated, and then the volume of the expansion tank is determined, in keeping with the maximum work temperature. Afterward, the tank of immediately superior size is chosen from Table 11, since the sizes of these tanks are standard [7].

G. *Heat exchanger*

While *the hydraulic circuit* connects *two components* of the mCCHP system which are *located far away* from each other, in order *to transfer* the heat from one component to other (similarly with the electrical *circuit*), *the heat exchanger* connects *two hydraulic circuits* which *are in contact* with one another in order

Table 11 Standard dimensions of expansion tanks

Lt	Diameter (mm)	Height (mm)	Connection (")
4	225	195	3/4
8	220	295	3/4
12	294	281	3/4
18	290	400	3/4
24	324	415	3/4
35	404	408	3/4
50	404	530	3/4
80	450	608	3/4
105	500	665	3/4
150	500	897	3/4
200	600	812	3/4
250	630	957	3/4
300	630	1,105	3/4
400	630	1,450	3/4
500	750	1,340	1
600	750	1,555	1
700	750	1,755	1

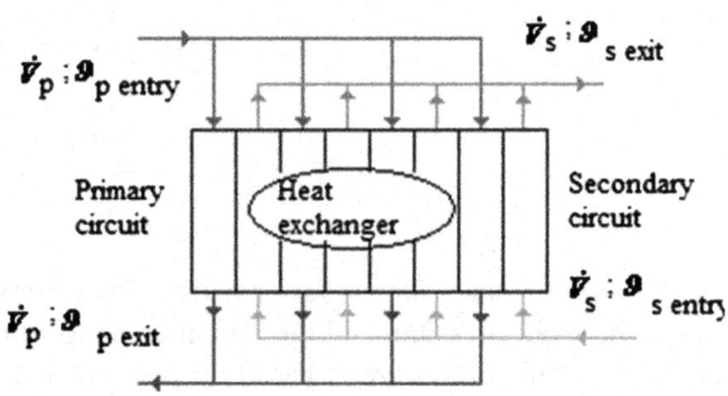

Fig. 15 Heat exchanger scheme

to transmit the heat from one circuit to other (similarly with the electrical *transformer*). Mathematical model of the heat exchanger (Fig. 15) is the following:

$$\dot{Q} = \rho_{p\cdot} \cdot c_p \cdot \dot{V}_p(\vartheta_{p_{\text{entry}}} - \vartheta_{p_{\text{exit}}}) = k \cdot A \cdot (\vartheta_p - \vartheta_s) \tag{49}$$

$$\dot{Q} = \rho_{s\cdot} \cdot c_s \cdot \dot{V}_s(\vartheta_{s_{\text{entry}}} - \vartheta_{s_{\text{exit}}}) \tag{50}$$

where:

$\vartheta_p = \frac{\vartheta_{p_{entry}} + \vartheta_{p_{exit}}}{2}$ is the average temperature of the thermal fluid for primary circuit (°C);

$\vartheta_s = \frac{\vartheta_{s_{entry}} + \vartheta_{s_{exit}}}{2}$ is the average temperature of the thermal fluid for secondary circuit (°C);

ρ_p, ρ_s—density of the thermal fluid for primary and secondary circuits (kg/m^3);

c_p, c_s—specific heat of the thermal fluid for primary and secondary circuits (kWh/kg °C);

$\dot{V}_p$, $\dot{V}_s$—volume flow of the thermal fluid for primary and secondary circuits (m^3/s);

k—overall heat transmission coefficient of the heat exchanger (kW/m^2 °C);

A—heat exchanger area (m^2);

$\dot{Q}$—heat flow transmitted by the heat exchanger (kW).

In heat exchanger selection, the required characteristics are k and A, while the rest of variables which appear in mathematical model are the entry data.

The heat exchanger may be of plate type that consists of exchanger package and casing. The exchanger package consists of aluminum or stainless steel plates with pressed-in spacers; condensate drainage is possible in every direction. The plates are connected by a fold, which gives a several fold material thickness at fluid entry and exit (Fig. 16).

Plate heat exchangers operate on the cross-flow principle. Heat is transmitted via the plates from the warm to the cold stream.

The heat flow transmitted is directly dependent on the exchanger surface area A. If the heat flow to be transmitted is known, then the area A is determined with relations:

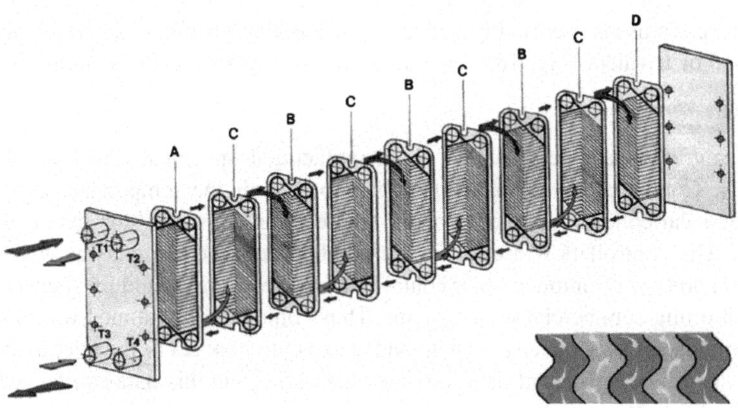

Fig. 16 Plate heat exchangers

Table 12 Typical overall heat transfer coefficients

	Hot fluid	Cold fluid	U [W/m^{2}C]
Heat exchangers	Water	Water	800–1,500
Coolers	Organic solvents	Water	250–750
Heaters	Steam	Water	1,500–4,000
Condensers	Aqueous vapor	Water	1,000–1,500
Vaporizers	Steam	Aqueous solutions	1,000–1,500

$$A = \frac{\dot{Q}}{k \cdot LMTD} \tag{51}$$

$$\text{or} \quad A = \frac{\dot{Q}}{U \cdot \Delta\vartheta} \tag{52}$$

where: *LMTD*—logarithmic mean temperature difference between the temperature of the primary and secondary circuits;

U—Value is the overall heat transfer coefficients (as example, the typical U-Value are given in Table 12).

By choosing the number of plates and their spacing, the heat flow transmitted by the heat exchanger is easily changed and even optimized to meet a particular specification.

H. *Hot water distribution*

The residential house may use one of the several types of heating systems. They can range from blowing hot air through ductwork to piping hot water, through floor. The most popular types of home heating systems are the following: forced air, radiant heat, hydronic (hot water base), and steam radiant. Hot water system uses heated water to heat a space by a combination of radiation and convection. Hot water in the buffer tank is piped to radiators mounted along walls. The fins increase the surface area of heat dissipation making the unit more efficient. Air is distributed by convection as air rises and is heated by the radiators. Buffer tank can be bivalent (Fig. 17) or trivalent (Fig. 18) depending on the number of heat sources.

I. *Solar controller*

Solar controllers are devices that are manufactured specifically to control solar systems. Controller command is generally based on the temperature difference between solar collector and heat storage. These commands start/stop the circulation pumps. The controllers will differ in the number and type of inputs as well as the number and type of outputs. Solar controllers are a convenient and inexpensive way for controlling commercial solar systems. They come preprogrammed with many of the common systems already in them and are easy to connect to the overall system. The units provide excellent data logging capabilities and this data can be accessed over the Internet.

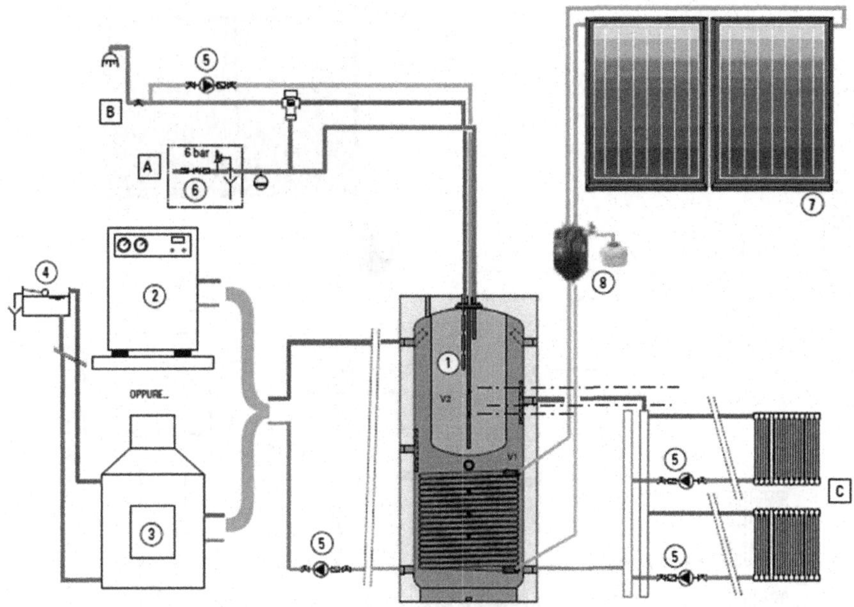

Fig. 17 Connection of the bivalent buffer tank

The standard solar controllers for fixed speed pumps will have 24VAC as an output and will require a magnetic starter, where as PLCs can be customized with several different voltage outputs, but their current output is quite small and will need to be run through a magnetic starter as well. A magnetic starter is similar to a relay, but intended for higher currents and they have overload protection on them to stop the motor if it exceeds its capacity.

The magnetic starter can also operate at two different voltages; the coil can be switched by 24VDC and the contacts can connect 240VAC. The correct motor starters should be chosen based on the power and voltage of the motor and the available switching voltage from the controller.

It is desirable in solar systems to vary the motor speed depending on the amount of energy the solar system is collecting. This will ensure the maximum output from the system and reduce the energy used to run the pumps within the system. A desired increase in temperature across the solar collector is set and the pump speed is varied to maintain this value. This is all handled in the controller. For larger systems, a Variable Frequency Drive (VFD) is required. A VFD is a device that connects between the power supply and the pump and varies the frequency of the current going to the pump motor to achieve different speeds. The VFD needs an input from an outside control as to how fast it should spin the motor. This is usually a 0–10VDC, or 4–12 mA signal. Some solar controllers can supply this output. Most Building Management System will have the option for an analog output and PLCs can be configured. There is a program in the inverter that will translate the signal from the controller into the motor speed that will need to be programmed, along with the motor parameters.

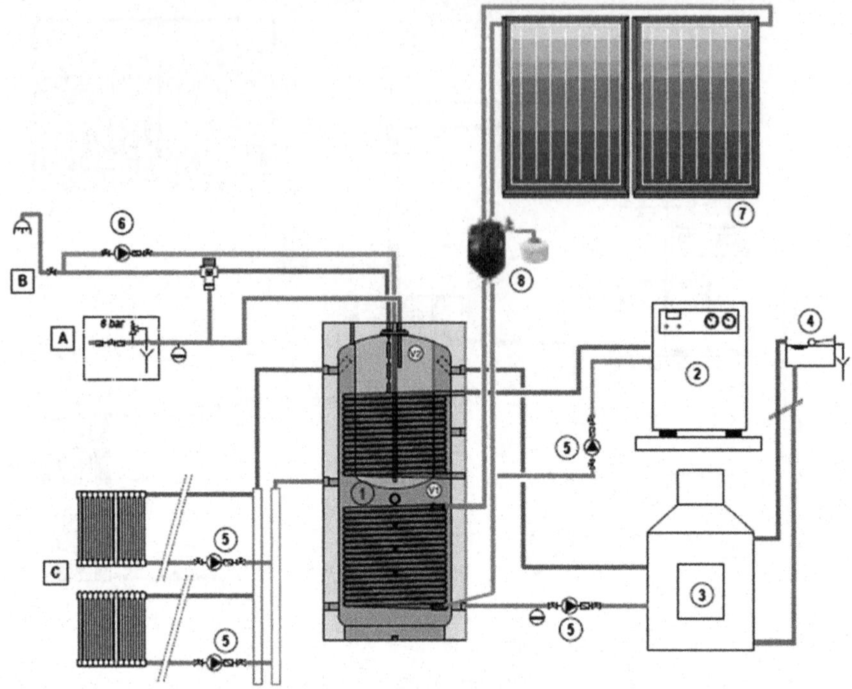

Fig. 18 Connection of the trivalent buffer tank [12]. *A* Domestic cold water inlet, *B* domestic cold water users, *C* heating system, *1* buffer tank, *2* HP unit, *3* additional burner, *4* opened expansion tank, *5* circulation pumps, *6* hydraulic safety group, *7* solar panels, and *8* solar circulation pump

3 System Operating and Control

Nicolae Badea, Emil Ceanga, Marian Barbu and Sergiu Caraman

3.1 Operating Modes of the CHP Unit

Depending upon requirements, the mCHP unit can operate using different types of operating modes (electricity and/or heat-driven). Time-variant operating modes may be implemented by means of an energy management system, which selects the optimum operating mode for the specific requirements.

- *Heat-driven operating mode*
  The heat demand is always the control variable for CHP operating in the heat-driven operating mode. The generated electricity has to be supplied for own use or to be fed into the grid.

- *Electricity-driven operating mode*
  For this operating mode, the electrical power demand is the control variable of the power output from the CHP unit. In principal there are two possible operating modes:
  - working in parallel with other systems; the CHP unit supplies the consumers until it reaches its maximum electrical output.
  - working independently; the CHP unit (very often in combination with a battery system) has to cover its own consumers' demand.
  - by taking additional measures, the CHP unit may also be used as stand-by power supplying system.

  The thermal energy, which is produced simultaneously by the CHP unit, should be used as well as possible. When it is appropriate, the heat storage tanks or other capacities may be used for heat storage.
- *Combined operating modes*
  It is also possible to apply combined operating modes, for example:
  - Heat-driven with peak-electricity function,
  - Maximum electricity and/or heat demand,
  - Minimum electricity and/or heat demand.

3.2 Control Strategy of the CCHP System

Designing the control system needs an analysis of all the elements that influence the performance level indicators of the mCCHP system: the thermal and electrical energy sources production, the loads which occur in summer and winter regime, as well as the state of the main thermal and electrical battery sources.

In a larger frame, this analysis is useful for achieving the following mutually dependent objectives:

- to establish the potential solutions set regarding the structure of a mCCHP system based on renewable energy sources and designed for residential consumers;
- to establish the potential solutions set regarding the system control, all these being viable from the point of view of the steady-state regime performances; and
- to analyze the system's operating through modeling and numerical simulation in dynamic regime. The purpose of this analysis is the functional design of the system, based on comparative evaluations, when a certain working solution is applied associated to a certain control solution.

In order to achieve these objectives, the modeling and numerical simulation of the mCCHP system is based on the aspects below mentioned.

1. The mathematical model of the system must be determined based on energetic balances, where the main energy accumulations existent within the system are considered, as follows:

   (a) the accumulation of electrical energy in a battery, through which the power transfers are achieved;
   (b) the accumulation of thermal energy in a heat accumulation tank, through which the thermal power transfer from the sources to the thermal consumers is also achieved.

2. The principle applied in the process control consists in using the voltage of the electrical energy accumulator and the temperature of the heat accumulation tank, as values sensitive to the misbalance between the produced power and the consumed one. In the case of both accumulations, if the power produced is less than the consumed one, then the electrical/thermal potential decreases and vice versa. The control system must maintain at constant (nominal) values the capacities through which the electrical/thermal potential (voltage and temperature) are evaluated, by adjusting the produced power. Thus, the equilibrium between production and consumption is achieved.
3. The dynamic components of the system's global model are generated by:

   (a) the energetic mass-balance equations, at the level of the two accumulators mentioned before,
   (b) the models of the control subsystems, and
   (c) the simplified models of the dynamics of those energy sources which have their own control system (such as the Stirling engine or the pellet boiler).

4. For some sources, such as the solar thermal collector or photovoltaic system, as well as for the thermal and electrical consumers, steady-state models can be used.
5. As models of the solar collector and photovoltaic system, the diurnal power graphs are adopted representing the powers delivered by these sources when they are located in the residence site. The graphs are built based on the statistical data regarding the site.
6. The diurnal power graphs (built on the basis of some plausible hypothesis regarding the consumption requirements over a 24 h interval at a residential level) are used as models of the electrical load components and of some of the thermal load components (such as the power consumed for heating the domestic water).
7. The electrical load has two components: the load used at the residential level and the load generated by the own consumption of mCCHP system equipments (such as pumps, ventilators etc.).
8. Designing the mCCHP system and analyzing its dynamic performances must be carried out in two distinct operating regimes: in winter regime, when the thermal energy is mainly used for heating the residence, and in summer regime, when the dominant thermal load is the air conditioning system. The thermal load in both operating regimes is designed in steady-state regime, but also for

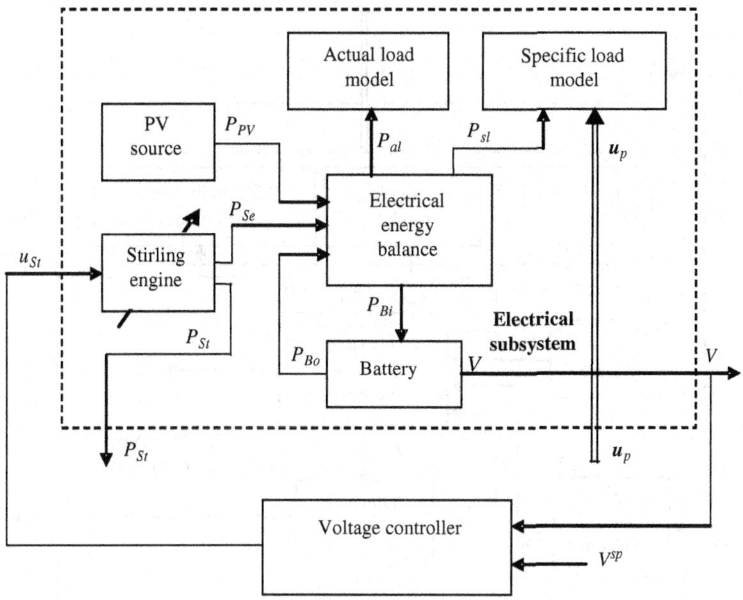

Fig. 19 Balance of both produced and consumed powers in the electrical subsystem

maintaining the possibility of considering any other plausible values of the weather conditions and for the temperature values imposed inside the residence.

9. In order to test the system's capacity to ensure the comfort requirements, even in the most difficult situations, the analysis of the dynamic performances (in winter regime, as well as in summer regime), not only the normal meteorological condition, but the extreme ones must be considered.
10. The control system must include monitoring functions, which can provide information regarding the quality standards of the mCCHP system.

Figure 19 presents the balance of produced and consumed powers in the electrical subsystem using the control loop for the battery voltage adjusting, V. The voltage controller, with a setpoint of V^{sp}, modifies the Stirling engine power, through u_{St} control.

The power delivered into the local electrical grid includes the following components:

- the power supplied by Stirling engine, P_{Se};
- the power of the PV (photovoltaic system) source, during the interval when it is active, P_{PV}; and
- the power delivered by the battery, P_{Bo}, when the sum of powers delivered by the sources mentioned above does not cover the whole consumption.

The electrical load contains the following components:

- the useful load of the residence, P_{al};

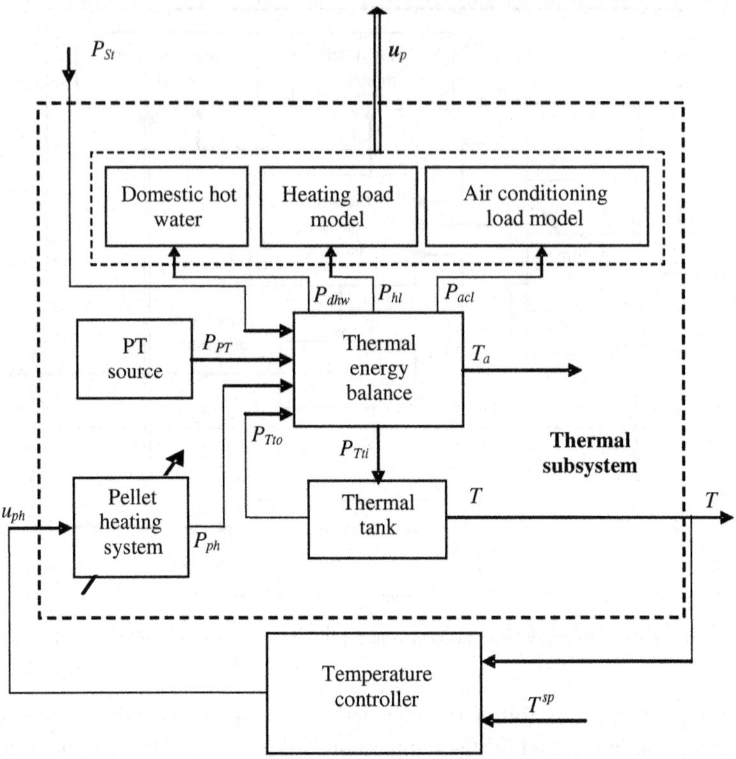

Fig. 20 Balance of both produced and consumed powers in the thermal subsystem

- the mCCHP system's own consumption, P_{sl}, mainly resulting from the air conditioning equipment; and
- the consumption for loading the battery, P_{Bi}, when the total power delivered by the sources is greater than the useful load summed up with its own consumption.

The thermal power, P_{St}, delivered to the thermal subsystem depends on the Stirling engine control through the voltage controller. In the same time, the on/off controls of the mCCHP system components (most part are components of the thermal subsystem) are included in the vector u_p and determine the own consumption of the system. It results that the electrical subsystem interacts with the thermal one through P_{St} and u_p values.

Figure 20 presents the production's balance with respect to the load, in the thermal subsystem.

In this case, the accumulation of energy takes place in a heat accumulation tank, where the temperature, T, of the thermal agent is maintained constant at the setpoint T^{sp}. The temperature controller modifies the power of the pellet boiler, P_{ph}, through the control variable u_{ph}. Based on the residence thermal balance, the total power which must be delivered by the thermal sources, so that the desired values of the temperature in the rooms (T_a) should be obtained, is determined in conditions of extreme values of the external temperature in winter and summer regimes.

In the thermal balance, the delivered power has the following components:

- the power delivered by the pellet boiler, P_{ph}, established by the temperature controller;
- the thermal power of the Stirling engine, P_{St}. In the thermal subsystem, it is a random variable, because in the operating regime of the Stirling engine the ratio P_{St}/P_{Se} is constant and the power P_{Se} modifies randomly, due to the fact that it is a control variable in the voltage control loop;
- the power of the solar thermal collector, P_{PT}, during the interval in which this source is active; and
- the power transferred from the thermal tank, P_{Tto}, when the sum of the powers delivered by the sources mentioned above do not cover the thermal load.

The thermal load has the following components:

- the power necessary for heating the residence during the cold season, P_{hl};
- the power consumed by the air conditioning equipment in the hot season, P_{acl};
- the power consumed in the domestic water circuit, P_{dhw}; and
- the power consumed for completing the energy accumulated in the thermal tank, P_{Tti}, that is the power used to increase the temperature T up to the setpoint, T^{sp}, when the total thermal power delivered by the sources is greater than the total thermal power used by the consumers.

From the interconnected operating mode of the two subsystems, it results that the electrical subsystem "subordinates" the thermal subsystem. The electrical energy variations of the Stirling engine, which take place in the process of balancing the produced power with the consumed one in the electrical subsystem (through the voltage controller), determine important variations of the thermal power produced by the Stirling engine. These variations must be compensated at the level of the temperature control system, through the adjusting of the pellet boiler power. To exemplify, let us consider the situation in which, at a certain moment, there is an important consumption of electrical energy in the electrical subsystem and the load is reduced in the thermal subsystem. In the above-mentioned situation, the control system previously presented imposes that the Stirling engine should produce the nominal electric power, through the temperature control loop. Simultaneously, it also delivers the nominal thermal power so that, if the energy accumulation in the thermal tank is already at the nominal level, the total produced thermal power is greater than the current consumption (if the solar thermal collector also operates). It is obviously that the excess of thermal energy must be dissipated, which reduces the economical efficiency of the mCCHP system. This situation occurs more frequently when monovariable controllers are used for the system control. There is another possibility of adjustment, which takes into account the interaction between the electrical and the thermal subsystems. In the situation presented above, when the electrical load is big and the thermal load is small, a multivariable controller

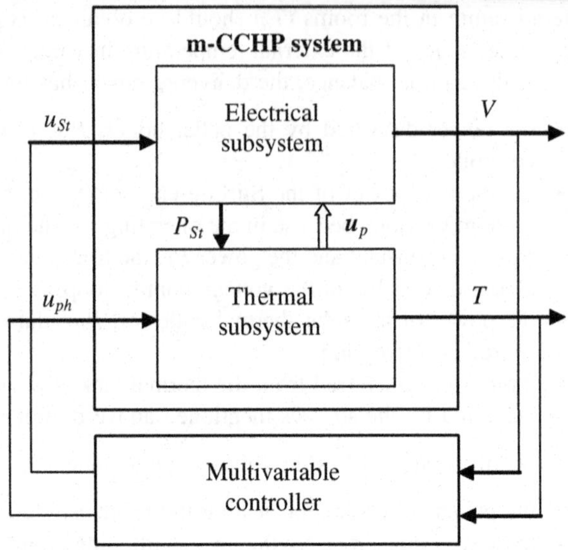

Fig. 21 Multi-variable control of a mCCHP system

(Fig. 21) may give a smaller control to the Stirling engine than a monovariable controller, so that the dissipated thermal energy should be as reduced as possible.

The electrical power delivered by the Stirling engine is not yet sufficient to compensate the load, but it is completed by the energy accumulated in the battery. Obviously, the completion is sufficient only if the electrical overload does not last very long. If this persists much longer, then, after the energy accumulated in the battery is reduced to the minimum allowed level, the Stirling engine operates at nominal power, even there is a risk of increasing the amount of the dissipated heat.

In the context of using a multivariable controller, the optimal control objective may be formulated as follows: minimize the loss of thermal energy when the voltage V and the temperature T are maintained in a given interval, around their nominal values.

4 System Dynamics Analysis

Emil Ceanga, Marian Barbu and Sergiu Caraman

The analysis through numerical simulation has the following objectives:

- to check the viability of different structural models for the mCCHP system, from the point of view of the performances in steady state and dynamic regimes;
- to select the best functional model of mCCHP from the technical and economical point of view that meet the requirements of ensuring the residence microclimate in winter and summer regimes.

Further on, the methodology of analysis through numerical simulation of a mCCHP is presented, considering several hypothetical scenarios regarding:

(a) the structure of mCCHP system models, considered in the simulation scenarios,
(b) the diurnal graphs of the useful electrical and thermal loads, and
(c) the nominal parameters of the main components which may be part of the system configuration (Stirling engine, pellet boiler, air conditioning equipment, solar thermal collector, photovoltaic systems).

4.1 The Structure of the mCCHP System Models Considered in Simulation

Model 1 In this functional model, the electrical energy is produced by two sources: the Stirling engine and the photovoltaic source (PV). Both sources recharge the battery and the local electrical grid is supplied by an inverter, coupled to the battery. The thermal energy is produced by three sources: Stirling engine, a pellet boiler, and a solar thermal collector, all being collected into an accumulation tank. The thermal energy accumulated in the tank is used for domestic water heating and for the residence heating in winter regime. It is also used for air conditioning and domestic water heating in summer regime. The structure of the conditioning equipment contains an accumulation tank for the cold agent.

Model 2 The electrical energy is produced by a single source: the Stirling engine. The thermal energy is produced by three sources, as well as in Model no. 1: Stirling engine, pellet boiler, and solar thermal collector.

Model 3 The difference as compared to the previous model consists in the fact that, in the summer regime, the solar thermal collector is missing.

Model 4 In this functional model, the electrical energy is produced by the Stirling engine and by a PV source. The thermal energy is produced only by two sources: Stirling engine and pellet boiler.

Model 5 The case in which the air conditioning uses electrical equipment is considered. In this functional model, the electrical energy is produced by the Stirling engine and a PV source and the thermal energy is produced only by two sources: Stirling engine and the pellet boiler.

4.2 Principles of Modeling and Numerical Simulation

(a) The simulation of the system in different functional models is distinctly carried out for the winter and summer regimes.
(b) The thermal powers necessary requirements for heating—in winter regime, and for air conditioning—in summer regime are determined through the

thermal balances in steady-state regime. In this case, the temperature values within the residence, according to the requirements of the thermal comfort and the extreme values of the external temperature, in winter and summer regimes, are imposed. The balances determine the powers necessary to ensure the temperature values imposed in the residence. They are achieved by two calculus program which, in what follows, will be called winter thermal balance (WTB) and summer thermal balance (STB), respectively.

(c) The electrical load of the system is made up of two components:

- the useful electrical load of the residence, generated by the consumers within the residence (illumination, refrigeration etc.);
- the load generated by the equipment's own consumption (the consumption of pumps, ventilators etc.).

The useful electric load is adopted as a diagram of the power consumed in 24 h, which must be as close as possible to the average diurnal consumption. Figure 22 presents the diagram of the diurnal useful power. It is the same for both winter and summer regimes. The electrical load determined by the equipments own consumption is a variable value, whose time evolution depends on the controls given to various equipments of the system.

(d) The thermal load of the system is also made up of two components:

– *a slow variation component* necessary to bring the residence temperature to the desired values. This component modifies on the temporal scale of the weather conditions evolution and it will be considered constant in the numerical simulation. In simulation, the component of the thermal load is supplied by the calculus programs, WTB and STB;
– *a fast variation component* representing the power consumed in the domestic water circuits (kitchen and bathroom). The graph of variation in 24 h for this component is adopted so that it should be close to the diurnal

Fig. 22 Graph of the useful electric power consumed in 24 h

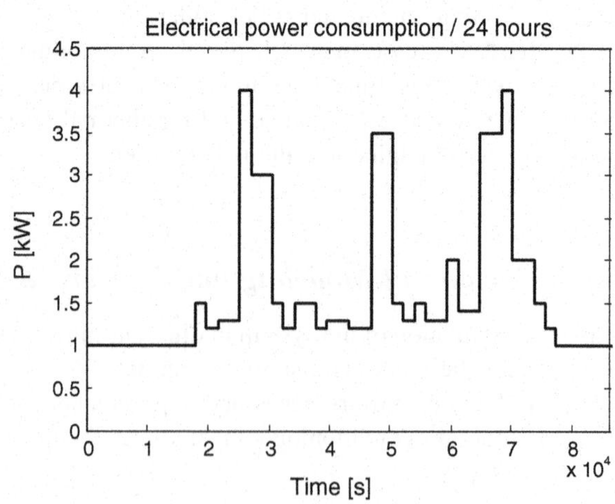

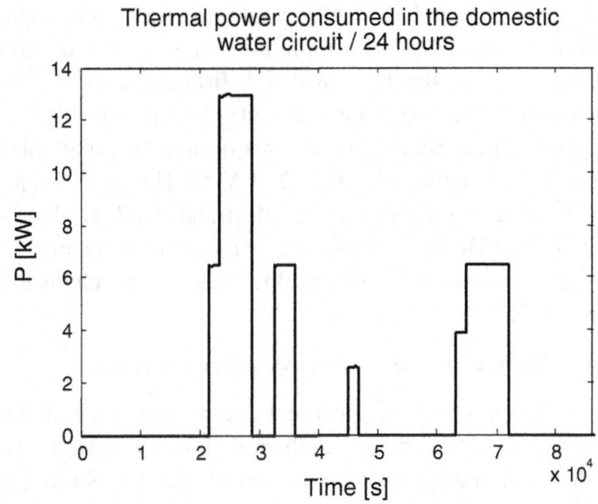

Fig. 23 Generic evolution of the thermal power consumed in the domestic water circuit in 24 h

consumption average of power in the circuit mentioned before (Fig. 23). This graph is considered the same in both winter and summer regimes.

(e) To obtain the permanent regimes, within the very slow thermal processes in mCCHP, the numerical simulation of the system must be carried out in enough extended interval. In what follows, the system simulation in a 3-day interval has been considered (3 day x 24 h x 3600 s, that means 259,200 s).

4.3 Simulation of the Dynamic Regimes

4.3.1 Electrical Subsystem

The electrical subsystem contains the following main components: the electrical battery, which represents the electrical energy accumulation subsystem in the whole system, the Stirling engine (the electrical part) and the electric load. The subsystem also contains a battery voltage control loop, having as control variable the power produced by Stirling engine. The simplified model of the battery is considered linear as:

$$V = V_m + \frac{V_M - V_m}{W_M - W_m}(W - W_m) \tag{53}$$

where V is the battery voltage at the current moment, when the accumulated energy in the battery is W; V_M and V_m are the voltage values at the maximum and minimum energies, W_M and W_m, respectively. A battery with a nominal voltage of 48 V was chosen. The relation (53) considers the following limit values of the voltage:

$V_M = 49.5$ V and $V_m = 47$ V. Admitting a battery capacity of 800Ah, it results that W_M = 138.24 MJ. The minimum value of the accumulated energy, at which the battery is practically considered discharged, is W_m = 17.28 MJ. If the battery is charged at the maximum capacity, that means W_M = 138.24 MJ, the battery may supply an electrical load of 1.4 kW in a 24 h interval. The transfer of the electrical energy from the source of 230 V-50 Hz, which represents the electrical part of the Stirling engine, to the electrical load is done through the chain: source 230 V-50 Hz → transformer → rectifier → battery → inverter 230 V-50 Hz. The analysis of the component efficiency of this chain leads to a global efficiency of $\eta = 0.7$.

A. *The model of the electrical energy sources*

The model of the electrical energy sources is defined by the graph of the time variation of the power delivered into the system. Within mCCHP system, two electrical energy sources are considered: the Stirling engine (its electrical component) and the PV system. Obviously, the first source is the main source and the second source is the complementary one. The electrical power given by the Stirling engine is controlled by the battery voltage controller, which achieves the balance between the produced power and the consumed one in the electrical subsystem. The power delivered by the PV source depends on the winter/summer regime of the system.

(a) *In winter regime* the daily operation of the photovoltaic system between 8 a.m. and 4 p.m. was considered in optimum operating conditions. For the rest of the day, the irradiance was considered equal to 0.
(b) Table 13 presents the values calculated for the irradiation and electrical powers during one winter day.

The relations from the Sect. 7 in Chap. "Renewable energy sources for the mCCHP-SE-RES systems" (where the PV sources are presented), considering an average efficiency of 10 % and a surface of 12 m^2 for the photovoltaic system were used in Table 13.

Figure 24a presents the profile of the power delivered by the PV source in 24 h (in winter regime). The average power corresponding to this profile is equal to 0.298 kW.

(c) *In summer regime* it was considered that the PV system operates between 6 a.m. and 9 p.m. Table 14 presents the values calculated for the irradiation and for the electrical one during one summer day and Fig. 24b shows the profile of the power delivered by the PV source in 24 h. An average power of 0.623 kW corresponds to this profile from Table 14 .

B. *Load model*

The useful electrical load model is given by the variation of the power consumed in 24 h that is presented in Fig. 22. The average diurnal power corresponding to this graph is of 1.58 kW.

Table 13 Daily operation of the photovoltaic system in winter regime

Interval within one day	Irradiance (W/m^2)	Irradiation (kW h)	Electrical power (kW h)
0–8^{00}	0	0	0
8^{00}–8^{30}	200	2.4	0.24
8^{30}–9^{00}	400	4.8	0.48
9^{00}–9^{30}	600	7.2	0.72
9^{30}–10^{00}	800	9.6	0.96
10^{00}–14^{00}	1,000	12	1.2
14^{00}–14^{30}	800	9.6	0.96
14^{30}–15^{00}	600	7.2	0.72
15^{00}–15^{30}	400	4.8	0.48
15^{30}–16^{00}	200	2.4	0.24
16^{00}–24^{00}	0	0	0

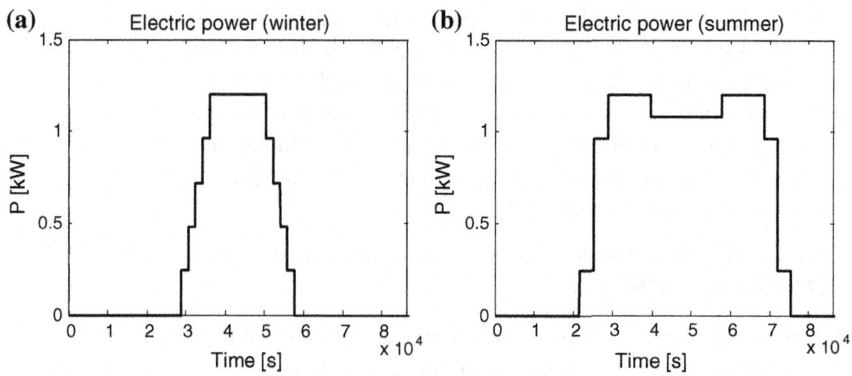

Fig. 24 Profile of the electrical power in 24 h generated by the PV source in winter (**a**) and summer (**b**)

The load generated by the equipment's own consumption is dependent on the summer/winter operating regime. In both regimes, the analysis has to take into account the difficult (extreme) meteorological conditions, for which the thermal balances of the residence were drawn up. These extreme situations generate the operating modes in which the system equipments are active most of the time, and therefore produce the highest equipment's own consumption.

(a) *In winter regime* the equipment's own consumption contains two components:

- the permanent regime consumption in difficult meteorological conditions, when the heating plant operates permanently. This consumption is given by: the electrical subsystem of the Stirling engine (0.09 kW), the electrical subsystem of

Table 14 Daily operation of the photovoltaic system in summer regime

Intervals within one day	Irradiance (W/m^2)	Irradiation (kW h)	Electrical power (kW h)	Electric power corrected by temperature (kW h)
$0–6^{00}$	0	0	0	0
$6^{00}–7^{00}$	200	2.4	0.24	0.24
$7^{00}–8^{00}$	800	9.6	0.96	0.96
$8^{00}–11^{00}$	1,000	12	1.2	1.2
$11^{00}–16^{00}$	800	9.6	1.2	1.08
$16^{00}–19^{00}$	1,000	12	1.2	1.2
$19^{00}–20^{00}$	800	9.6	0.96	0.96
$20^{00}–21^{00}$	200	2.4	0.24	0.24
$21^{00}–24^{00}$	0	0	0	0

the pellet boiler (0.09 kW), the cooling pump of the Stirling engine (0.071 kW), the recirculation pump of the pellet boiler (0.071 kW), the circulation pump of the pellet boiler (0.311 kW), the cooler ventilator of the Stirling engine (0.340 kW), and the pump used in the hot water circulation in convectors (0.6 kW). It results a permanent consumption of 1.573 kW;

- the consumption of the domestic water recirculation pump (0.202 kW), when the respective circuit operates. Considering the operating intervals for the pump given by Fig. 23 and the pump operates at the nominal power, it results an average consumed power of 0.051 kW.

As a conclusion, the total average consumption in winter is: $P_c^w = 1.58 + 1.573 + 0.051 = 3.204$ kW.

(b) *In summer regime* the equipment's own consumption differs from the one used in winter regime through its permanent regime component, since the equipments which operate are different. The electrical load is made up of the following components:

- the permanent load of 0.902 kW contains: the pellet boiler (0.09 kW), the cooling pump of the Stirling engine (0.09 kW), the recirculation pump for the pellet boiler (0.071 kW), the circulation pump for the pellet boiler (0.311 kW), and the cooler ventilator of the Stirling engine (0.340 kW);
- the consumption of the cooling subsystem, with a total power of 2.218 kW (pump for the thermal agent in the chiller—0.26 kW; the cold water pump for the chiller circuit—accumulation tank—0.27 kW; the pump for the external circuit of the chiller—0.474 kW; recooler engine—1.2 kW; adsorption chiller—0.014 kW). The cooling subsystem operating regime (that is of the adsorption chiller) is scheduled in the time intervals 10 pm–4 am and 10 am–4 pm. It results an average consumed power of 1.109 kW;

- the consumption of the pump for cold water circulation into the convectors (0.6 kW). The pump is scheduled to work between the 4 am and 10 pm, therefore it results an average consumed power of 0.15 kW;
- the consumption of the pump for domestic water recirculation (0.202 kW), when the respective circuit is fully operational. An average power of 0.051 kW corresponds to this circuit;
- the consumption of the pump for the water circulation in the solar thermal collector (0.2 kW). The pump is scheduled to operate between 6 am and 7 pm, therefore it results an average power of 0.108 kW.

As a conclusion the equipment's own average consumption in summer regime is: 0,902 + 1,109 + 0.15 + 0.051 + 0.108 = 2.32 kW and the total average power of electrical energy consumed during summer is $P_c^s = 1.58 + 2.32 = 3.9$ kW.

C. *Energy balance checking*

The average power consumed in winter regime is $P_c^w = 3.204$ kW and in summer regime $P_c^s = 3.9$ kW. Based on these results, the minimal power of the Stirling engine may be chosen. The inferior limit value of this power is $P_{St,e} = 5$ kW. Taking into account a global efficiency $\eta = 0.7$ of the power produced by the Stirling engine, it results a useful power delivered by the Stirling engine of $P_{Stu} =$ 3.5 kW.

In winter regime the average power generated by the PV source, 0.298 kW, may also be counted on. Since this source discharges directly into the battery, an efficiency of its use of 0.85 is also admissible and the average power available in the load from the PV source is $P_{PVu}^w = 0.2533$ kW. The total average power available in this regime is $P_d^w = P_{Stu} + P_{PVu}^w = 3.753$ kW, that is greater than the average power consumed, $P_c^w = 3.204$ kW.

In summer regime the general average power generated by the PV source is equal to 0.623 kW and the useful average power in the load is $P_{PVu}^s = 0.529$ kW. The total average power available in this regime is $P_d^s = P_{Stu} + P_{PVu}^s = 4.029$ kW, that is greater than the average power consumed, $P_c^s = 3.9$ kW.

It results that the average powers of the sources in winter and summer regimes are greater than the average consumptions in the respective regimes, in difficult meteorological conditions admissible in achieving the energetic balances (external temperature is −17 °C in winter and +35 °C in summer).

Further on, the main issue is to determine if the control loops, which ensure the balance between the produced power and the consumed one in the electrical and thermal subsystems, operate with admissible errors in the dynamic regime. Essentially, it is necessary to check if:

- the dynamic error of the voltage loop does not reach excessive values, for which the battery is practically discharged;
- the dynamic error of the temperature loop in the thermal tank does not exceed the limits of a correct operating regime of the thermal subsystem.

These tests are achieved through simulations of the whole system in dynamic regime.

D. *Numerical simulation of the electrical subsystem*

The analysis of the dynamic regimes is achieved on the time scale corresponding to the evolution of the physical variables of the Stirling engine. In the following, a time constant of 60 s was taken into consideration for the Stirling engine, to which corresponds a transitory time of about 300 s. Obviously, on this time scale, the fast dynamic regimes of the electrical processes are neglected.

The control law of the battery voltage loop, according to the principle scheme given by Fig. 21, may be of 0/1 or continuous control type. Further on, it is considered the case in which the Stirling engine allows an on/off control, that corresponds to a power in the range of 0 – $0.45P_n$, and a continuous control in the range of $0.45P_n$– P_n (P_n denotes the nominal power). For smaller powers, under 0.45 P_n, the control of the Stirling engine is achieved through 0/1 pulse width modulation (PWM), the minimum period of these impulses being 30 min (the minimum duration between two successive starts of the Stirling engine being imposed by the engine characteristics). The controller used in the battery voltage control loop is of PI type. Taking into account the global efficiency of using the electrical energy delivered by the Stirling engine, $\eta = 0.7$, the controller will provide controls of PWM type in the range of 0–1.6 kW and continuous controls in the range of 1.6–3.5 kW.

In these conditions, the numerical simulation scheme of the electrical subsystem is given by Fig. 25 where the block "Stirling engine" is detailed in Fig. 26. Figures 25 and 26 include also the Simulink blocks "Electrical process controller" and "PWM Subsystem," respectively. These blocks are further presented in Figs. 89 and 91b in Sect. 4.1. The variation of the power consumed in 24 h, with the form given in Fig. 22, is generated by the block of the from File type, named "var2." The simulation scheme allows the recording of the main variables of the subsystem: battery voltage, useful power delivered by the Stirling engine, difference between

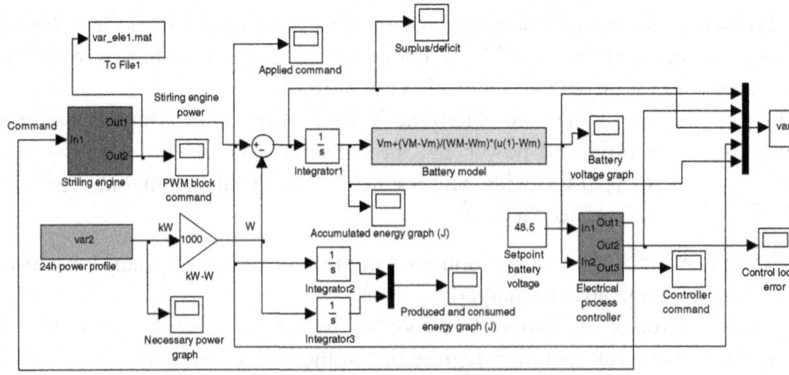

Fig. 25 Power subsystem simulation scheme

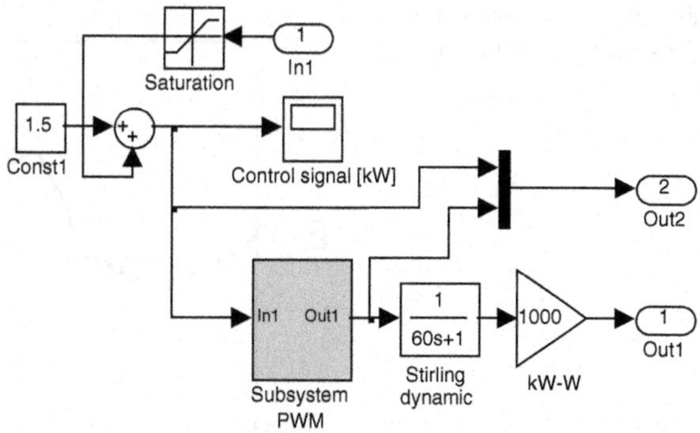

Fig. 26 "Stirling engine" simulation block

the useful produced power and the consumed one, energy accumulated in the battery, control error etc. In what follows, considering that at the initial moment the energy accumulation in the battery is 90 MJ, the results of the numerical simulation over a 24 h interval will be presented.

Figures 27 and 28 present the evolutions of the battery voltage and of the voltage control error. The evolution of the difference between the produced power and the consumed one is given by Fig. 29. The Stirling engine is controlled by the voltage controller, whose power has the evolution given by Fig. 30. It must be noticed that, for useful produced powers, less than a threshold of 1.6 kW, the power supplied has a 0/1 variation, with pulse width modulation. The period of the impulses is of 30 min. The evolution of the energy accumulated in the battery is given by Fig. 31.

Remark The presented results illustrate the principle of electrical subsystem simulation, without considering the real load (the equipment's own consumption has been neglected).

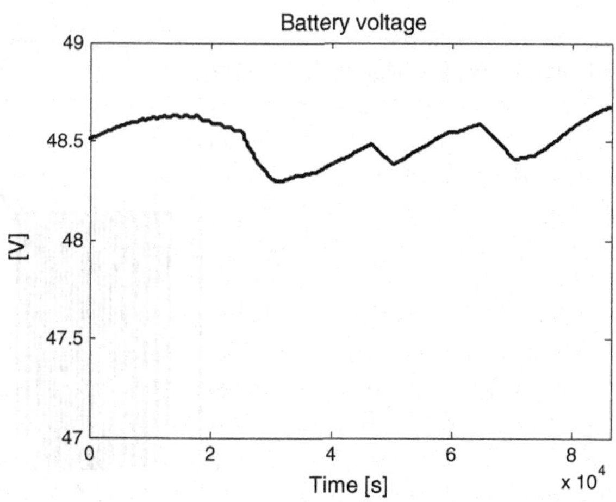

Fig. 27 Battery voltage evolution in electrical subsystem

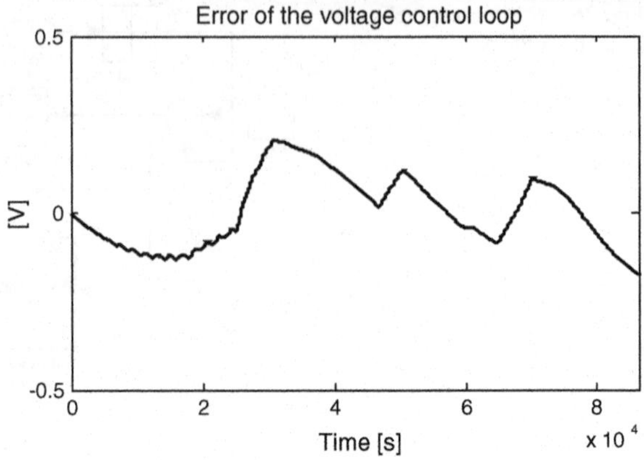

Fig. 28 Error evolution of the voltage control loop

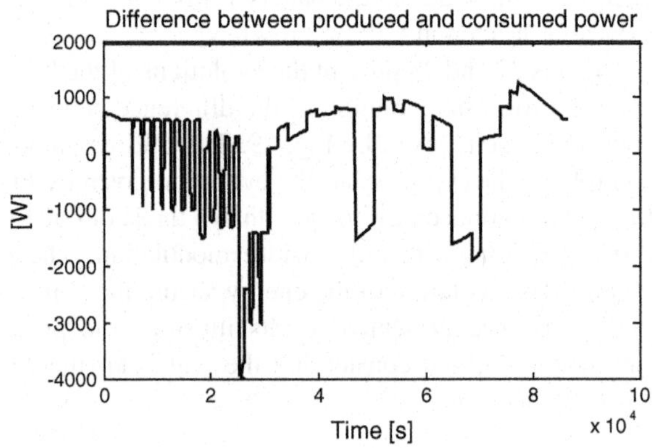

Fig. 29 Evolution of the difference between produced and consumed power

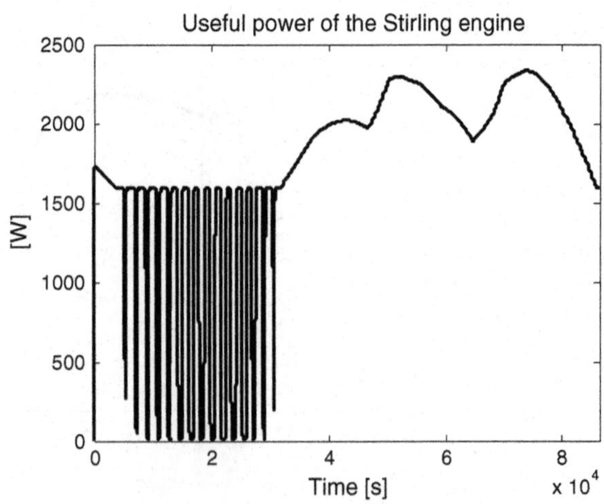

Fig. 30 Useful power evolution of Stirling engine

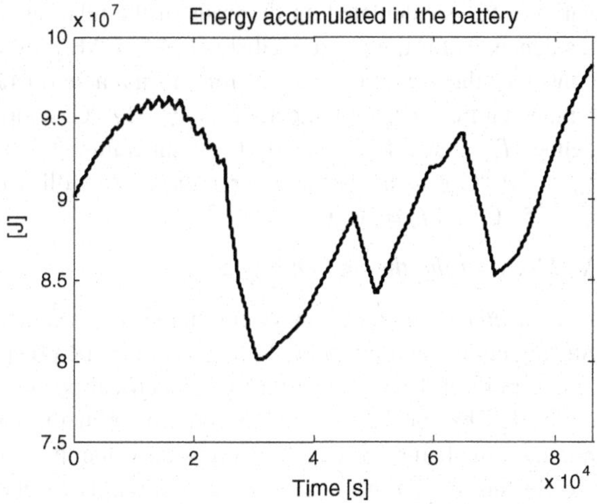

Fig. 31 Evolution of the energy accumulated in the battery

4.3.2 Thermal Subsystem

The mathematical model is determined based on the thermal balances of the accumulation tank and the residence. The equation of the thermal balance of the thermal accumulator is:

$$m_w c_w \frac{dT_c}{dt} = P_{Stir} + P_{pel} - P_{tv} - P_{am} \tag{54}$$

where m_w, c_w are the mass, the specific heat of the water accumulated in the tank, respectively, T_c–the temperature of the water in the tank, P_{Stir}–the thermal power of the Stirling engine, P_{pel}–the power delivered by the pellet boiler, P_{tv}–the power consumed in the residence aiming to cover the losses through transmission and ventilation, and P_{am}–the power consumed in the domestic water circuit.

The balance between the thermal power delivered by the sources (Stirling engine and pellet boiler) and the power consumed for the residence heating and for the domestic hot water is achieved by the control of the water temperature in the accumulation tank (setpoint equal to T_c^{ref}). Since the thermal power of the Stirling engine is a random value (given by the voltage controller), the temperature control of the thermal agent in the accumulation tank is done by the power control of the pellet boiler.

The second thermal balance equation is:

$$m_a c_a \frac{dT_a}{dt} = k_c(T_c - T_a) - P_{tv} \tag{55}$$

where m_a, c_a are the mass and the specific heat of the air in the residence, respectively, k_c–transfer coefficient through convection, and P_{tv} is the power consumed in the residence to compensate the losses through transmission and ventilation. The variable P_{tv} is calculated on the basis of the relations in the thermal balance of the residence, that is done using a distinct MATLAB program. This MATLAB program considers the following values for the main temperatures, corresponding to the winter regime: outside temperature $t_e = -17$ °C, temperature of the air in the residence, $T_a = 293$ K ($t_a = 20$ °C), temperature of the domestic cold water, and $t_{ar} = 10$ °C etc. In the summer regime, the following values were considered: $t_e = 35$ °C and $t_a = 26$ °C.

A. *Model of the thermal energy sources*

In winter regime the subsystem includes two sources of the thermal energy: the Stirling engine and the pellet boiler. The thermal power of the Stirling engine, $P_{St,t}$, is in a ratio of 3:1 with respect to its electrical power. Since the electrical power is controlled by the battery voltage controller in the electrical subsystem, so as to maintain the battery voltage at a constant value, it results that the thermal power of the Stirling engine has a random evolution. The pellet boiler is controlled by the temperature controller, which maintains constant the temperature of the thermal agent in the accumulation tank. This control has two objectives: bringing the temperature to the setpoint value, T_c^{ref}, and compensating the power variations of the Stirling engine. In the simulation scheme of the thermal subsystem, the pellet boiler was modeled as a simple dynamic system, with a time constant of 100 s.

The two sources transfer the thermal energy to the accumulation tank. Finally the energy is transferred to the thermal load. The temperature in the accumulation tank has a value sensitive to the difference between the produced power and the consumed one. Through the control of this temperature, the balance between the two powers is achieved.

In summer regime a third energy source is available: the solar thermal collector. It was considered that this source is active in the thermal subsystem circuit between 6 am and 6 pm. Using the mathematical model of the solar thermal collector and considering a day when the maximum air temperature reaches 38 °C, the graph of the diurnal power given by Fig. 32 is obtained. Integrating this power over a 24 h

Fig. 32 One day profile of the solar collector power

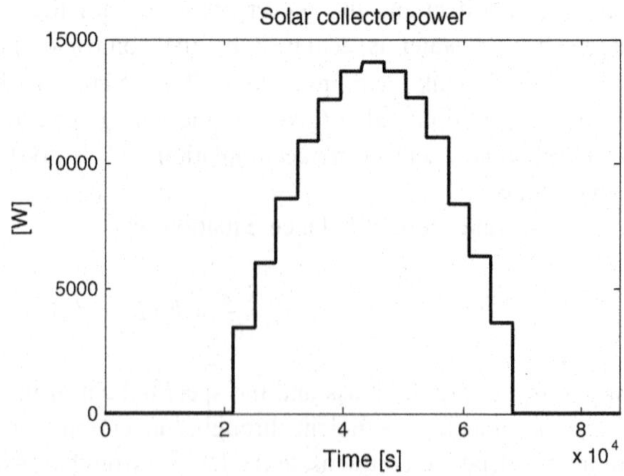

interval, an energy of 4.48×10^8 J is obtained, to which an average diurnal power of $P_{\text{collector}} = 5.18$ kW does correspond.

B. *Load model*

The thermal load differs depending on the season (winter/summer). In winter regime, the thermal load includes two components:

- the power necessary to the residence;
- the power consumed for heating the domestic water. The graph of this power in a 24 h interval is given by Fig. 22. If the power given by the graph presented in Fig. 22 is integrated, the energy consumed in this circuit, equal to $1.98 \cdot 10^8$ J, is obtained.

In summer regime the main component of the thermal load is the power necessary for the air conditioning. The second component, for the domestic water circuit, is considered the same as in the case of winter regime.

Both powers are independent variables, which play the role of disturbances in the control loop of the thermal agent temperature in the accumulation tank.

C. *Energetic balance checking*

In both operating regimes, winter/summer, the main components of the load are given by the power necessary for the residence heating and the one necessary for the air conditioning equipment. The values of the powers mentioned before are obtained with WTB and STB energetic balances, for the two regimes.

In winter regime, the thermal balance of the residence indicates that the power necessary to compensate the heat losses through transmission and ventilation is 14.8 [kW]. If the average power consumed by the domestic water circuit is added, it results a total average consumed power equal to $P_s^w = 17.1$ kW.

In the summer regime, the thermal balance of the residence provides a necessary power of about 30 kW for the air conditioning. The total average power necessary in this case, together with the consumption of the domestic water circuit, is 32.3 kW.

With these data, the necessary power of the pellet boiler may be chosen, the tank being controlled by the temperature controller. The selection criterion is the maximum value of the average load. Based on this value, of 32.3 kW, a boiler with a nominal power of 30 kW is chosen. In this case, the average thermal power, in kW, available in summer regime, is $P_t^s = 30 + P_{collector} + P_{St,t} > 32.3$ [kW].

The difference between the available average thermal power and the average thermal load may seem excessive, especially in winter regime. It is important to notice that the instantaneous values of the consumed power in the domestic water circuit are much higher than the average value. Under these conditions, there may be intervals in which the instantaneous value of the load exceeds sensitively the instantaneous value of the produced thermal power. Thus, *only the simulation of the system in dynamic regime may show whether the deviation of the collector temperature from the setpoint value remains into an admissible limits interval.*

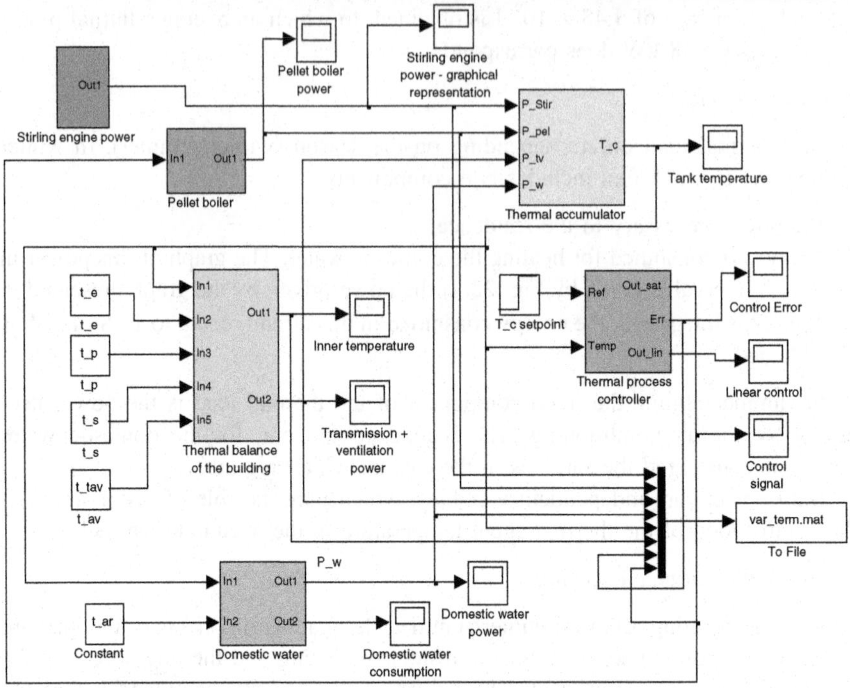

Fig. 33 Thermal subsystem simulation scheme for the winter mode

D. *Numerical simulation of the thermal subsystem*

As in the case of the electrical subsystem, the analysis done through the simulation of the dynamic regimes is shown for a 24 h interval, considering the winter regime.

The simulation scheme of the thermal subsystem is shown in Fig. 33. The simulation of the thermal energy sources is achieved through the following blocks:

- "Stirling engine power" which gives a random power variation $P_{\_Stir}$. In reality, the thermal power variations of the Stirling engine are generated as a result of the action performed by the battery voltage control loop. Since the simulated thermal subsystem is decoupled from the electrical one, a variation such as the one given by Fig. 34 of the variable $P_{\_Stir}$. has been imposed.
- "Pellet boiler" which is controlled by the temperature controller. This block is a simple dynamic element, with a time constant of 100 s.

The "Thermal accumulator" block has as output value the water temperature from the tank, $T_{\_c}$, and as inputs the thermal power of the Stirling engine, $P_{\_Stir}$, the power delivered by the pellet boiler, $P_{\_pel}$, the power consumed in the residence for covering the transmission and ventilation losses, $P_{\_tv}$, and the power consumed in the domestic water circuit, $P_{\_w}$. This block has the model given by the Eq. (54) and it is implemented by the scheme shown in Fig. 35.

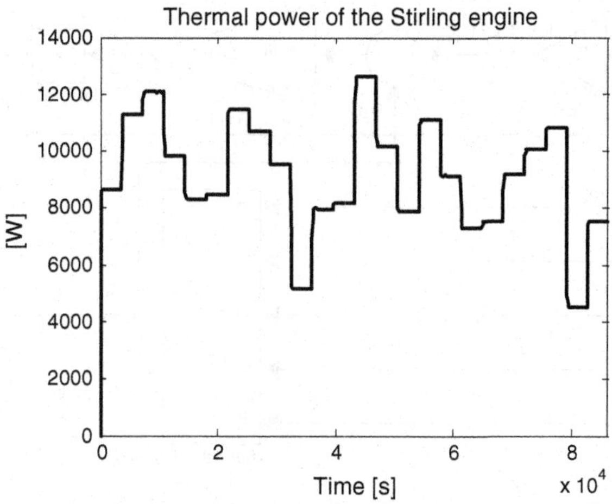

Fig. 34 Thermal power evolution of the Stirling engine

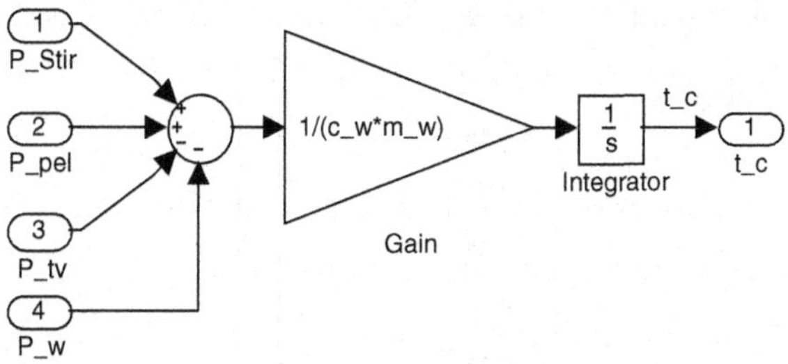

Fig. 35 "Thermal accumulator" simulation block from Fig. 33

The "Thermal balance of the building" is implemented through the scheme presented in Fig. 36 and it provides two variables:

- The temperature inside the residence, $t_{\_a}$. This variable is obtained according to the Eq. (54);
- the power lost through transmission and ventilation, $P_{\_tv}$.

The "Thermal balance" block from Fig. 36 achieves the thermal balance calculus in order to obtain the power $P_{\_tv}$, based on the following input data: the temperature of the air inside the residence ($t_{\_a}$), the temperature of the external environment ($t_{\_e}$), the temperature of the cellar ($t_{\_p}$), the temperature of the ground ($t_{\_s}$), and the temperature of the ceiling ($t_{\_tav}$).

The results of the thermal subsystem simulation are shown in Figs. 38, 39 and 40, when the variable $P_{\_Stir}$ (power delivered by the Stirling engine) is the one from Fig. 34 and the power graph of the domestic water circuit is the one given by Fig. 37. Figure 38 presents the power evolution of the pellet boiler. Figure 39 presents a very good performance of the temperature control system of the thermal

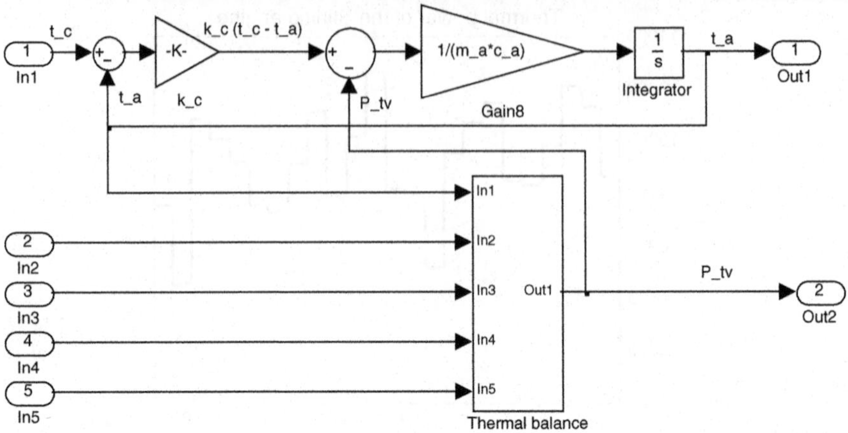

Fig. 36 "Thermal balance of the residence" simulation block from Fig. 33

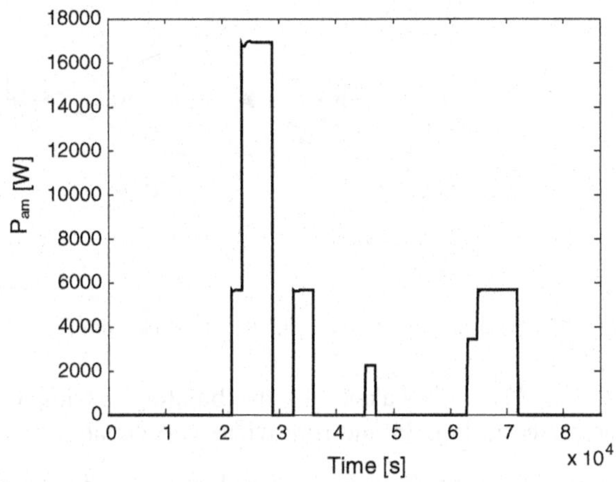

Fig. 37 Graph of the thermal power consumed in the domestic water circuit in 24 h

tank, the setpoint being equal to 72 °C. The evolution of the temperature inside the residence is shown in Fig. 40. The obtained results show (as in the case of the electrical subsystem) the simulation principle of the thermal subsystem, in a single operating regime and in the conditions when the two subsystems are decoupled.

4.4 Simulation and Analysis of mCCHP System—Model 1

In this functional model of mCCHP system, the electrical energy is produced by two sources: the Stirling engine and a PV source. Both sources discharge in a battery. The battery supplies the local electrical grid through an inverter, coupled to

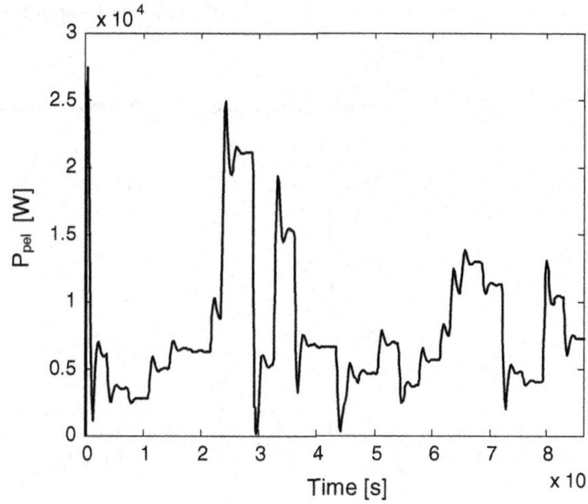

Fig. 38 Pellet boiler power evolution in 24 h in the thermal subsystem

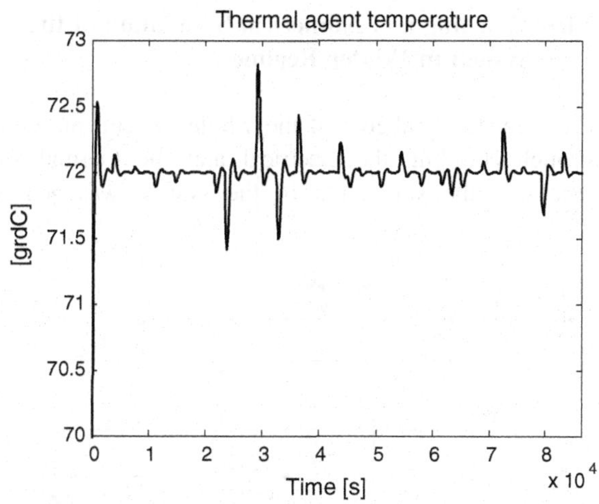

Fig. 39 Thermal agent temperature evolution in the accumulation tank in the thermal subsystem

the battery. Thermal energy is provided by three sources: the Stirling engine, a pellet boiler, and a solar collector, all being collected in an accumulating tank. The accumulated thermal energy is used for heating the domestic water and the residence, in winter regime, as well as for air conditioning, in summer regime. In the structure of the conditioning equipment, there is an accumulating tank for the accumulation of the cold reserve. This is not included in the model, since the conditioning equipment was characterized in the global model of the system through the total power consumed, without given details about the dynamics of the internal processes of the conditioning equipment.

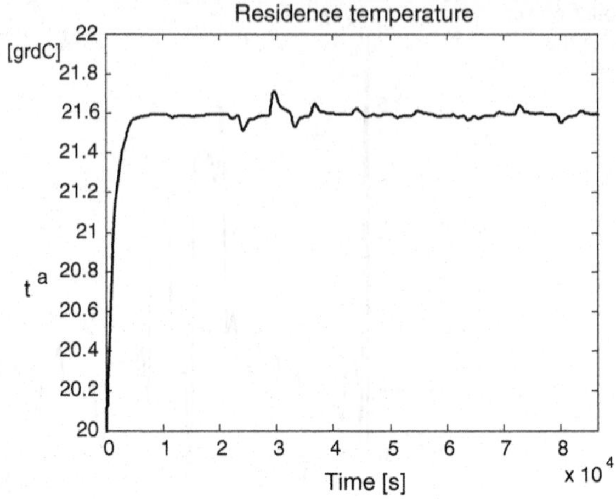

Fig. 40 Temperature evolution inside the residence

4.4.1 Modeling and Numerical Simulation of the System in Winter Regime

The simulation scheme of the whole system, presented in Fig. 41, was obtained through coupling the electrical and the thermal subsystems. As indicated in Sect. 4.2, the simulation of the system was achieved over a 3-day interval.

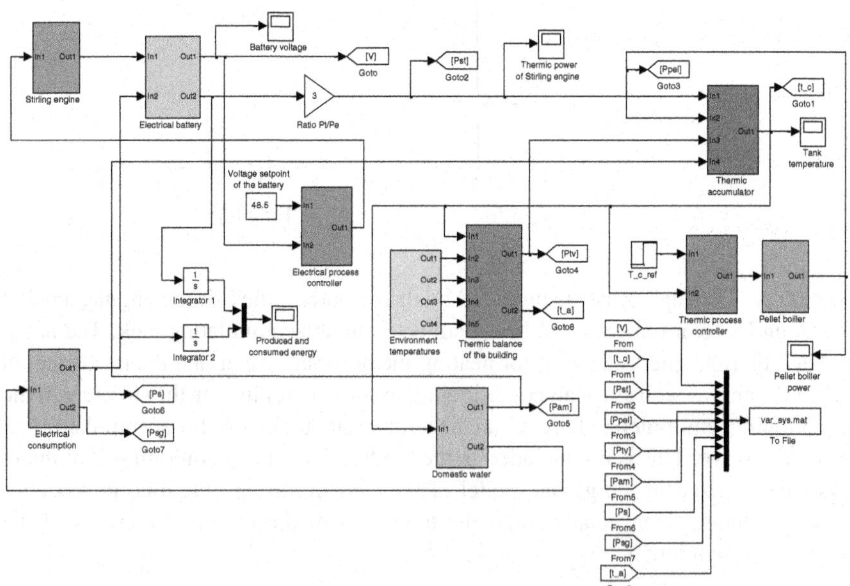

Fig. 41 mCCHP system simulation scheme for the winter regime (*functional model* 1)

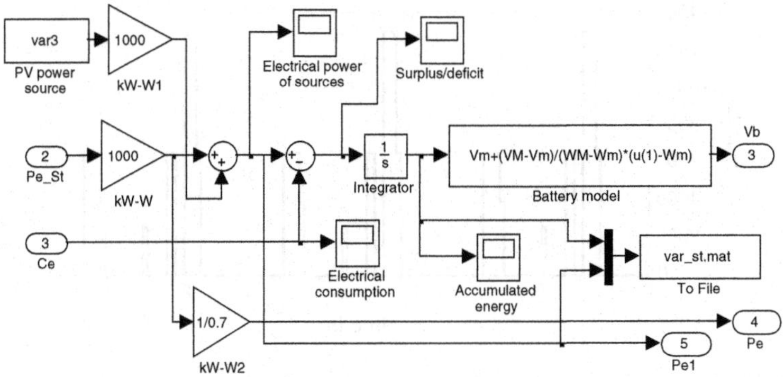

Fig. 42 "Electrical battery" simulation block

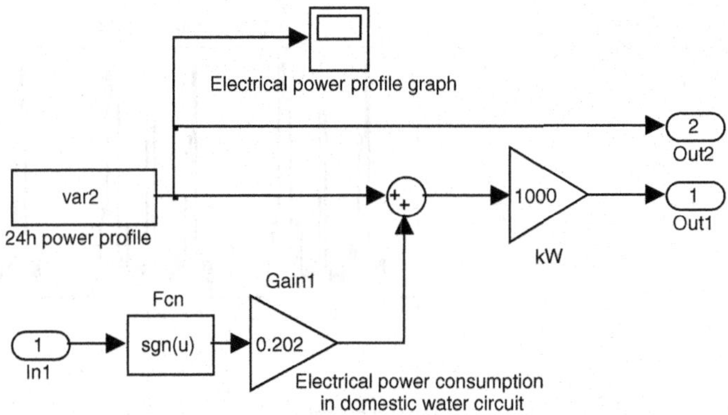

Fig. 43 The detailed scheme of the "electric consumption" block

Choosing such a long interval was justified by the necessity to obtain the permanent regime of the system, that is, to exhaust the transitory component, which is influenced by the initialization of the system (for example, the initial charging/discharging regime of the battery).

Figures 42 and 43 present in detail the "Electrical battery" and the "Electrical consumption" blocks. The operating conditions are given by the following imposed requirements:

- the thermal power consumed in the domestic water circuit is given by Fig. 44;
- the consumed electrical power is given by Fig. 45. The evolution of this power results based on a permanent load graph, given by Fig. 46, to which the consumption of the pumps from the domestic water circuit is added, when these start to operate.

Figure 47 gives the time evolution of the battery voltage and Fig. 48 shows the energy accumulated in the battery. It can be noticed that, after the transitory regime, these evolutions practically indicate a permanent one. That means the electrical

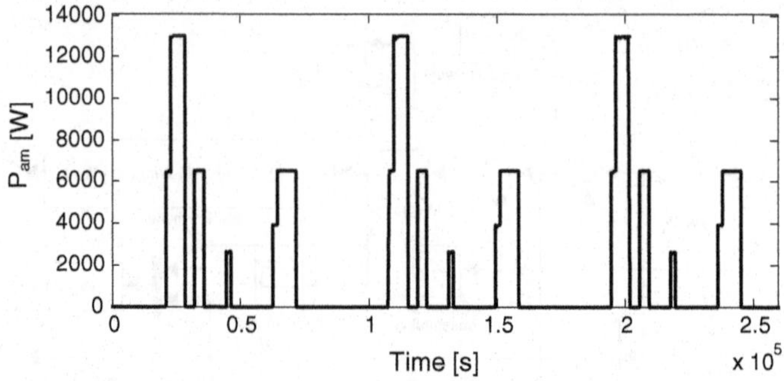

Fig. 44 Thermal power consumed in the domestic water circuit in winter regime, model 1

Fig. 45 Total consumed electrical power in winter regime, model 1

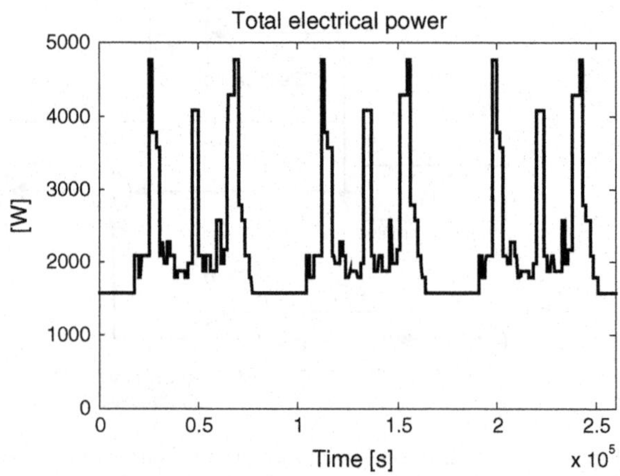

Fig. 46 Evolution of the permanent electric load in winter regime, model 1

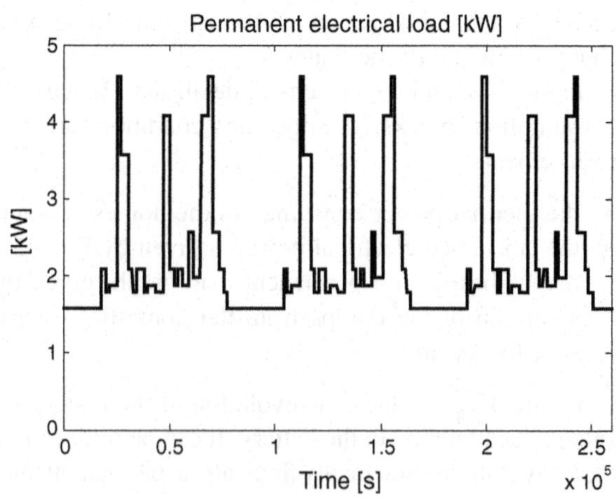

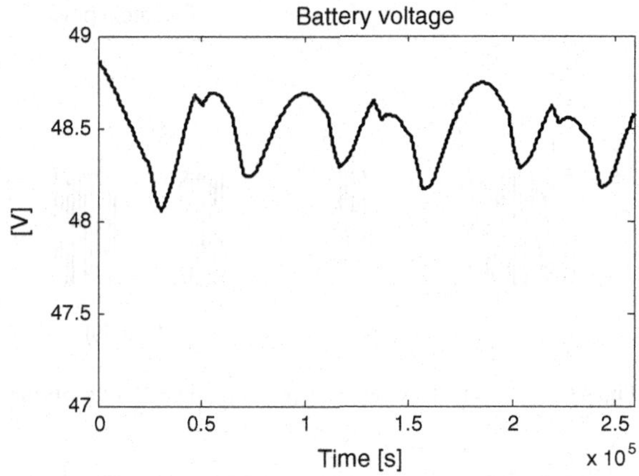

Fig. 47 Battery voltage evolution in winter regime model 1

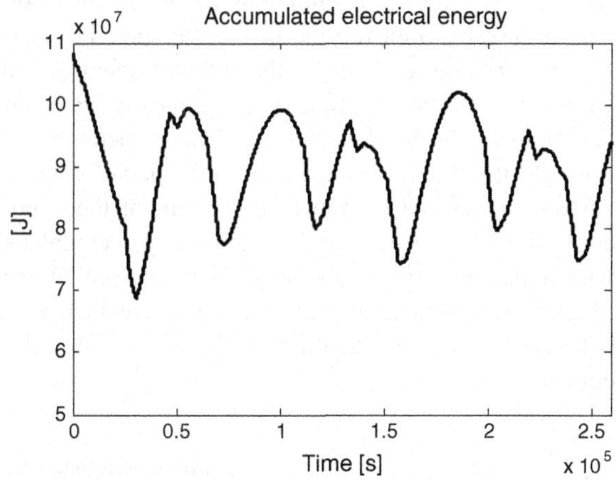

Fig. 48 Accumulated electrical energy evolution in winter regime, model 1

energy sources can compensate the consumption, in the conditions of energy accumulated in the battery. The existence of the subsystem of the electrical energy accumulation (the battery) allows the equality between the average energy production and the average consumption in the system, if the instantaneous values of the consumed power have very important variations (see Fig. 45). The voltage control is done through the control of the Stirling engine power. Figure 49 gives the variation of the electrical power delivered by the Stirling engine. One can notice its discontinuous operating regime when the delivered power values are less than 1.6 kW. In this case, the delivered power varies under the pulse width modulation, with a 30 min period. The thermal power of the Stirling engine is in ratio of 3:1 with respect to the electrical one.

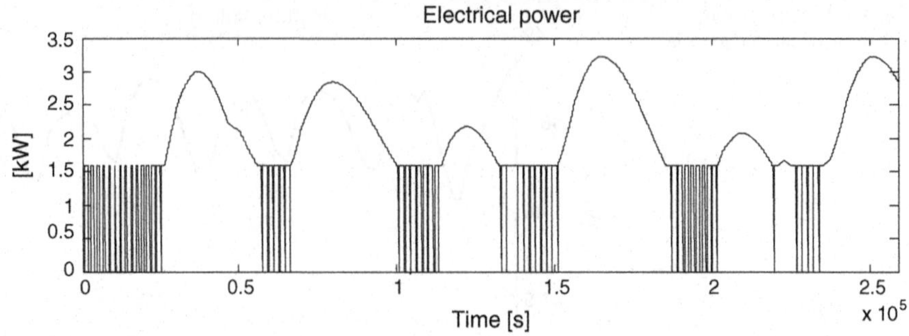

Fig. 49 Evolution of the electrical power produced by the Stirling engine in winter regime, model 1

The performances of the temperature control system of the thermal agent in the accumulating tank are illustrated in Fig. 50 (the set point is equal to 72 °C). The temperature oscillation corresponds to the moments when the Stirling engine operating regime is discontinuous (compare Figs. 49 and 50). When the thermal load necessary to ensure the desired temperature in the residence is constant, the temperature control loop of the thermal agent is influenced by two disturbance values: the power variation in the domestic water circuit and the thermal power variations of the Stirling engines. The temperature control of the thermal agent is achieved through a control given to the pellet boiler. Figure 51 presents the power variation of the pellet boiler, as a result of the control given by the temperature controller. One can notice that the energy accumulated in the tank varies within a wide range, almost up to the nominal power. When the outside temperature is constant, the temperature variations in the residence are very small (Fig. 52), being determined by the dynamic deviations of the thermal agent temperature in convectors.

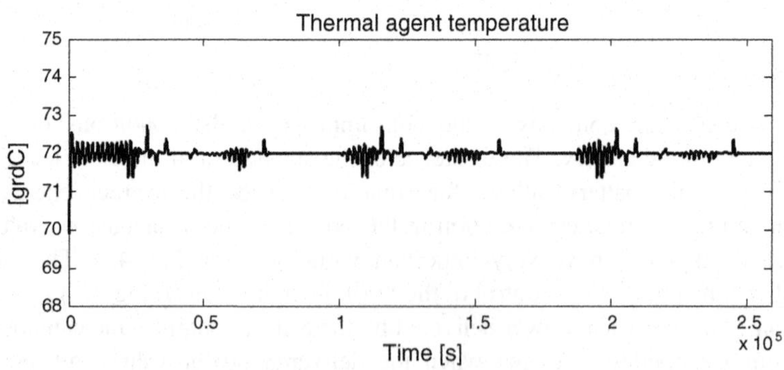

Fig. 50 Thermal agent temperature evolution in the accumulation tank in winter regime, model 1

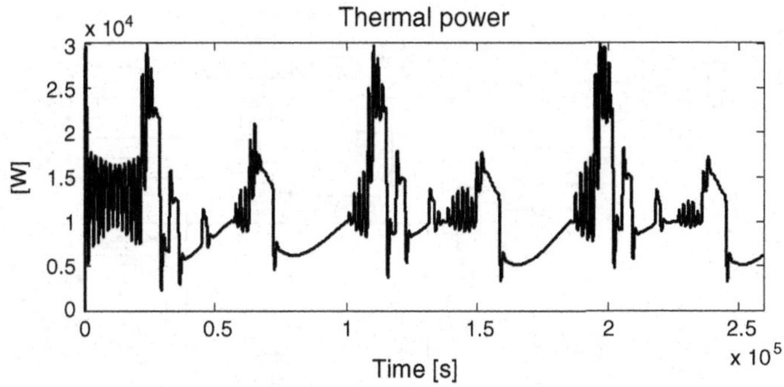

Fig. 51 Pellet boiler power evolution in winter regime, model 1

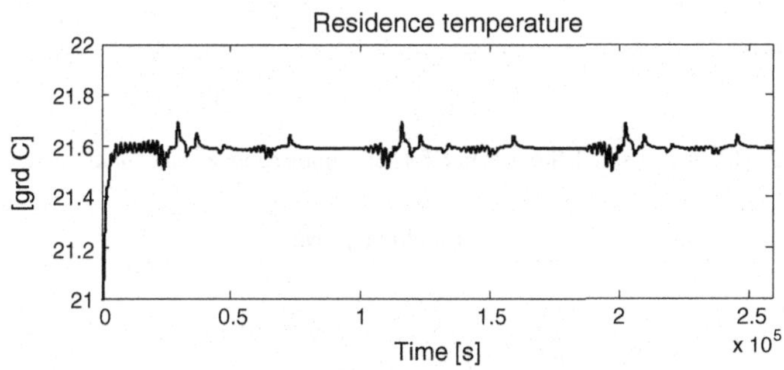

Fig. 52 Temperature evolution inside the residence in winter regime, model 1

4.4.2 Modeling and Numerical Simulation of the System in Summer Regime

The simulation scheme of the system is presented in Fig. 53. Two new blocks are added:

- the block for the calculus of the power necessary to the operation of the air conditioning equipment. Depending on the temperatures imposed on the external environment, this block achieves the energetic balance of the building and provides the power necessary to the air conditioning equipment;
- the block for the calculus of the power provided by the solar collector.

The sources which supply the electrical subsystem are: the Stirling engine, controlled by the voltage controller, and the PV source. The graphs of the power variation of the two sources are given in Figs. 54 and 55, respectively.

The electrical load has the components presented in Sect. 4.3.1.B. The graph of the consumed electrical power took into consideration that some equipments operate according to a daily schedule as follows:

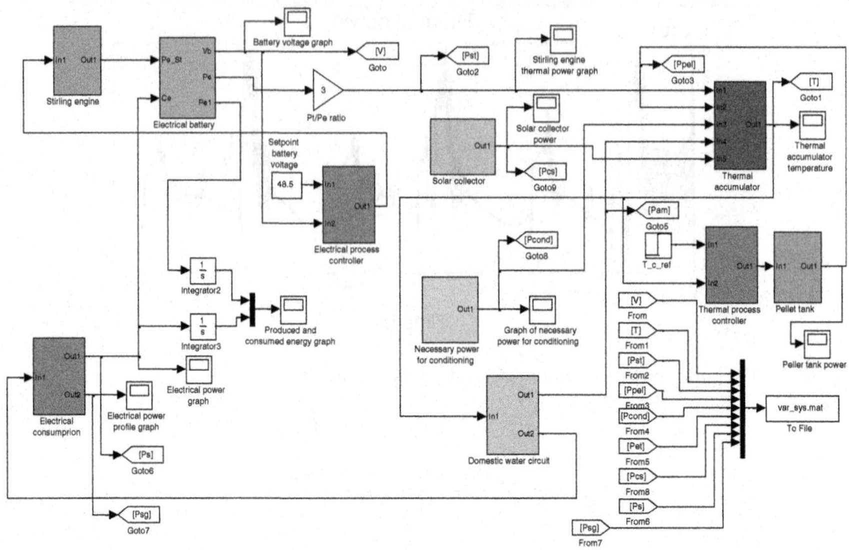

Fig. 53 mCCHP system simulation scheme for the summer mode (*functional model* 1)

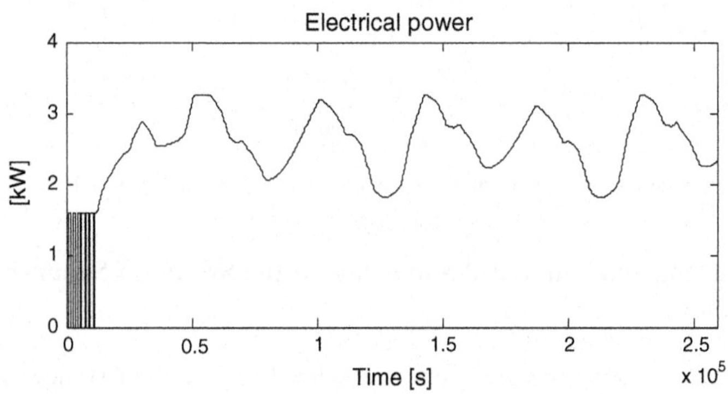

Fig. 54 Evolution of the electrical power produced by the Stirling engine in summer regime, model 1

- the cooling subsystem (adsorption chiller) is scheduled to operate between 10 pm–4 am and 10 am–4 pm.
- the pump for the cold water circulation in the convectors, with a power of 0.6 kW, is scheduled to operate between 4 am and 10 pm.
- the pump for the water circulation in the solar collector, with a power of 0.2 kW, is scheduled to operate between 6 am and 7 pm.

The graph of the consumed electrical power is given by Fig. 56.

In the conditions presented before, regarding the evolutions of the source and load powers, the battery voltage and the energy accumulated in the battery are given

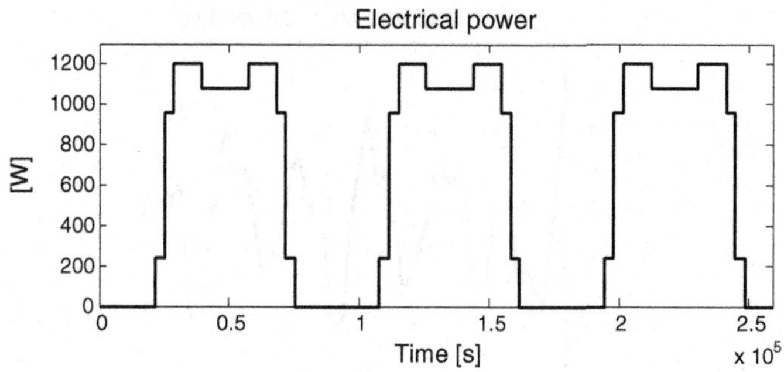

Fig. 55 Profile of the power generated by the PV source in summer regime, model 1

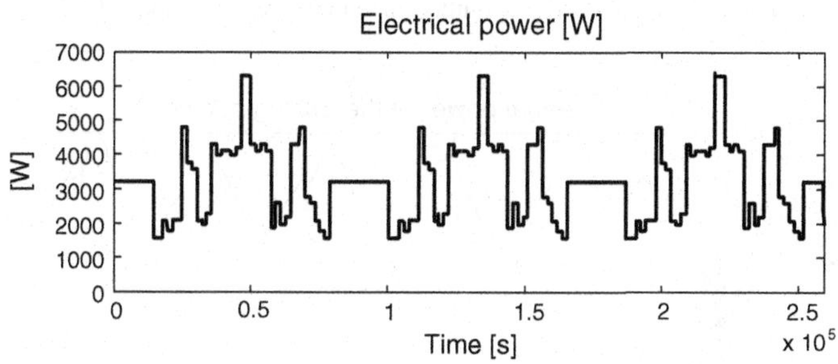

Fig. 56 Total consumed electrical power in summer regime, model 1

Fig. 57 Battery voltage evolution in summer regime model 1

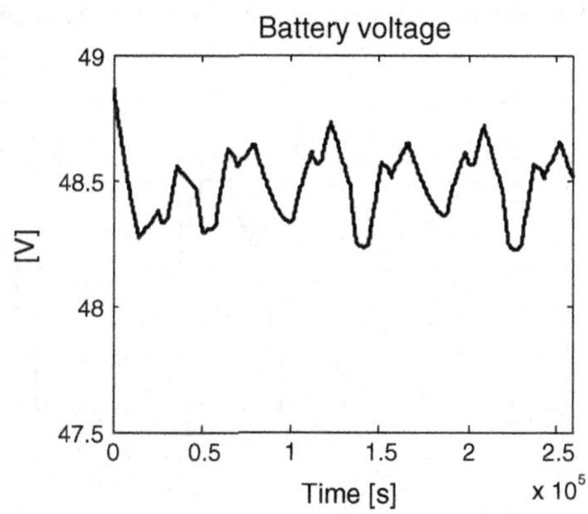

by Figs. 57 and 58, respectively. It can be noticed that a permanent regime is obtained, so the operation in dynamic regime of the electrical subsystem takes place with respect to the requirement of the battery voltage control at the imposed set-point (48.5 V).

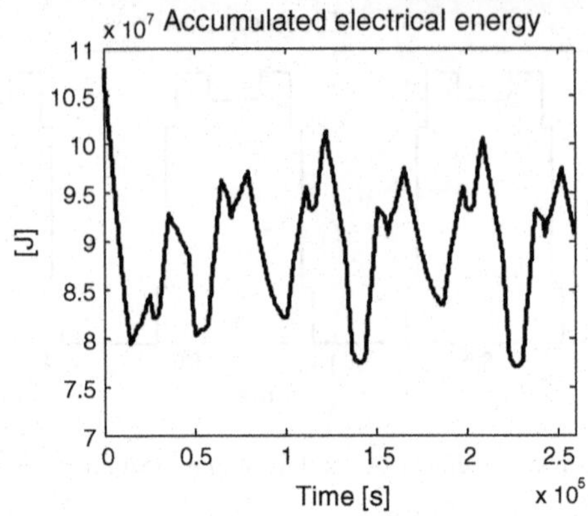

Fig. 58 Accumulated electrical energy evolution in summer regime, model 1

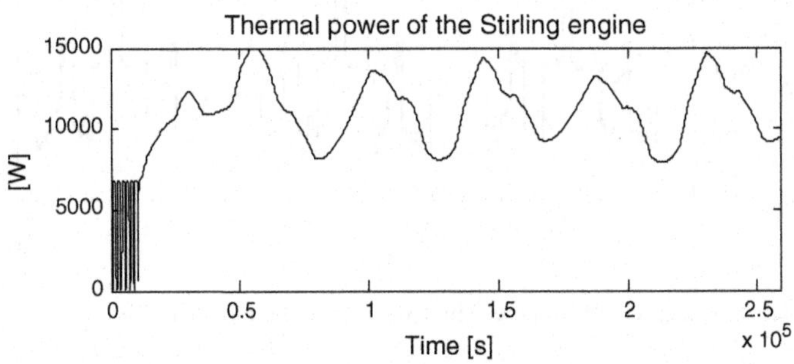

Fig. 59 Thermal power evolution of Stirling engine in summer regime, model 1

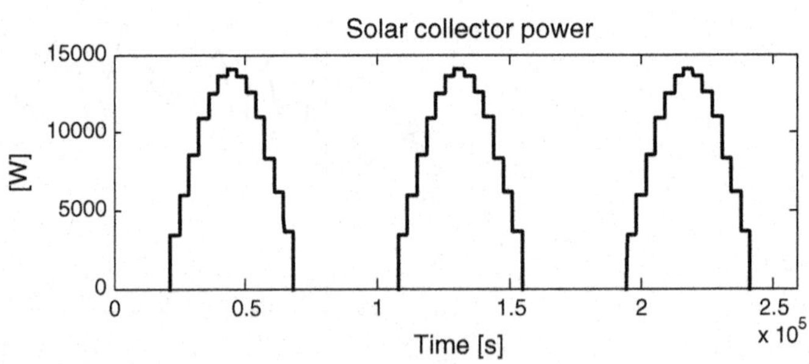

Fig. 60 Profile of the solar collector power in summer regime, model 1

In the thermal subsystem, there are three sources of energy: the Stirling engine, the solar collector, and the pellet boiler which is controlled by the temperature controller of the thermal agent in the accumulation tank. The evolutions of the powers of these sources are presented in Figs. 59, 60 and 61, respectively.

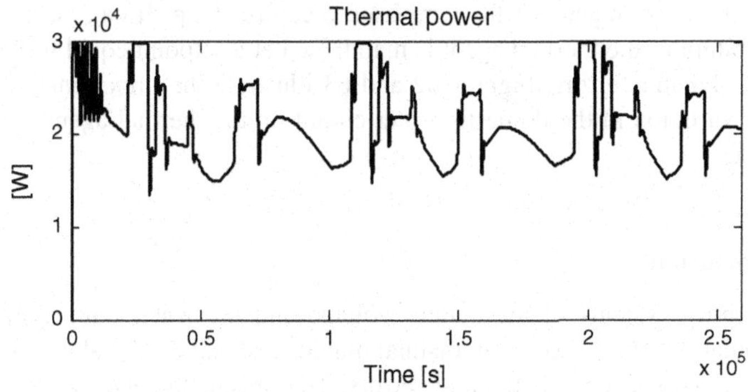

Fig. 61 Pellet boiler power evolution in summer regime, model 1

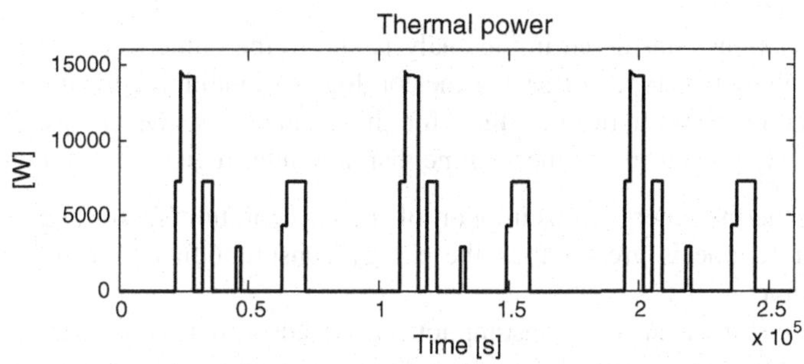

Fig. 62 Thermal power consumed in the domestic water circuit in summer regime, model 1

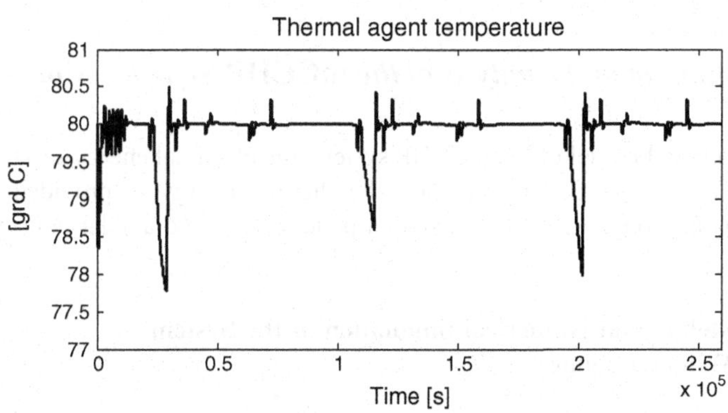

Fig. 63 Thermal agent temperature evolution in the accumulation tank in summer regime, model 1

The thermal load contains two components: the power absorbed by the conditioning equipment, which is considered constant at about 30 kW, and the variable power from the domestic water circuit, which is given by Fig. 62. Figure 63 illustrates the performance of the temperature control loop. It can be noticed that the temperature of the thermal agent is maintained at a setpoint equal to 80 °C with admissible dynamic errors. Figures 62 and 63 illustrate the important influence of the power variation in the domestic water circuit on the thermal agent temperature in the tank.

4.4.3 Conclusions

- The control systems of the battery voltage and the water temperature in the accumulation tank allow the maintaining of the adjusted values around the setpoints with acceptable dynamic errors. That means the energy sources allow the balance in the dynamic regime of the electrical and thermal loads, in the condition of preestablished variations of the electrical and thermal energy consumption.
- The situations considered in the analysis and in the numerical simulation have had in view two types of severe meteorological conditions, for the winter and summer regimes. It resulted that, for these situations, the energy demand is sensitively greater in summer regime than in winter regime, out of two reasons:
  - the thermal energy consumption of the air conditioning equipment in summer regime is greater than the energy consumption for heating in winter regime;
  - the electrical energy consumption of the adsorption equipment for the air conditioning sensitively increases the total electrical energy consumption of the system.

4.5 Simulation and Analysis of the mCCHP *System—Model 2*

In this functional model of the mCCHP system, the electrical energy is produced by a single source: the Stirling engine. The thermal energy is provided by three sources, as in model 1: Stirling engine, a pellet boiler, and a solar collector.

4.5.1 Modeling and Numerical Simulation of the System in Winter Regime

The numerical simulation scheme is given by Fig. 41, in which the elimination of the PV source was done by modifying the "Electrical battery" block. In the detailed scheme of this block, the element "From file," denoted as "PV power source," which provides the power delivered by PV source over the simulation interval, was removed.

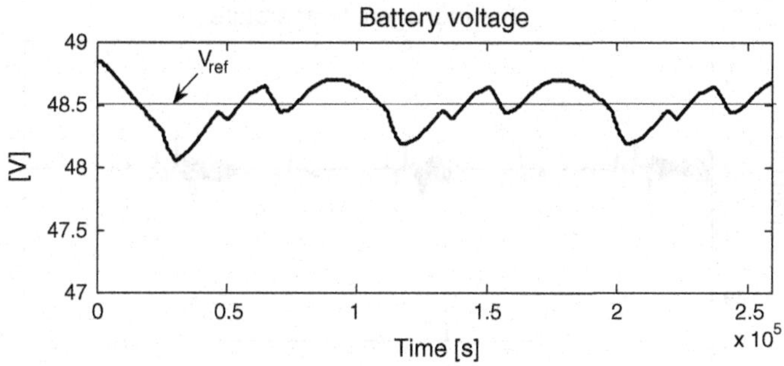

Fig. 64 Battery voltage evolution in winter regime, model 2

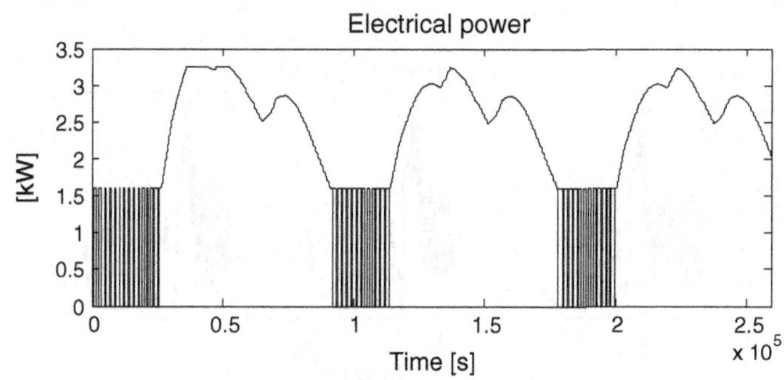

Fig. 65 Evolution of the electrical power produced by Stirling engine in winter regime, model 2

The evolutions of the battery voltage and of the electrical power delivered by the Stirling engine are presented in Figs. 64 and 65, respectively. The control loop within the electrical subsystem manages to maintain the voltage average value at a setpoint equal to 48.5 V. In the same time, the control loop within the thermal subsystem maintains the temperature of the thermal agent in the accumulation tank at a setpoint equal to 72 °C (Fig. 66). The electrical power of the Stirling engine is greater than the one mentioned in the case of model 1 (see Fig. 52), due to the fact that the PV source is missing. Consequently, the average value of the pellet boiler power (Fig. 67) is smaller than in the case of model 1 (see Fig. 51).

4.5.2 Modeling and Numerical Simulation of the System in Summer Regime

The numerical simulation scheme is given by Fig. 53, in which the PV source was removed from the "Electrical battery" block. The evolutions of the controlled values and of the powers controlled by the two controllers are presented in Figs. 68, 69, 70, and 71. One can notice that a permanent regime is obtained in the battery

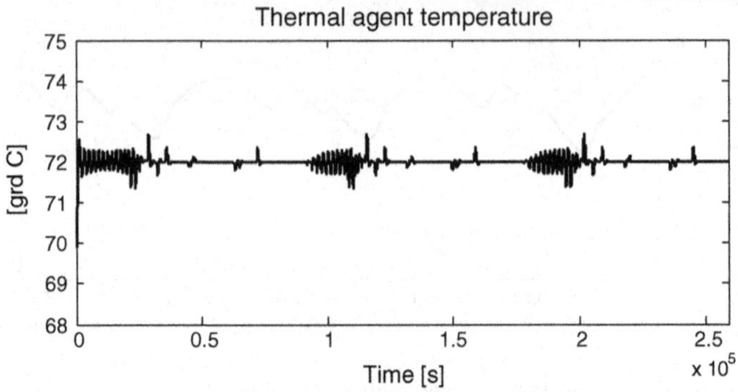

Fig. 66 Thermal agent temperature evolution in the accumulation tank in winter regime, model 2

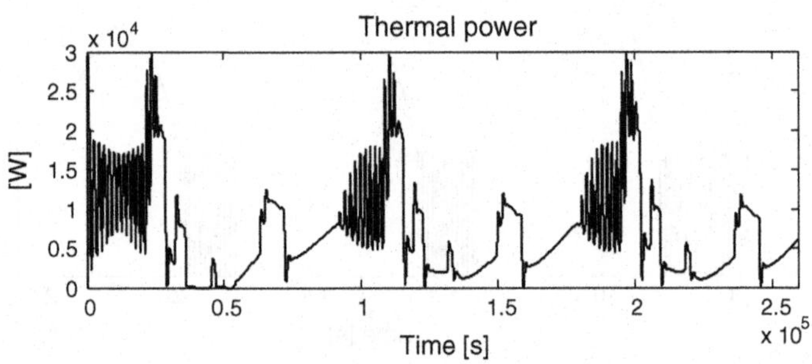

Fig. 67 Pellet boiler power evolution in winter regime, model 2

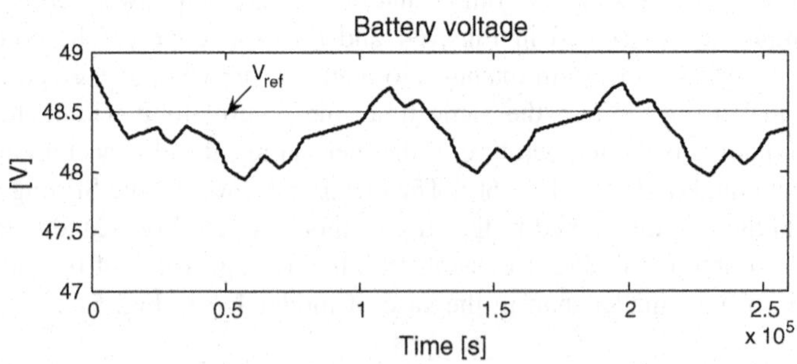

Fig. 68 Evolution of the battery voltage in summer regime, model 2

circuit, but the controller from the electrical subsystem does not manage to maintain the average value of the voltage at the setpoint value (48.5 V). In the greatest part of the time, in permanent regime, the Stirling engine delivers a maximum power. In these conditions, a very small increase of the useful electrical load may lead to the impossibility of achieving the equality between the produced electrical energy and

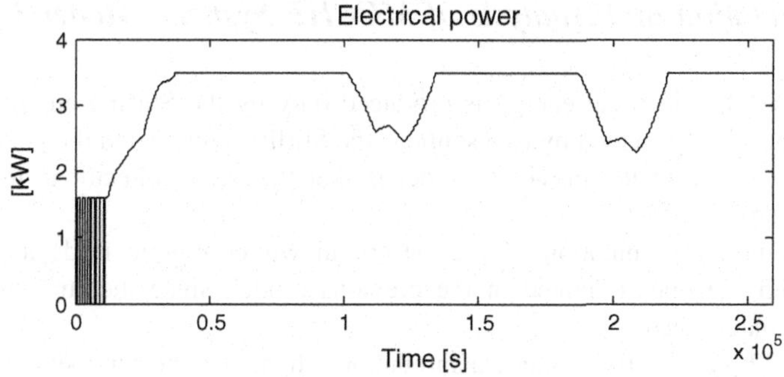

Fig. 69 Evolution of the electrical power produced by Stirling engine in summer regime, model 2

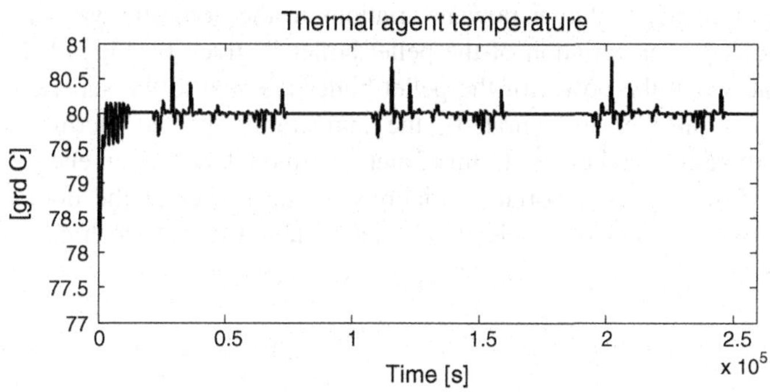

Fig. 70 Thermal agent temperature evolution in the accumulation tank in summer regime, model 2

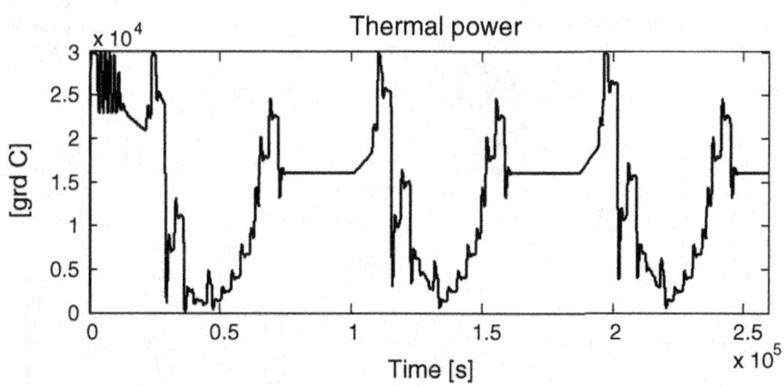

Fig. 71 Pellet boiler power evolution in summer regime, model 2

the consumed one. It is obviously that the removing of the PV source is not an appropriate solution even if the analyzed structure of mCCHP system may barely be accepted. The main drawback of this model consists in achieving in a small measure of the electrical energy demand during summer regime.

4.6 Simulation and Analysis of MCCHP System—Model 3

In this case, the electrical energy is produced only by the Stirling engine and the thermal energy is provided by two sources: the Stirling engine and the pellet boiler. The difference from the previous model is that the solar collector is missing in summer regime.

The numerical simulation of the system in winter regime leads to identical results with the ones obtained in the previous model, since the structure of the system is maintained.

Figure 72 presents the numerical simulation scheme for the summer regime. The evolution of the thermal agent temperature in the accumulation tank is given by Fig. 73. It can be noticed the existence of a number of important dynamic deviations, which are still acceptable, of the temperature comparing with the setpoint (80 °C), caused by the larger power variations in the domestic water circuit. The instantaneous power variation of the pellet boiler is given by Fig. 74. In the time intervals in which the power of the pellet boiler is equal to the superior limit (the temperature controller is saturated), the temperature of the thermal agent has important dynamic deviations. In this functional model, a thermal energy deficit in summer regime, when important variations of the power in the domestic water circuit occur, is added to the electrical energy deficit in the electrical subsystem, noticed also in model 2.

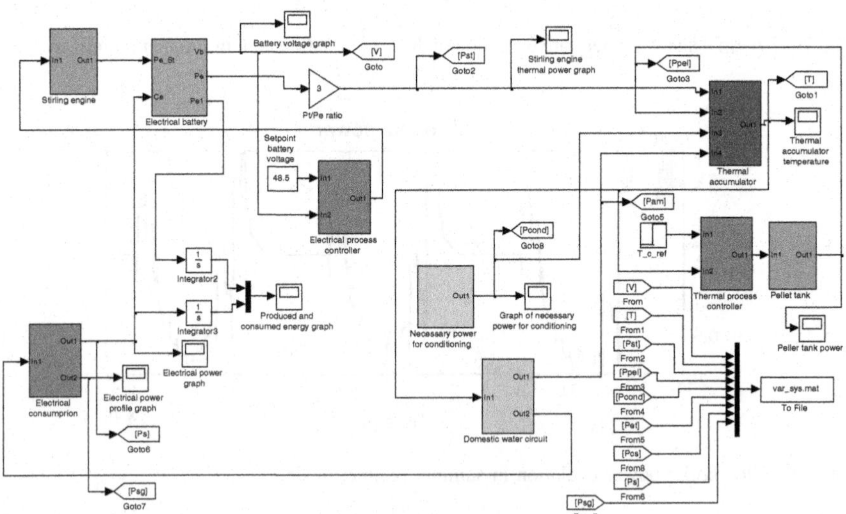

Fig. 72 mCCHP system simulation scheme for the summer regime (functional model 3)

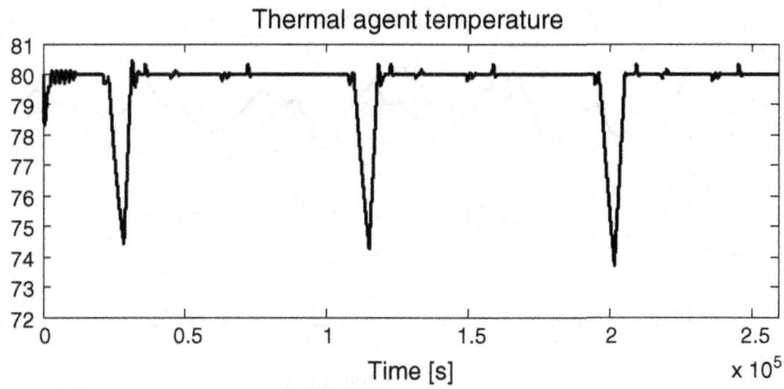

Fig. 73 Thermal agent temperature evolution in the accumulation tank in summer regime, model 3

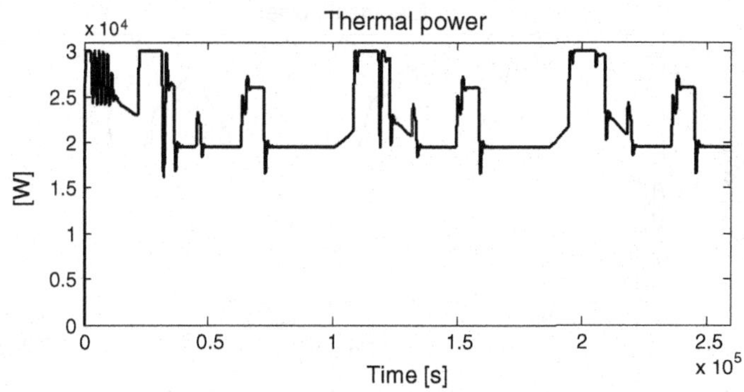

Fig. 74 Pellet boiler power evolution in summer regime, model 3

4.7 Simulation and Analysis of MCCHP System—Model 4

In this functional model, the electrical energy is produced by the Stirling engine and a PV source. The thermal energy is provided only by two sources: Stirling engine and the pellet boiler.

The numerical simulation of the system in winter regime leads to identical results with the ones obtained in model 1, since the structure of the system is the same.

Figure 72 presents the numerical simulation scheme of the system in summer regime, in which the "Electrical battery" block has the form given by Fig. 42. The operating regime of the electrical subsystem is illustrated by Figs. 75 and 76, where the evolutions of the battery voltage and the electrical power delivered by the Stirling engine are shown. These variations are similar to the ones obtained in functional model 2 (see Figs. 57 and 54, respectively), since there is only a small difference between the two models: the consumption of the pump in the thermal agent circuit of the solar thermal collector.

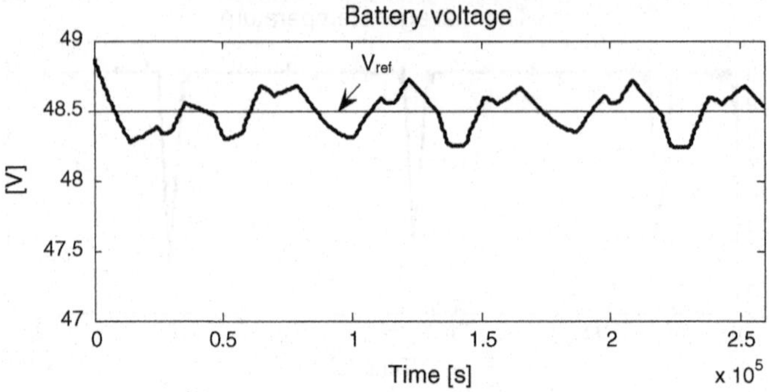

Fig. 75 Battery voltage evolution in summer regime, model 4

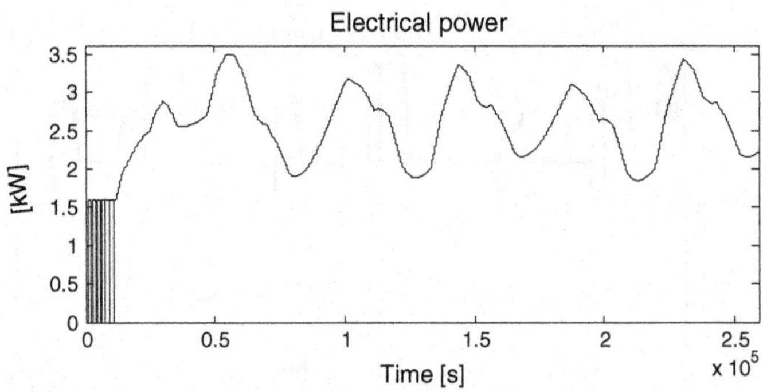

Fig. 76 Evolution of the electrical power produced by Stirling engine in summer regime, model 4

The thermal subsystem is affected by the lack of the thermal collector. The operating regime of the temperature control loop is illustrated in Fig. 77. Figure 78 gives the power variation of the pellet boiler, controlled by the temperature controller. It is noticed that in the time intervals in which the power consumed in the domestic water circuit is high, the pellet boiler is controlled at a maximum power (the control is limited) and there are important dynamic deviations of the temperature from the setpoint (80 °C).

4.8 Simulation and Analysis of MCCHP System—Model 5

A structure with a major difference compared to the ones previously analyzed is considered: air conditioning is done using electrical equipments (mechanical compression chiller). In this functional model, the electrical energy is produced by the Stirling engine and a PV source and the thermal energy is provided by two sources: the Stirling engine and a pellet boiler.

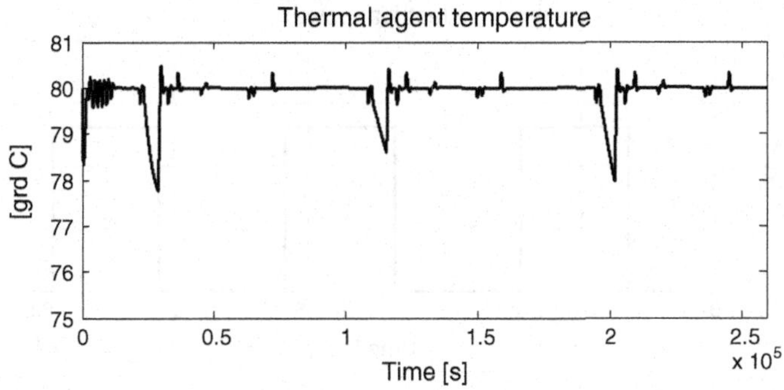

Fig. 77 Thermal agent temperature evolution in the accumulation tank in summer regime, model 4

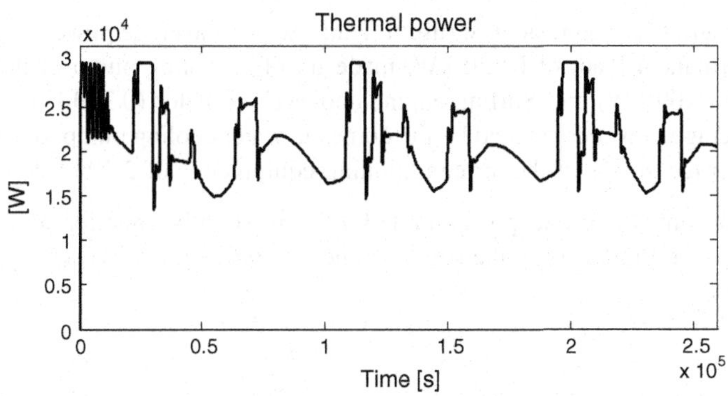

Fig. 78 Pellet boiler power evolution in summer regime, model 4

In winter regime, the numerical simulation scheme and the results are identical to the ones from obtained in model 1.

In summer regime, the electrical and thermal loads are radically modified. For the air conditioning, six equipments were adopted, as follows: one of 12,000 BTU, with a consumed electrical power of 1.12 kW, and five of 9,000 BTU, each one having a consumed electrical power of 0.82 kW. In summer regime, an extreme situation regarding the outside temperature was considered. It is admitted that the air conditioning equipments operate 11 h out of 24, according to the graph presented in Fig. 79.

The analysis of the energetic balance in the electrical subsystem, carried out in relation to the average powers, highlights the following:

A–the electrical energy sources: the Stirling engine, with a maximum useful average power of $P_{Stu} = 3.5$ kW, and the PV source, with a useful average power of 0.623 kW. It results a useful average power of 4.123 kW;

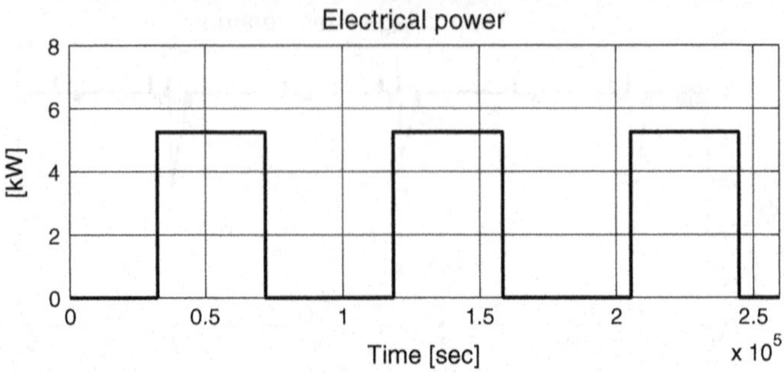

Fig. 79 Profile of the consumed power by the air conditioning system

B–the electrical load that is composed of:

- consumers from the residence (useful load), with an average power of 1.58 kW,
- the permanent load, of 0.430 kW, made up of a cooling pump of the Stirling engine (0.09 kW) and Stirling engine cooler's ventilator (0.340 kW);
- the average power consumed by the domestic water cooling pump, of 0.051 kW;
- the average power of the air conditioning equipments, of 2.393 kW.

The consumed average power (4.433 kW) is slightly superior to the useful average power, produced by the sources. The system's simulation scheme is given by Fig. 80.

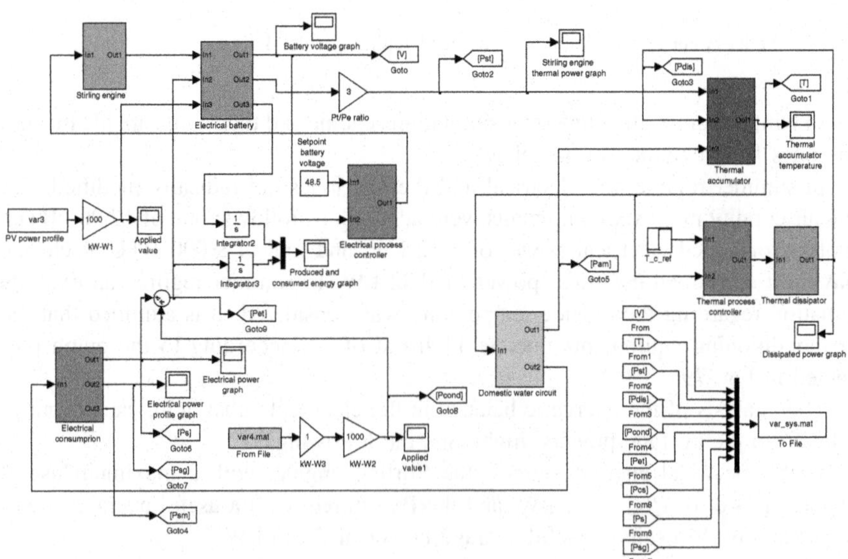

Fig. 80 mCCHP system simulation scheme for the summer regime (*model* 5)

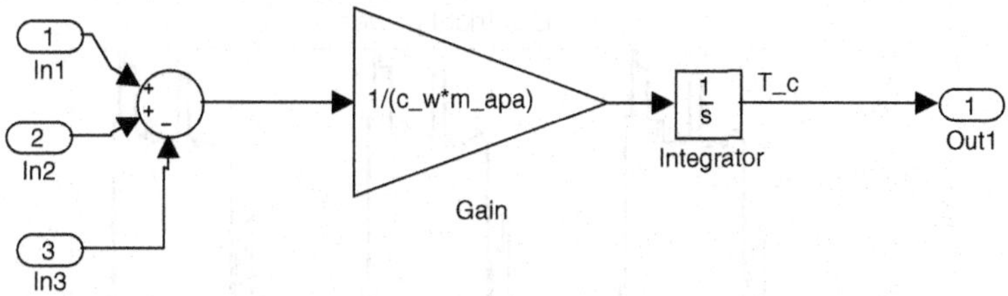

Fig. 81 Scheme of the "Thermal accumulator" block in functional model 5

Most of time, the used blocks are identical to the ones presented in the previous functional models. The new elements of the scheme are:

- the block From file "var 3," which provides the power generated by the PV source, having the profile presented in Fig. 55;
- the block From file "var 4," which gives the time evolution of the power consumed by the air conditioning equipments, according to Fig. 79.

The "thermal accumulator" block was modified, compared to the previous functional models; Fig. 81 presents the new scheme.

The following results in the electrical subsystem were obtained: Fig. 82 shows the total power provided, Fig. 83 shows the total consumed power, and Fig. 84 shows the battery voltage evolution. It is noticed that, at the beginning of the dynamic regime, the battery voltage is greater than the setpoint and the voltage controller brings the power of the Stirling engine to 0. The PV source begins to deliver energy after 6 h from the beginning of the dynamic regime, so that, in the initial period of this regime, the total electrical power provided by the sources is null and the battery discharges. In what follows, the voltage controller controls the Stirling engine to deliver its maximum power, but the average of the total load is superior to the average of the produced total power and the average power of the battery voltage has a decreasing evolution. The main conclusion consists in the fact that the system cannot operate in a permanent regime.

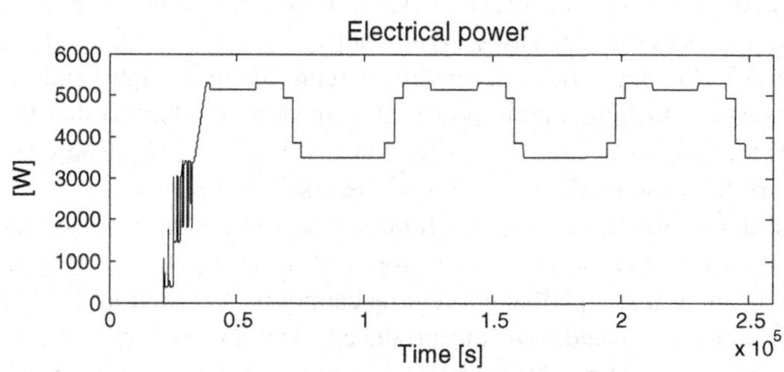

Fig. 82 Total produced electrical power in summer regime, model 5

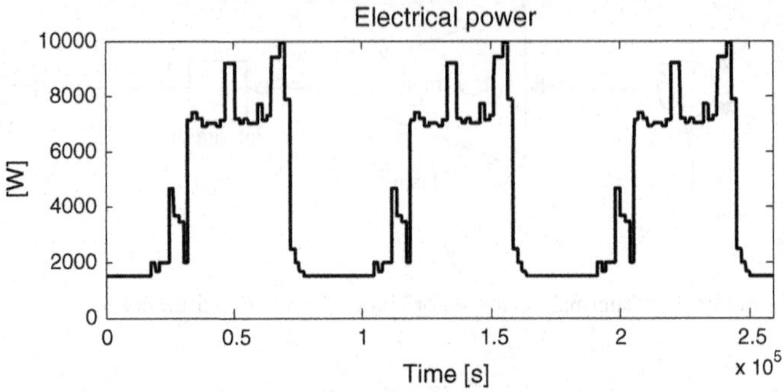

Fig. 83 Total consumed electrical power in summer regime, model 5

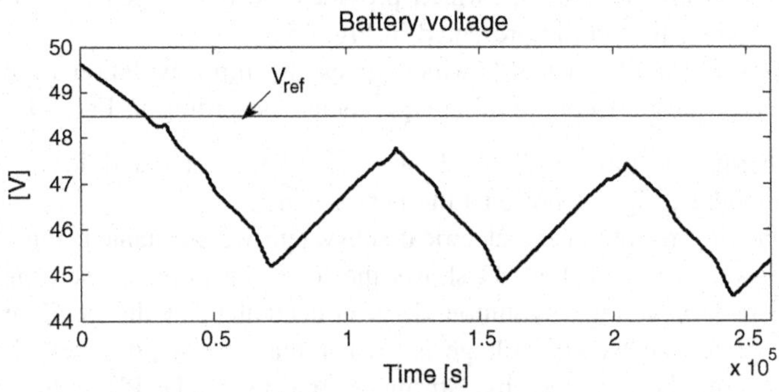

Fig. 84 Battery voltage evolution in summer regime, model 5

Analyzing the operating regime of the thermal subsystem, it can be noticed that–in summer regime–there is only one source (the Stirling engine) and a single load (the domestic water circuit). The temperature evolution of the thermal agent inside the accumulation tank is presented in Fig. 85. In the initial period of the dynamic regime, the thermal and electrical powers of the Stirling engine are null and the thermal energy consumption in the domestic water circuit is achieved from the energy saved in the accumulation tank. Figure 85 shows that in the initial period, the temperature decreases significantly. In the following, the energetic consumption in the domestic water circuit is reduced significantly and the voltage controller sets the Stirling engine power at a maximum value, so that the thermal energy produced is in excess. This situation involves the use of an energy dissipater. Figure 86 presents the evolution of the dissipated power.

Overall, the results obtained through numerical simulation does not validate the proposed solution, because the power produced in the electrical subsystem is not sufficient, meaning that a permanent regime cannot be achieved in the evolution of the battery voltage. In addition, the produced electrical energy is insufficient in summer regime, but the produced thermal energy is in excess, which imposes the use of a thermal energy dissipater.

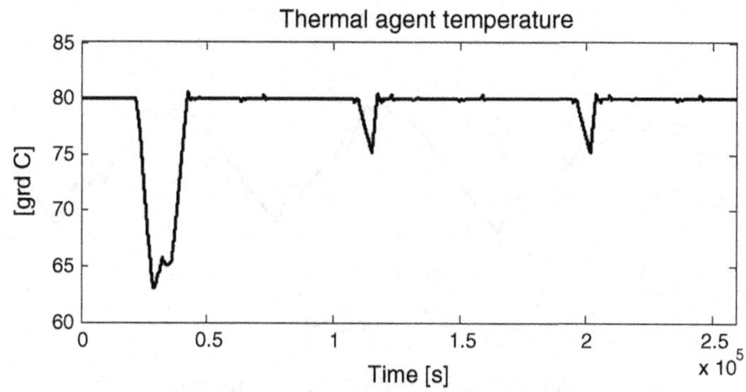

Fig. 85 Thermal agent temperature evolution in the accumulation tank in summer regime, model 5

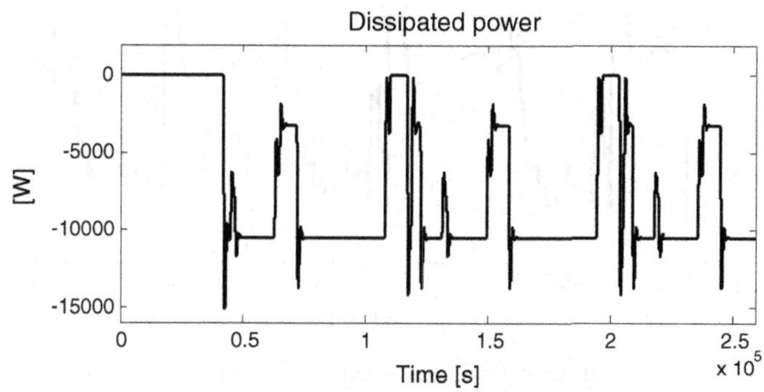

Fig. 86 Dissipated power evolution in summer regime, model 5

The analyzed functional solution can be considered only if the energy consumption parameters are modified in summer regime. Therefore, if it is considered that all the air conditioning equipments are identical, with an electrical energy consumption of 0.82 kW, and they are scheduled to operate daily, between the hours: 10 pm–1 am; 12.30 pm–2.30 pm; 3 pm–5 pm; and 5.30 pm–7.30 pm, then the battery voltage, given by Fig. 87, represents the existence of a permanent regime, acceptable in the electrical subsystem.

4.9 Conclusions

1. A first conclusion is referring to the choice of the type of the air conditioning equipment. When using the adsorption device, the total thermal power consumed in summer regime has the evolution presented in Fig. 88. In the hypothesis that the air conditioning equipments with electrical supply are used, a correct operating regime of the electrical subsystem requires in summer regime a

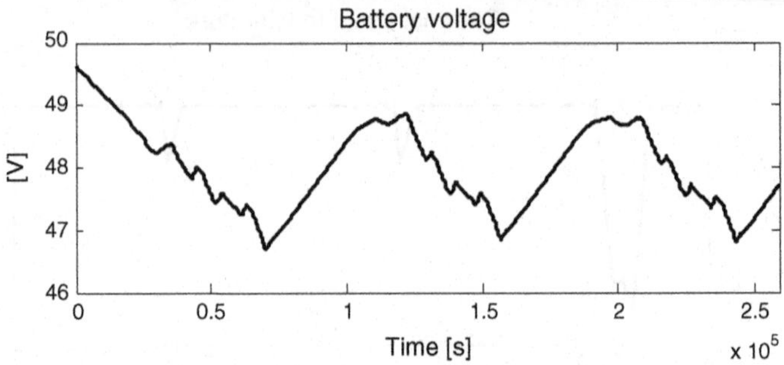

Fig. 87 Battery voltage evolution in a modified refrigeration regime

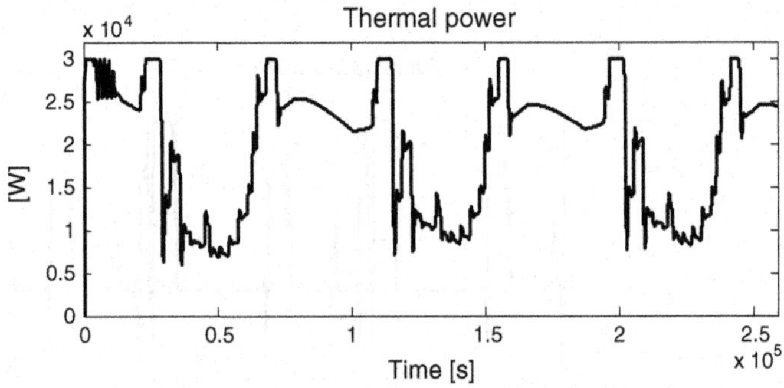

Fig. 88 Graph of the total thermal power consumed in the summer regime

power given by the Stirling engine greater than 5 kW (for 5 kW, the graph representing the battery voltage is the one shown in Fig. 84). This will lead to an average value of the thermal power in summer regime of about 18–20 kW. It is obvious that the thermal energy demand in summer mode is much lower in Model 5, but using the Stirling engine to produce electrical energy for air conditioning leads to dissipation of large quantities of thermal energy. Although air conditioning installations using electrical equipment are more advantageous in terms of energy consumption, it is important to consider the qualitative aspects of the air conditioning process. The achievement of the adsorption air conditioning equipment has obviously qualitative advantages, such as: the thermal transfer is done through convectors, which have the advantage of a thermal comfort as compared to the classic conditioning equipments; the cold accumulating tank included in the system allows the balancing of the temperature level inside the residence etc.

2. The first functional model satisfies the electrical and thermal power requirements in winter and summer regimes and therefore it is recommended, compared to the other models, due to the following reasons:

- the lack of a free renewable energy source, as the solar collector or the PV source, will determine the decrease of the safety margin in compensating the electrical and thermal loads;
- the renewable energy sources mentioned before diminishes the fuel consumption of the system (biogas and pellets);
- the potential of the energy deficit is more important in the electrical subsystem than in the thermal subsystem. Therefore, the use of a PV source is necessary. On the other hand, the solar collectors are very convenient with respect to the technical and economical aspects, so that the conclusion reached in this chapter is that the functional model 1 is the most rational.

5 Design of the Control Sub-System

Emil Ceanga, Marian Barbu and Sergiu Caraman

Designing the control subsystem involves the achievement of the following steps:

- designing the monitoring and control system includes choosing the field equipments (transducers, actuators etc.), the designing of the connection device of these equipments to the data acquisition system, and the design of the monitoring software of the whole system;
- Designing the numerical controllers in the thermal and electric subsystems. These controllers are implemented in the data acquisition numerical system and the process monitoring, either by using the interface equipments of a PC, or by using a programmable logic controller.

5.1 Design of the Numerical Controllers

The control system of a building can be divided into at least two hierarchical levels, as follows:

- lower level that ensures the balance between the produced thermal/electrical power in respect with the consumed power;
- higher level, which should provide the comfort parameters inside the building, especially the reference values for temperature in winter and summer regimes.

5.1.1 The Lower Level of the Hierarchical Control System

The objective followed in this section is to design numerical controllers which intervene in the control loops aiming to perform the balance between the produced power and the consumed one in the thermal and electrical subsystems.

In the electric subsystem, the balance between the produced power and the consumed one is achieved by the control loop of battery voltage (Fig. 19). The control u_{St} given by the controller is applied to the Stirling engine. As a rule, it can work in two control modes:

- on/off control;
- heterogeneous control: on/off type, in 0,..., P_1 operating domain, where P_1 is a fraction of nominal power, and continuous control in $P_1,\ldots,P_{max}$ operating domain.

The on/off control is the simplest solution, its software implementation being an elementary issue. For example, if the system operates with a 48 V battery and the imposed setpoint is $V^{ref} = 48.4$ V, occurs the problem of determining the control switching thresholds: $V^{ref} + \varepsilon_1$, at switching $1 \to 0$, and $V^{ref} - \varepsilon_2$ at switching $0 \to 1$. These thresholds could be chosen $\varepsilon_1 = 0.2$ and $\varepsilon_2 = 0.4$. The parameters ε_1 and ε_2 determine the amplitude and the period of the temperature oscillation in on/off control regime. It is recommended that the amplitude of the temperature oscillation should be reduced, if the oscillation period is less than a certain limit, imposed by the characteristics of the Stirling engine (e.g. more than 20 min). The width of the hysteresis, $\varepsilon_1 + \varepsilon_2$ of the on/off controller is experimentally established, in the system's commissioning.

The continuous control is supplied by a *PI* controller equipped with an antiwind-up mechanism. For illustrating the controller operating regime, it is considered the situation when the nominal electric power of the Stirling engine is 3 kW. The Simulink scheme from Fig. 89 presents the algorithm for the numerical implementation of the *PI* controller. The error signal, *err*, must control the power of the Stirling engine, which can vary in a certain domain. Let us consider this domain: 0, $\ldots,P_{max}$, where $P_{max} = 3$ kW. Parameters T_i and K_p of the controller are determined at the tuning of the voltage control loop.

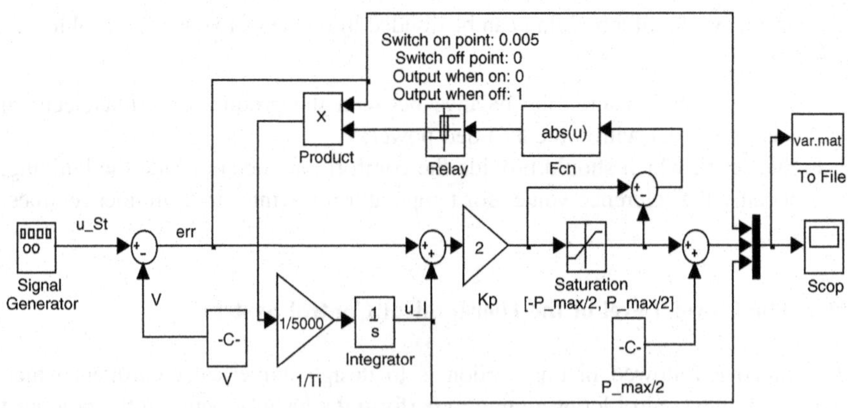

Fig. 89 PI controller structure for the voltage

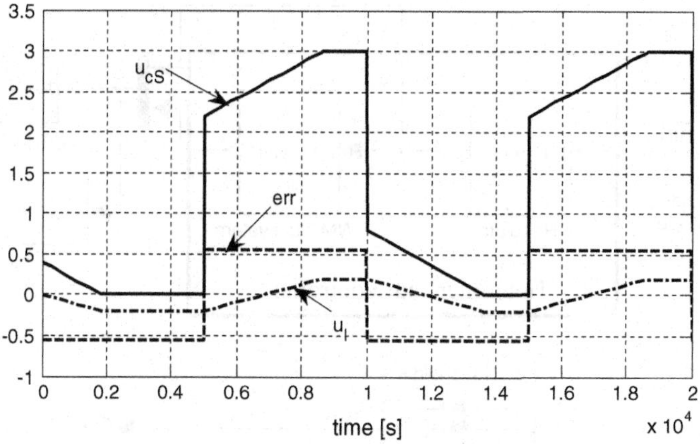

Fig. 90 Evolution of: error, *err*, control, u_{cS} and integral component, u_I

Figure 90 presents the controller operation, when the input signal (the error from the control loop of the battery voltage, *err*) is a rectangular wave. The control given by the controller, u_{cS}, varying between 0 and 3, is equal to the electrical power (kW) imposed to the Stirling engine. To illustrate the anti wind-up mechanism, Fig. 90 also presents the evolution of the integral component u_I. It can be noticed that when u_{cS} control reaches the saturation value, the component u_{cI} remains constant.

Most Stirling engines are endowed with heterogeneous control. They may be controlled by 0/1 controls when the electric power discharged in the system is in the range of $[0, P_1]$ and by continuous controls, when the discharged electric power is in the range of $[P_1, P_{max}]$. To achieve the control of the Stirling engine within the battery voltage control loop, two variants are possible:

- using an on/off controller when the load is under the threshold P_1 and continuous control for the rest;
- using the PWM (Pulse Width Modulation) technique, when the load is in the domain $P_1,\ldots,P_{max}$.

The second solution is more efficient, even if it involves a more complex algorithm. In what follows the second solution will be presented in detail.

In accordance with this solution, the PI control is applied to the input of a PWM subsystem which offers to the output the same value with the one of the input signal, if the level of the control exceeds the threshold P_1, and a PWM signal, with an average value equal to the input signal, if the signal is under this threshold. Figure 91 presents the controller structure. The numerical values illustrated in this scheme correspond to the situation in which $P_1 = 1.6$ kW and $P_{max} = 3$ kW. In Fig. 91.b it is considered a sawtooth wave signal given by the signal generator with an amplitude of $P_1[\text{kW}]/2 = 0.8$ and a period equal to the minimum duration imposed for two successive 0/1 controls. This was supposed to be equal to 30 min (1,800 s). The operation of the heterogeneous controller is illustrated in Fig. 92, when the input signal (the error *err*) is a rectangular wave.

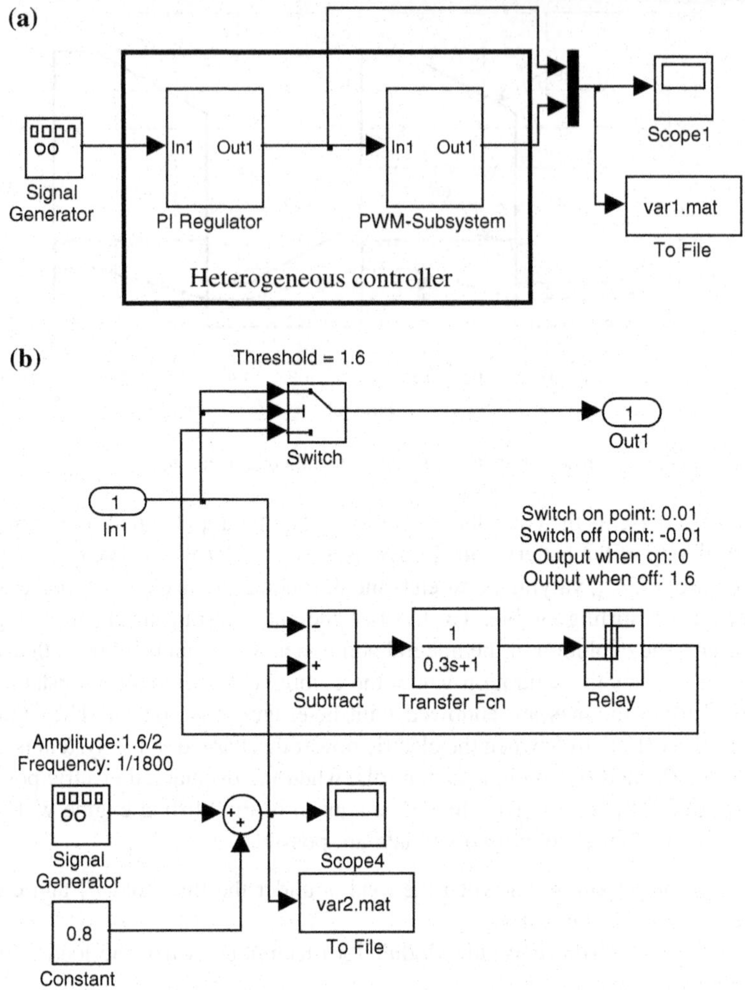

Fig. 91 Structure of a PI heterogeneous controller: **a** block scheme of PI controller; **b** PWM subsystem

A potential solution for improving the performance of the battery voltage control loop is the use of an additional anticipatory control, depending on the power consumed in the system, according to the principle scheme from Fig. 93. The controlled process is made up of Stirling engine and the electric subsystem of the battery (*SM*), plus the proper battery (*B*). The disturbance variable which affects the battery is the power *P*, consumed in the whole electric subsystem. If the consumed power is measured at the level of the monitoring system, then a feed-forward control may easily be achieved, through the anticipation controller, *AC*. In this case, the control received by the Stirling engine, u_{cS}, has two components: the control of the voltage controller—*VC*, u_c^{fb}, and the anticipative control, u_c^{ff} . Regarding the balance of the

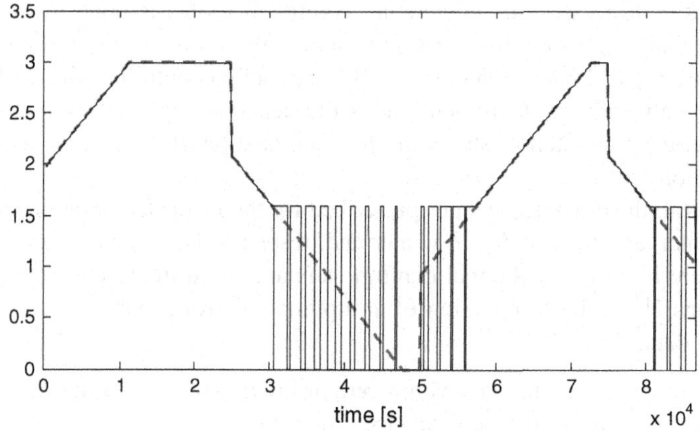

Fig. 92 Evolution of PI control (*dot*) and of heterogeneous control (*solid line*)

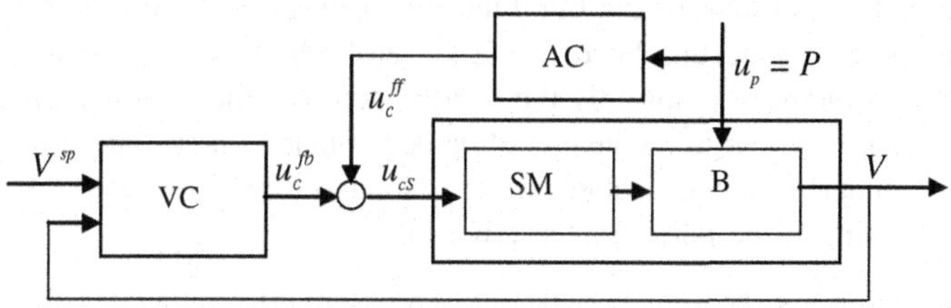

Fig. 93 Feed-forward control of the battery voltage

produced electric power with the consumed one, the feed-forward control has a reduced efficiency, so that the use of this solution is not an important problem.

In the thermal subsystem, the balance between the produced power and the consumed one is achieved through the temperature control loop for the thermal tank (see Fig. 20). The u_{ph} control given by the controller from this loop modifies the power of the pellet tank. The design of the control law in the loop mentioned above is done depending on the control possibilities offered by the pellet tank: discrete 0/1 control, by imposing a minimum period between two successive starts of the tank; continuous control using a *PI* controller and, eventually, the heterogeneous control. The last control is necessary when the loads are under a specified threshold and the control should be of 0/1 type. All these solutions could be similarly applied in the cases presented for the control of the Stirling engine.

5.1.2 The Higher Level of the Hierarchical Control System

The fundamental objective of this level is to control the temperature inside the building. In modern control systems of buildings an optimal control structure is implemented, in relation to a criterion that provides a trade off between quality requirement regarding the thermal comfort and energy saving requirement.

In the simplest case, the temperature control in rooms is done using a on/off controller that commands the electromagnetic valves which feed the convectors. Less commonly is the solution to use continuous action controllers that controls the three-way valve. These controllers varies the heat/cooling agent flow throughout the convertors, simultaneously with the recirculation flow to the heat/cooling accumulation tank.

In general, the references values prescribed for the inside temperatures may vary in some intervals, D_w and D_s, in winter and, respectively, summer regimes. These intervals are imposed by thermal comfort standards used in current design HVAC applications [13]. The formulation of an optimal control problem is based on the following data:

- inside temperature can vary within certain limits within an admissible domain, in order to optimize the performance criterion;
- there are many variables in the system that can predicted. These relate to powers of the renewable energy sources or the consumed powers (especially at the change of thermal load of the building, due to weather conditions);
- there is an energy accumulated in the system, which can be used in the optimization problem. It is not only the energy stored in the building structure, but also the energy stored in the heat/cooling accumulation tank. This energy can be operated under a strategy called in [15] "load shifting" or "active storage" in order to optimize the performance criterion.

In these conditions, the higher level control problem is formulated in accordance to the methodology "Model Based Predictive Control" (MBPC), which is widely used today [13–17]. A schematic diagram illustrating the MBPC is given in Fig. 94 [17]. MBPC provides the control signal that is transmitted to the control system of the inside temperature. Sometimes this signal is the reference of the temperature control loop, which can be adjusted in the intervals D_w and D_s, depending on the operating regimes.

The process modeling is based on the analogy with the electrical circuit resistance-capacitance (RC). Regardless of the operating regime (summer/winter), the process dynamics is described by the differential equation [15]:

$$C\frac{\mathrm{d}\theta_{\mathrm{a}}}{\mathrm{d}t} = u + P_d + (\theta_{oa} - \theta_a)/R \tag{55}$$

where θ_a and θ_{oa} are the air temperature inside and, respectively, outside the building, R and C are resistance and, respectively, heat capacity, P_d is the disturbance caused by building occupants (including the power variation in the domestic water circuit), the direct sunlight and electrical appliances in the house, and u is the command sent by the MBPC. In summer regime, $u \leq 0$, and in winter regime $u \geq 0$. By discretizing the Eq. (55) it results:

$$x(k+1) = Ax(k) + Bu(k) + d(k) \tag{56}$$

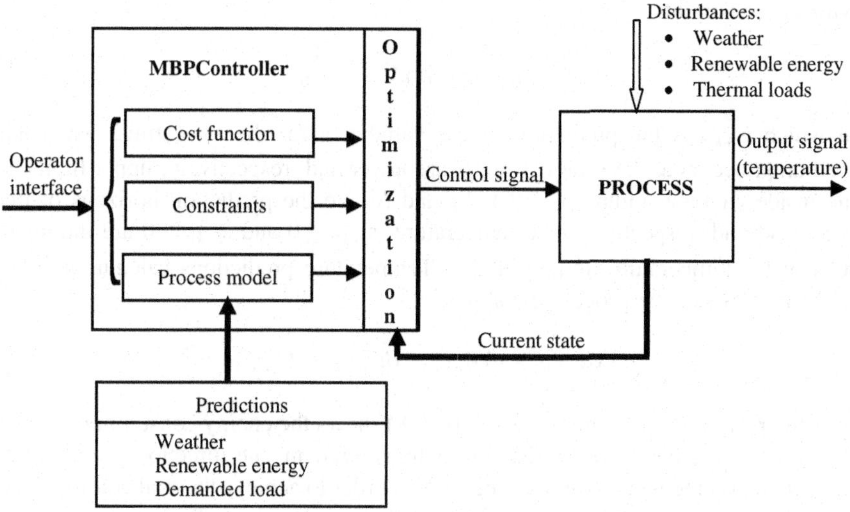

Fig. 94 MBPC structure for the higher level of the hierarchical control system [17]

in which $x \equiv \theta_a$ is the process state variable,

$$A = 1 - \frac{T_s}{RC}, \quad B = \frac{T_s}{C}, \quad d(k) = \frac{P_d(k)T_s}{C} + \frac{\theta_{oa}(k)T_s}{RC} \tag{57}$$

and T_s is the sampling time.

The process model, of the form (55), is the simplest form possible, in which it is considered only one heat accumulation. In reality, the process model is more complex and includes thermal energy storage tanks used in winter and summer regimes for "load shifting." Adopting the model process (55) is justified by the need of a simple presentation of the MBPC principle.

A key step in the formulation of the optimal control problem is the choice of the performance criterion. It must reflect a compromise between the requirements of the thermal comfort and of reducing the energy consumption. There are many concrete ways of formulating this compromise. In the following is exemplified by [15] the selection of criteria and constraints. At each sampling step of the MBPC is required to solve the following optimization problem:

$$\begin{aligned} \min_{\substack{u_{t|t}, u_{t+1|t}, \ldots, u_{t+N_u-1|t} \\ \underline{\varepsilon}_{t+1|t}, \ldots, \underline{\varepsilon}_{t+N_y|t}; \overline{\varepsilon}_{t+1|t}, \ldots, \overline{\varepsilon}_{t+N_y|t}}} & \left\{ \sum_{k=0}^{N_u-1} \left| u_{t+k|t} \right| \cdot T_s + \kappa \max\left[\left| \mathrm{u}_{t|t} \right|, \ldots, \left| \mathrm{u}_{t+N_u-1|t} \right| \right] \right. \\ & \left. + \rho \sum_{k=1}^{N_x} \left[\left| \overline{\varepsilon}_{t+k|t} \right| + \left| \underline{\varepsilon}_{t+1|t} \right| \right] \right\} \end{aligned} \tag{58}$$

with the constraints:

$$\underline{x} - \underline{\varepsilon}_{t+k|t} \leq x_{t+k|t} \leq \overline{x} + \overline{\varepsilon}_{t+k|t} \quad (59)$$

in which $v_{t+k|t}$ is the prediction of the variable v for the $t+k$ time, when the prediction occurs at the time t; $\underline{x}$ and $\overline{x}$ are lower and, respectively, upper limits of the inside air temperature ($x \equiv \theta_a$); N_u and N_x are the prediction horizons of the command and, respectively, the temperature; $\underline{\varepsilon}_{t+k|t} \geq 0$ and $\overline{\varepsilon}_{t+k|t} \geq 0$ are the overruns of the temperature limits $\underline{x}$ and $\overline{x}$. Temperature predictions calculus is done using Eq. (56) of the process dynamics:

$$x_{t+k+1|t} = Ax_{t+k|t} + Bu_{t+k|t} + d_{t+k|t} \quad (60)$$

The first term of the criterion given by (58) shows the energy consumption, since the control variable u shows the power transferred in one direction or the other (depending on the winter/summer regime), in order to achieve thermal comfort. The second term, which has the weighting factor κ, penalizes high values of the transferred thermal power, in order to ensure a feasible control. The third term, with the weighting factor ρ, ensures thermal comfort requirement. It is obvious that, for the criterion (58) with the constraints (59), the intervals D_w and D_s for the temperatures evolution in winter and summer regimes are not established a priori, but results from solving the optimization problem. The weighting factor ρ is higher as these intervals are narrower, but this leads to a possible higher energy consumption.

The optimization problem (58) and (59) is solved at the current sampling step, result the command $u(t) = u^*_{t|t}$, which is applied to the process. In the next sampling step, the temperature is measured $x(t+1) = x_{t+k|t+1}$ and again the optimization problem is solved, for calculus of the next command $u(t+1)$ etc.

6 Interface System

Marian Barbu

The graphical user interface (*HMI—Human–Machine Interface*) is the software component of the monitoring and control system that realize the interaction between the human operator and the control system. It is designed to be used by people who, in most cases, do not necessarily have knowledge on programming or automatic control.

Modern solutions on monitoring and control of physical systems are usually implemented using Programmable Logic Controllers (PLC). Using PLCs involves the design and development of two software components: *PLC Application* and *HMI-Human–Machine Interface*. These components are usually developed using the software tools provided by the hardware manufacturer. In this paragraph are presented the main general elements that should be considered in the design and development of the two software components mentioned above. The practical elements of the implementation are given in Chap. "Experimental case study", for the real application considered as case study.

The *PLC Application* is structured tasks which, in turn, contain program modules and routines. In designing the *PLC Application* one must consider a number of elements, which leads to a general structure of subfolders, such as:

- Controller Application—contains tags (variables) visible at the PLC level (Controller Tags), the logic of handling major errors from PLC (Controller Fault Handler), and control logic for powering the PLC (Power-Up Handler);
- Tasks—include tasks forming the process control logic. Each task consists of programs which in turn contain variables (tags) and routines;
- Motion Groups—contains instructions for coordinating movements on multiple axes;
- Add-On Instructions—contains instructions written by the user;
- Data Types—contains predefined and user defined data types;
- Trends—contains charts for online monitoring of variables;
- I/O Configuration—contains information about the project hardware configuration. PLC hardware structure contains a set of input–output modules for the acquisition of process variables and sending of commands to process.

The *PLC Application* uses a number of variables for acquisition and processing of input–output process variables, process parameterization, and implementation of logical control sequences, and to ensure the communication with higher levels of supervision of the installation. These variables can be of two types:

- Global—declared at the higher level of the application (Controller Tags);
- Local—declared within each program module of the application tasks (Program Tags).

Local variables are visible only to that program module, while global variables are visible inside any task, program or routine. Also, only global variables may be transferred outside of PLC, for example, to the Human–Machine Interface or to the higher control level.

Human–Machine Interface should provide a schematic representation of the plant (block diagram) and should offer for users clear options regarding the control and monitoring of the system elements. HMI runs on the engineering station and can have the following functions:

- allows real-time visualization of the main variables from the plant;
- allows selecting the operating mode of the plant;
- allows the operator to apply direct commands to the plant;
- allows graphical display of the evolution of certain variables from the plant;
- allows to store the values of the key variables for possible further processing.

The HMI application is usually done using the development tools provided by the hardware manufacturer, but there are other general software tools for developing HMIs (e.g., IntegraXor, IGSS).

Typically a HMI is organized under the following subfolders structure:

- System—contains an command line utility for sending commands to the system;
- HMI Tags—contains the database with the system inputs and outputs;
- Graphics—contains graphic screens and related instruments to achieve the Graphical User Interface (GUI);
- HMI Tag Alarms—defines the alarm display and storage options;
- Logic and Control—definition of variables, events, macros, and attached shortcut keys;
- Data Log—defines the data sets to be acquired and stored on disk.

HMI is an independent application with respect to the control application and needs a prior definition of the used input/output variables. HMI connects to the PLC using a specific communication driver. In this way occurs in both directions the transfer of the interest process variables.

HMI is divided into several sections, each trying to capture different aspects of the monitored and controlled plant. There are some important aspects to be considered in developing a HMI. These aspects lead to the following sections:

1. General block diagram

The first section of HMI provides an overview of plant, including all the system elements and process variables. All the plant components must be displayed according to their position in the physical plant. Thus, this section must allow to readily seeing the global state of the system and that of its various elements. For each active object of the plant it is displays its state (*ON/OFF*) or the value at that moment. If the element is stopped, the status/value will be displayed using the usually the red color, otherwise it will be displayed using the green color.

2. Graphical monitoring of process variables

In this screen, the operator can view real-time evolution of the process variables. A screen can display simultaneous several variables and the operator must be able to select the desired variables from configuration menu of the screen.

3. Alarms monitoring

This screen contains a list of alarms occurred in the application, representing alerts about overcoming certain limits, malfunction of the system equipments or in case of damage. This screen must contain a button "Confirm" for the operator. This button has the meaning of "acknowledgment," the acknowledged alarms being removed from the current list. However, the alarms should not be removed completely from the system; they are stored and can be viewed later.

4. Analog inputs/outputs monitoring;

In this screen can be viewed in real-time the values of all analog inputs and outputs from the system.

5. Digital inputs/outputs monitoring;

In this screen can be viewed in real-time the values of all digital inputs and outputs from the system.

6. Plant setup screen

The setup screen is designed to offer the possibility of modifying certain parameters the controlled plant. This screen usually can also provide different working regimes (e.g., manual, automatic), with the possibility of switching between them.

7. Users menu

User menu must contain a set of options on connecting/disconnecting the operator from application, change the password of the current user or application exit.

References

1. Günter S (2006) Simader-Micro CHP systems:state-of-the-art-European Commission Deliverable 8 (D8) of Green Lodges Project (EIE/04/252/S07.38608)
2. http://batteryuniversity.com/learn/article/how_to_restore_and_prolong_lead_acid_batteries
3. https://archive.org/stream/Home_Power_Magazine_009/Home_Power_Magazine_009_djvu.txt
4. http://energy.gov/energysaver/articles/balance-system-equipment-required-renewable-energy-systems
5. EN 50160 (1999) Voltage characteristics of electricity supplied by public distribution systems
6. IEC 038 (1999) IEC standard voltages
7. Complete Solar Thermal Solutions Kingspan Solar (2013) http://www.kingspansolar.co.uk/pdf/UK%20Solar%20Manual%20Feb%2013_LR.pdf. Accessed 20 Feb 2013
8. Adsorbtion chillers'—SorTechh AG product
9. http://productfinder.wilo.com/en/COM/productrange/0000001b0000accb00010023/fc_range_description/
10. http://utahbiodieselsupply.com/heatexchangers.php
11. http://www.engineeringpage.com/technology/thermal/transfer.html#coil
12. http://www.cordivari.it/product.aspx?id=1&gid=2720&prd=2775&lng=0
13. Freire RZ, Oliveira GHC, Mendes N (2008) Predictive controllers for thermal comfort optimization and energy saving. Energy Buildings 40:1353–1365
14. Verhelst C (2012). Model predictive control of ground coupled heat pump systems for office buildings. PhD Thesis, Katolieke Universiteit Leuven, Belgium
15. Ma Y, Kelman A, Daly A, Borelli F (2012) Predictive control for energy efficient buildings with thermal storage. IEEE Control Syst Mag 2012:44–64
16. Barata FA, Igreja JM, Neves-Silva R (2012) Model predictive control for thermal house comfort with limited energy resources. In: 10th Portuguese Conference on Automatic Control, pp.146–151. Funchal, Portugal, 16–18 July 2012
17. Sakellariou F (2011) Model predictive control for thermally activated building systems. MSc. Thesis, Eindhoven, Netherlands

Experimental Case Study

Nicolae Badea and Marian Barbu

Abstract In order to be shown in detail, the design process algorithm (see Fig. 2 in Chap. "Structural Design of the mCCHP-RES System") was divided into two stages, namely structural design and functional design. In Chap. "Structural Design of the mCCHP-RES System" was presented the first stage while in Chap. "Functional Design of the mCCHP-RES System" was presented the second. Here it is shown a concrete example, which consists in the design of mCCHP system for a familiar type housing. The design was done in accordance with the algorithm, which is shown in Fig. 2 in Chap. "Structural Design of the mCCHP-RES System" and described in Chaps. "Structural Design of the mCCHP-RES System" and "Functional Design of the mCCHP-RES System" in detail. The 12 paragraphs of this chapter contain data resulting from application of the 12 steps of the algorithm. Testing of the real mCCHP system that has been achieved through the practical application of this algorithm confirmed the algorithm validity.

1 Conceptual Framework

Nicolae Badea

The case study that is presented here consists in developing an experimental mCCHP-SE-RES system, designed according to the algorithm presented in Chaps. "Structural Design of the mCCHP-RES System" and "Functional Design of the mCCHP-RES System", and its integration into an experimental residential building located on the campus of "Dunarea de Jos" University of Galati, as well as the assessment of the experimental couple building—mCCHP-SE-RES system thus obtained, in order to prove the design algorithm validity.

N. Badea (✉) · M. Barbu
"Dunarea de Jos" University of Galati, Galati, Romania
e-mail: nicolae.badea@ugal.ro

M. Barbu
e-mail: marian.barbu@ugal.ro

N. Badea (ed.), *Design for Micro-Combined Cooling, Heating and Power Systems*,
Green Energy and Technology, DOI 10.1007/978-1-4471-6254-4_7

In achieving this assemble, *the conceptual framework* is composed of all the restrictions and requirements imposed by the location, host, and financial supporter. Development of the experimental couple building—mCCHP-SE-RES system, using renewable energy available in the southeastern region of Romania has two global requirements:

- to be inhabited, offering all comfort conditions of a house;
- to become experimental basis for masters and Ph.D. students.

The conceptual framework includes the following main components:

- Local needs analysis for the experimental building with mCCHP system;
- Obtaining legal conditions to permit the experimental building construction;
- Local needs analysis for mCCHP system implementation.

1.1 Local Needs Analysis for the Experimental Building

This component of the framework has aimed to establish the location for building on the campus of "Dunarea de Jos" University of a residential building in the upper energy class. Once the location is established, the construction of the experimental building remains dependent on the building permit.

1.2 Obtaining Legal Conditions to Permit the Experimental Building Construction with MCCHP System

Under this component of the conceptual framework, it was identified the standard procedure for obtaining the legal conditions for providing the experimental building and legislative framework to be complied with. Buildings may be constructed only based on a building permit. The building permit is the authority document of the local public administration, which ensures the enforcement of the legal measures regarding the layout, design, execution, and functioning of buildings. The building permit is issued according to, abiding by the provisions of urbanism documentation, fully authorized, and approved according to the laws in force. The mayors of cities or towns issue the building permit for buildings and works of any kind inside and outside the limits. The building permit is issued within 30 days at the latest from the date of the application entry, based on the documentation submitted to the authorities, which shall include:

(a) the urban certificate;
(b) proof of ownership of land and/or buildings;
(c) the project for building permit;
(d) the necessary legal notices and agreements, established through the urban certificate;
(e) proof of legal tax payment.

A. *The urban certificate* is the information document by means of which the authorities, in compliance with the related provisions of the urban plans (and related regulations) and land development plans (notified and approved according to the law),

- make known to the applicant the elements regarding the judicial, economic, and technical status of the lands and buildings existent at the date of the request and
- establish the planning requirements to be met according to the specific site and list of legal approvals and agreements necessary to authorize.

B. *Technical documentation for building permit* is a fragment of the technical project and is drawn up in accordance with the frame-content, requirements of the town—planning certificate and the notices and agreements thereto; it is drawn up, signed, and verified according to the law.

1.3 Local Needs Analysis for mCCHP System Implementation

The purpose of this local needs analysis for mCCHP system implementation is identifying the location inside or outside house of mCCHP system and the topology of its components, in accordance with the general principles on which the concept of trigeneration is based, and considering the energy resources available in the region. The analysis was performed from three points of view:

- in terms of structures (charts), at block diagram level and of energy distribution system integrated to building;
- from the functional point of view, represented by the degree of complexity (limited by certain barriers) and energy efficiency (EFF);
- from the architectural limitation imposed by building permit.

The comparative analysis in terms of structure, at block diagram level, leads to a conceptual scheme for the mCCHP system based on the use of renewable energy (solar energy and energy from the combustion of biomass) and thermal energy distribution by hot/cold water to ventilator-convectors.

From the functional point of view, the analysis performed of the appropriate systems for the production of cold, the following conclusions can be drawn:

- both absorption and adsorption systems can be used successfully in generating cold within the trigeneration systems;
- the use of refrigerating systems with high temperature heat flows requires cumulative absorption systems (in several steps), relatively complex and hence of the low reliability;
- adsorption refrigeration system does not emit toxic gases and can be mounted inside the building.

From the architectural limitation imposed by building permit, the analysis performed on the house project for placing inside the house of the mCCHP system without affecting the requirements of building permits, is a necessary technical basement in the building.

The execution of a building may only be done based on the technical project and the execution details. Technical project was developed by specialized technical teams, and signed by technical staff having higher education degrees in architecture, construction, and related facilities.

2 Manufacturer Business Plan

Nicolae Badea

The second step in achieving a mCCHP system is the analysis of manufacturer business plan in order to select the appropriate approach. We have considered the case in which, for a given residence, the design of a dedicated mCCHP system is required, using nondedicated (or general use) components, available on the market today. Once the manufacturer businesses plan, have been completed, the design should be continued with the collection of all initial data that form the hypothesis of the design problem.

3 Initial Data Collection

Nicolae Badea

Initial data refers to: (a) the residence building features, (b) the customer needs and requirements, (c) the residence functional needs, and (d) the residence energetic environment.

3.1 Experimental Residence Features

3.1.1 Residence Location

The Galati city is situated in the southeast of Romania, 45°47″ latitude to the north and 28°2″ longitude to the east, on the left bank of the Danube at the confluence of the rivers Siret (to the west) and Prut (to the east), near Lake Brateş, the biggest lake in this part of the country (Fig. 1). It is the capital of Galaţi County which is situated in the "Lower Danube" Euroregion, the eastern border of the EU in proximity to the Republic of Moldavia and Ukraine. With a population of 249,432 inhabitants, Galaţi is the seventh largest city in Romania. The higher education institution within the municipality boundaries is "Dunarea de Jos" University with 15,330 students. The experimental residence is situated within the built-up area of the city

Fig. 1 Building location [1]

of Galați, Domnească Street no. 155. The location is in the university campus, having the cafeteria to the north (building D + P + E), the C hostel to the east (building P + 3E), the green in front, toward hostel D (building P + 3E) to the south, and the thermal station (building ground floor) to the west.

3.1.2 Building Drawings

The experimental residence has the following height state: underground floor + ground floor + first floor. The residence is individual, representative for the eastern area of Romania. The building is made up of ground floor plus first floor and attic (Fig. 2). Figures 3 and 4 present the overviews for the ground floor and for the first floor with geometric characteristics (lengths and widths) of rooms.

3.1.3 Characteristic Elements of the Architecture

Characteristic elements of the architecture related to each floor are described below as follows:

Fig. 2 *Frontal view* of the building (South facade)

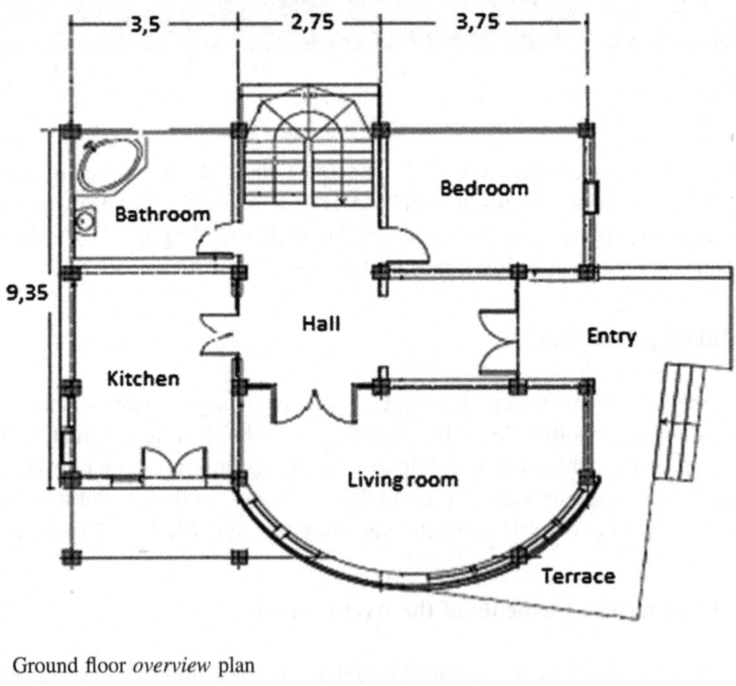

Fig. 3 Ground floor *overview* plan

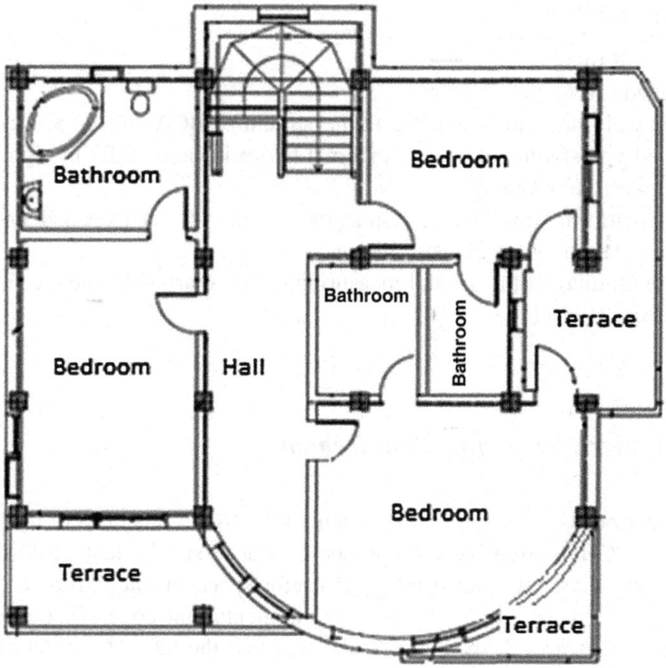

Fig. 4 First floor *overview* plan

Total floor area (TFA) is 299 m^2.

Floor:

- heated surface: 100 m^2;
- heated space average height: 3.05 m;
- exterior wall structure: concrete frame structure, BCA filler, thermo-isolation expanded polystyrene of 15 cm;
- platform on top of the ground floor: steel concrete platform of 15 cm, 15 cm extruded polystyrene thermo-isolation, 3 cm blanket;
- on top of the floor there is a nonheated loft, covered in steel plate tile type, mineral wool thermal isolation of 10 cm.

Ground floor:

- heated surface: 100 m^2;
- heated space average height: 2.95 m;
- exterior wall structure: concrete frame structure, BCA filler, thermo-isolation expanded polystyrene of 15 cm, not in contact with the ground;
- platform on top of the underground floor: steel concrete platform of 15 cm, 15 cm extruded polystyrene thermo-isolation, 3 cm blanket.

Underground floor:

- Surface: 99 m^2;
- space medium height: 2.75 m;
- exterior wall structure: concrete frame structure, BCA filler, thermo-isolation expanded polystyrene of 15 cm, external hydro isolation 2.05 m high, being in contact with the ground;
- ground platform: low steel concrete of 15 cm, 15 cm extruded polystyrene thermo-isolation, 3 cm blanket, mosaic;
- concrete plinth over the ground measuring 0.7 m, marginal concrete sidewalk at 0.7 m from 0 level.

3.2 Customer Needs and Requirements

The customer needs and requirements mainly refer to the functions of the residence and to the mCCHP system, which must be easy accessible, silent, easy to operate, and its components must not occupy all useful space in underground level. The mCCHP system must satisfy all these requirements and cover all the functional needs of the residence. A specific demand was that the mCCHP system to be off-grid type with thermal compression chiller (TCC), given the fact that the requirement of heat is much higher than electricity. At the same time in the house, the electric installations and general purpose outlets have total installed power, $P_n = 12$ kW. Lighting circuits are made from copper conductors (phase, neutral) and conductor (electrical protection) protected in protective tube mounted flush in building elements.

3.3 Residence Functional Needs

Energy is used in buildings for various purposes as heating and cooling, ventilation, lighting, and the preparation of domestic hot water (DHW). In the location where the building is, there are available electricity grid with 3 × 380 V, 50 Hz and natural gas distribution network, operated by local distribution system operator. Also in the building site there is cold water and sewerage network.

3.3.1 Distribution of Thermal Energy

The heating system chosen for the building runs on hot water, on its own thermal station—nonconventional energy, mixed bi-tubular distribution and heating enameled steel plate radiators or equivalents in baths and kitchens, ventilator-convectors in technical spaces and living spaces.

The horizontal distribution starts from the distributor/collector from the thermal station, the turn pipes and the return ones circulate in parallel circuits under the roof of the underground floor. A turn-return circuit using ventilator-convectors, runs in the cold season with hot water of 80/60 °C for heating the residence, and in the hot season with cold water of 5/15 °C for cooling the residence. The commutation of the cold/hot operation statuses is done by electronic control with electro-valves from the thermal station.

3.3.2 Sanitary Appliances

The cold water input comes from the local potable water network, based on the junction notification issued by the utilities provider. The residence will be branched to the public network through a fire proof polypropylene branch pipe for appliances from the existent cold water distribution pipe.

The hot water input is done through the residence's own thermal station, equipped with a hot domestic water preparation appliance.

The local distribution is done through an underground network into the PEHD pipe, assembled by being buried in a sand bed, at minimum 1.2 m from the surface. The hot and the cold water are distributed through a network assembled in the underground floor, into the water shafts that supply consumers vertically.

3.3.3 Electric Energy Input Appliance

In the house the electric installations and general purpose outlets have installed power, $P_n = 12$ kW. The mCCHP system is equipped with an accumulator/inverter 400 V, 50 Hz batteries system. The distribution to all consumers is achieved through all interior layouts in a copper (Cu) core conductor isolated with PVC for the internal isolations, type FY, protected by a protection tube embedded in the construction elements. A reinforced cable with Cu core isolated with PVC is used for the external layouts, assembled underground into a sand bed at a depth of minimum 0.8 m.

Illumination source placement in the rooms has been done so that a maximum degree of illumination and good surface uniformity is ensured.

The control of electric sources is done locally, through switches and knobs assembled into protection dozes embedded in the construction elements or through switches with motion sensors, depending on the area.

In the distribution electric panels, for the protection of the illuminating circuits, automatic bipolar switches are foreseen, with the thermal current according to the necessities of each circuit (in general 10 A). The illumination circuits will be composed of two FY 1.5 mm conductors (phase, null) and a 2.5 mm conductor (electric protection) covered in a protection tube embedded in the construction elements. The plug installation is divided into mono-phase circuits, with a maximum of eight plugs per circuit, grouped so that the power in the circuit does not exceed 2 kW. All plugs are of the null contact type (simple or double), inbuilt. The plug circuit for the general use plugs is achieved with three Cu FY 2.5 mm^2 conductors protected by a tube built in the construction elements.

The metal carcasses of the thermal stations and all the metallic structural elements are connected to the leak plug. The lead-in is done with a copper conductor, minimum 2.5 mm^2 in section. Automatic bipolar switches are fitted into the electric distribution panels, for the protection of the plug circuits; their thermal current is dimensioned in keeping the necessities of every circuit (usually of 16 A) and a differential protection of 30 mA.

3.4 Residence Energetic Environment

3.4.1 Local Climate

The energy consumption of a building depends on external factors what are climatic parameters specific to the location: air temperature, air humidity, wind speed, and sunshine. According to Romanian standard, Romania is divided into four climatic areas. Galaţi city is in area III, where conventional external temperature ϑ_e is of −18 °C.

To calculate the annual heat necessary for building and respectively, the amount fuel for heating, the *monthly medium external temperatures* are used, by means of which, the medium temperature during the period of heating $\vartheta_{e,\mathrm{hs}}$ and the number of degrees—days (HDD) are determined (Table 1).

The sizing of the ventilation-air conditioning installations for summer conditions and establishing the heating-cooling charge, the average outdoor temperature specific to the month of July is used (Table 2).

Wind speed values were adopted on statistic basis, regarding the wind-temperature concomitance, which determine four wind areas on the country territory. The wind speed conventional values in Galaţi are given in Table 3.

Table 1 Period of heating (according to Romanian standard) [1]

Climatic area	ϑ_e (°C)	$\vartheta_{e,\mathrm{hs}}$ (°C)	HDD (deg.-days)	Period of heating (days)
III	−18	2,9	3,510	205

Table 2 Daily average outdoor temperature, and amplitude of daily oscillation, A_z, during summer for the county of Galaţi [2, 3]

Locality	Daily average outdoor temperature ϑ_{em} (°C)		Amplitude of daily oscillation, A_z
	a	b	
Galaţi	25.8–25.6	24.6–24.4	6

Table 3 Wind speed value in Galaţi [4]

Individual settlements	Wind speed (m/s)	
	Inside the settlements	Outside the settlements
Galaţi	8	10

Table 4 Average monthly sunshine duration (d_s) in hours/month

Galaţi	Month											
	Jan	Feb	Mar	Apr	May	Jun	Jul	Aug	Sep	Oct	Nov	Dec
d_s (h/m)	76	82	138	193	251	294	307	293	230	185	85	63

The climatic data regarding sunshine (the duration of sunshine and the intensity of solar irradiation) is of interest both for the warm period of the year and for the cold one. They are used for dimensioning the air-conditioning installations during the warm season, establishing the solar inputs, which must be taken up. In addition, the climatic data regarding sunshine are used to adjust the heat necessary for heating as long as the building is designed accordingly to capture the solar energy during the cold season. The average sums of sunshine duration, in hours per month, for Galaţi are given in Table 4.

The Galati city is located in a temperate continental climate zone, slightly modified by the proximity of the Black Sea and the intrusion of warm wet air from the Mediterranean. Climatic seasons are clearly defined with a short and soft low-snow winter and a long summer which can be very hot and dry. Monthly air temperature and relative humidity of the external air has an important role in ventilation and acclimatization. The monthly climatic parameters for Galati city are shown in Table 5.

The land surface characteristics, where house is located in the Galati city, have great importance in defining the specific climate condition. Landscape features in the area, vegetation characteristics, the influence of the Danube river and Brates lake, and construction types (presence of C and D hostels with 15–20 m height) determine the local climatic particularities.

3.4.2 Local Energy Sources and Resources

A. *Solar energy potential of the Romanian territory*

Renewable energy sources have a large potential in Romania and moreover, each development region benefits from at least two renewable energy sources that might be exploited (Fig. 5).

Galaţi city is situated in southeast region with a solar energetic potential of (5,002–5,113) MJ/m^2/year or of (1,350–950) kWh/m^2/year (Fig. 6).

Following the building's location in the urban area, the wind energy cannot be used.

Table 5 Climate data for the city of Galati

Month	Air temperature °C	Relative humidity (%)	Horizontal solar daily irradiation (kWh/m^2/d)	Wind speed (m/s)	Earth temperature (°C)	Heating degrees days (°C-d)	Cooling degrees days (°C-d)
January	−1.7	87.0	1.47	5.7	−1.5	611	0
February	−0.6	84.0	2.31	5.7	0.0	521	0
March	5.0	79.0	3.44	5.7	5.9	403	0
April	11.1	74.0	4.94	6.2	13.6	207	33
May	16.7	73.5	5.94	4.6	20.0	40	208
June	20.0	75.0	6.69	4.6	23.4	0	300
July	21.7	72.5	6.56	4.1	26.2	0	363
August	21.7	71.5	5.75	4.6	26.1	0	363
September	17.8	73.5	4.39	4.6	20.6	6	234
October	11.7	77.0	2.97	4.6	13.7	195	53
November	5.0	86.0	1.64	5.1	5.3	390	0
December	0.6	89.0	1.19	5.7	−0.4	539	0
Annually	**10.8**	**78.5**	**3.95**	**5.1**	**12.8**	**2,912**	**1,554**

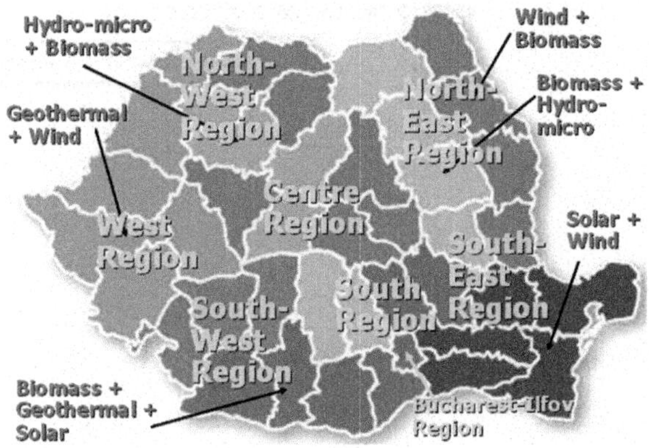

Fig. 5 Map of renewable energy potential in Romania [5]

B. *Solar energy potential for the building*

The simplified model of the residence meets the current standards for a living area and space volume needed for a four-member family while keeping the particulars of the southeast region of Romania. From climatic condition of the Galati city and location of the building using PVGIS data [7], we obtain the following monthly variation of the horizontal irradiation (Fig. 7).

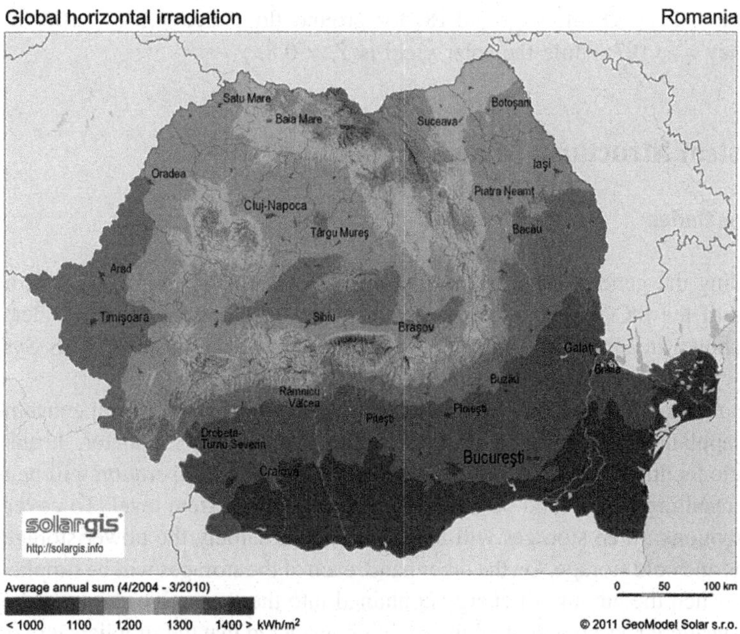

Fig. 6 Solar potential of Romania [6]

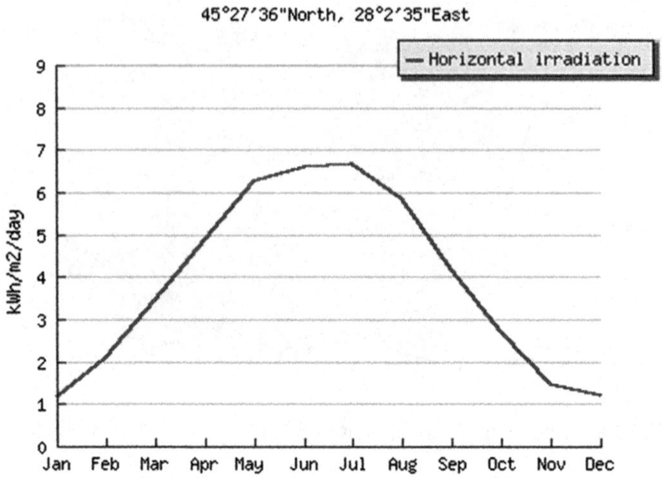

Fig. 7 Monthly variation of the *horizontal irradiation* [7]

In the Table 6 are presented the values of monthly power and heat production, both those specific for the panels as well as those specific for the experimental building.

These values were calculated with relations (7)–(10) in Chap. "Structural Design of the mCCHP-RES System", for the experimental building which has on its roof both the photovoltaic panel with area $A^r_{PV} = 12.3\ m^2$ ($k^r_{PV} = 0.123$ for ground floor area $A_{GFA} = 100\ m^2$) and efficiency $\eta_e = 0.15$, as well as the solar thermal panel with area $A^r_{ST} = 18\ m^2$ ($k^r_{ST} = 0.18$, for ground floor area $A_{GFA} = 100\ m^2$) and efficiency $\eta_t = 0.7$, while the solar yield is $Y_r = 0.8$.

4 System Structural Modeling

Nicolae Badea

Regarding the general structural model, it is shown in Fig. 9 in Chap. "Structural Design of the mCCHP-RES System". Here, the functional needs of residence are those shown in Fig. 1 in Chap. "Structural Design of the mCCHP-RES System", namely lighting, ventilation, and cooling, heating, and DHW.

The needs will be met by incorporating into the system of several consumers as home appliance, chiller with thermal compression (TCC), radiator, heater, and domestic facility. Consumptions occasioned by the system operation will be aggregated, resulting in heat and power consumption at the system level. To cover these consumptions, three storages will be incorporated, namely the power storage, heat storage, and cold storage. On the other hand, each of the storages will be supplied with energy when the amount of energy contained into the respective storage is below a level considered acceptable. In this case it is considered that the suppliers will only be the PV panel, Stirling engine as CHP unit, additional boiler, and ST panel.

Table 6 Monthly specific power and heat production

Month	Horizontal solar daily irradiation, H $(\mathrm{kWh/m^2/d})$	Monthly specific power and heat production of the panels		Monthly specific power and heat production of the building	
		e^r_{PV} $\left(\mathrm{kWh/m^2_{panel}}\right)$	q^r_{ST} $\left(\mathrm{kWh/m^2_{panel}}\right)$	e_{PV} $\left(\mathrm{kWh/m^2_{TFA}}\right)$	q_{ST} $\left(\mathrm{kWh/m^2_{TFA}}\right)$
Jan	1.47	5.47	25.52	0.23	1.53
Feb	2.31	7.76	36.22	0.33	2.17
Mar	3.44	12.80	59.72	0.54	3.58
Apr	4.94	17.78	82.99	0.75	4.98
May	5.94	22.10	103.12	0.93	6.19
Jun	6.69	24.08	112.39	1.01	6.74
Jul	6.56	24.40	113.88	1.02	6.83
Aug	5.75	21.39	99.82	0.90	5.99
Sep	4.39	15.80	73.75	0.66	4.43
Oct	2.97	11.05	51.56	0.46	3.09
Nov	1.64	5.90	27.55	0.25	1.65
Dec	1.19	4.43	20.66	0.19	1.24
Annually	**3.95**	**172.97**	**807.18**	**7.26**	**48.43**

5 Consumption Estimation

Nicolae Badea

At residence level, the consumption could be global or specific. Consumption is global when it refers to the whole residence. The global consumption divided to the TFA of the residence building, becomes specific. Moreover, the consumption is estimated for each form of energy consumed and each function of the residence covered.

5.1 Estimation of the Heat for Global Heating and the Cold for Air-Conditioning

For determining the heat for heating and the cold for air-conditioning, the overall heat transfer coefficient of the building will be used. The building heat loss coefficient (UA) is found by identifying every route of heat loss from the building and adding these together. The routes for heat transfer between interior and exterior are through the building envelope, air exchange between inside and outside (infiltration) and through the ground (perimeter). Landscape features in the area, vegetation characteristics, and the influence of the nearby buildings determine local climatic particularities given by the solar irradiation, and by the wind. This is why information regarding the shading in residence location, architectural orientation (Figs. 8 and 9), and building characteristics should be available.

Fig. 8 Southeast façade-3D of the experimental residence

Fig. 9 Southwest façade-3D of the experimental residence

The most important heat losses are those transferred through the walls, the roof, the windows, and the ventilation system. Thermal insulation of the building is ensured by the envelope. Thus, the exterior walls are extruded polystyrene over masonry brick structure and the floor above the basement, and most of the top floor and attic are insulated.

For the method application should identify for all sides, the walls and windows of the building and conductive heat transfer through all elements of the building envelope from one side to the other: walls, floor, roof, windows, doors, etc.

Table 7 presents the materials thickness and thermal proprieties for all elements of the building envelope. Based on these data, using the relationship 12 in Chap. "Structural Design of the mCCHP-RES System", the overall heat loss coefficient or U—value was calculated.

Taking into account the architectural details of the building and the materials used (for which the U—values are given in the Table 7), the building heat loss coefficient (UA) is calculated, by identifying every route of heat loss from the building (Table 8).

The annual energy need for heating was determined using Eq. 23 in Chap. "Structural Design of the mCCHP-RES System". and UA value from the Table 8. It results:

$$Q_{\text{heating, year}} = 0.024 \cdot \text{UA} \cdot \text{HDD} = 24{,}855\ \text{kWh/year} \tag{1}$$

Similarly, the annual energy need for cooling was determined using Eq. 24 in Chap. "Structural Design of the mCCHP-RES System", and it results:

Table 7 Indicated materials for the residence building

Layer type	Thickness (mm)	Conductivity (W/m °C)	*R* value (m^2 °C/W)	*U*-value (W/m^2 °C)
Foundation			**2.21**	**0.45**
Compact ground	1,500	0.83	1.81	
Reinforced Concrete dry	100	0.75	0.13	
Floor screed	15	0.055	0.27	
Walls			**4.58**	**0.21**
Exterior plaster	10	0.43	0.02	
Polystyrene insulation	150	0.03	3.33	
Reinforced concrete	400	0.33	1.21	
Interior plaster	10	0.43	0.02	
Ceiling and roof			**2.62**	**0.38**
Ceiling			*1.3*	
Air wedge	150	5.56	0.03	
Glass fiber plates	50	0.04	1.25	
Interior plaster	10	0.43	0.02	
Roof			*1.32*	
Ceramic plates (tiles)	30	1.59	0.02	
Air wedge	150	5.56	0.03	
Glass fiber plates	50	0.04	1.25	
Internal plaster	10	0.43	0.02	
Windows and Door				**1.4**

$$Q_{\text{cooling,year}} = 0.024 \cdot \text{UA} \cdot \text{CDD} = 13,263 \quad \text{kWh/year} \tag{2}$$

5.2 Analytical Estimation of the Heat Consumption for Domestic Hot Water

Assuming a temperature difference of $\Delta\vartheta = 50$ °C between the hot and cold water sides, and four persons, we determined, with the Eq. 39 in Chap. "Structural Design of the mCCHP-RES System", the monthly quantity of heat consumed for DHW, as follows:

$$Q_{\text{DHW}} = c_w \frac{m \cdot n \cdot N_d}{3{,}600} \Delta\vartheta = 4{,}245 \text{ kWh/year} \tag{3}$$

Table 8 UA—value for building

Envelope heat transfer coefficient (UA conduction)					
Building element		*U*-value	Area	*U* × Area	Description
Orientation	Element	W/m² °C	m²	W/°C	
South	Walls	0.21	37.2	7.81	
	Window	1.4	32	44.8	
East	Walls	0.21	33	6.93	
	Window	1.4	6.2	8.68	
North	Walls	0.21	61.8	12.97	
	Window	1.4	6.2	8.68	
West	Walls	0.21	41.1	8.63	
	Window	1.4	3.9	5.46	
	Ceiling/Roof	0.38	240	91.2	
$UA_{envelope} = U \times A$ (W/°C)				195.16	
Infiltration heat Loss ("UA" infiltration)					
Volumetric heat capacity (Air) (Wh/m³ °C)			1.16		
Building volume (m³)			597.2		
Air changes per hour (ACH) (1/h)			0.2		
"UA" infiltration = VHC × Volume × ACH (kW/°C)				138.56	
Perimeter heat loss ("UA" perimeter)					
Heat loss coefficient (*F*) (W/m °C)			0.48		
Length of building perimeter (*P*) (m)			45.65		
"UA" perimeter = *F* × *P* (W/°C)				21.91	
UA (W/°C)				355.63	

where:

> Q_{DHW}—the yearly energy needed to produce DHW (kWh/year),
> $c_w = 4.187$ (kJ/kg °C)—specific heat capacity of water,
> $m = 50$ (kg/capita)—daily quantity of water consumption,
> $n = 4$—number of building occupants,
> $N_d = 365$ is the number of days.

5.3 Analytical Estimation of the Global Power Consumption for Domestic Facilities

Using Eq. 41 in Chap. "Structural Design of the mCCHP-RES System", the lighting energy (E_{Lt}) required to fulfill the illumination function is:

$$E_{L,t} = \frac{(P_n \cdot F_c) \cdot [(t_D \cdot F_o \cdot F_D + t_N F_o)}{1,000} = 11,191\ \text{kWh/year} \tag{4}$$

where:

Installed power, P_n = 12 kW
Constant illuminance factor F_c = (1 + MF)/2 = 0.9
Maintenance factor MF is = 0.8
Daylight time usage $t_D = t_o - t_N$ = 2,250 h
Annual operating time t_o = 2,500 h
Occupancy dependency factor $F_O = F_{OC} + 0.2 - F_A = 0.5$
F_{OC} = 0.8 (from Table 19 in Chap. "Structural Design of the mCCHP-RES System")
Absence factor, F_A = 0.5
Daylight dependency factor $F_D = 1 - (F_{DS}\ F_{DC}\ C_{DS}) = 0.81$
Daylight supply factor, $F_{DS} = a + b \cdot \gamma_{site} = 0.81$
F_{DC} = 0.3, from Table 21 in Chap. "Structural Design of the mCCHP-RES System"; C_{DS} = 1.

Values of the coefficients a and b for determining the daylight supply factor was adopted from the Table 20 in Chap. "Structural Design of the mCCHP-RES System". The parasitic energy required to provide charging energy for emergency lighting and standby energy for lighting controls in the building was considerated zero ($E_{P,t}$ = 0).

5.4 Determination of the Specific Consumption

The specific heat, cooling, and power annual consumption for building is obtained by dividing the annual consumption to the TFA of the residence (299 m^2) thus:

- for heating, the specific heat annual consumption is obtained by dividing the annual consumption (Eq. 1) to the TFA and resulting:

$$q = 83.2\ \left(\text{kWh/m}^2\ \text{year}\right) \tag{5}$$

- for cooling, the specific cooling heat annual consumption is obtained by dividing the annual consumption (Eq. 2) to TFA and resulting:

$$q_C = 44.35\ \left(\text{kWh/m}^2\ \text{year}\right) \tag{6}$$

- for DHW, the specific DHW annual consumption is obtained by dividing the annual consumption (Eq. 3) to the TFA and resulting:

$$q_{DHW} = 14.2\ \left(\text{kWh/m}^2\ \text{year}\right) \tag{7}$$

Table 9 Specific energy consumption less RES

Annual specific consumption	Energetic class		Specific energy of building	Energetic class of building (less RES)
	A	B		
Heating (kWh/m^2 year)	70	117	83.2	**B**
Cooling (kWh/m^2 year)	20	50	44.35	**B**
Hot water (kWh/m^2 year)	15	35	14.2	**A**
Illumination (kWh/m^2 year)	40	49	37.42	**A**

- for power, the specific electricity annual consumption is obtained similarly by dividing the annual consumption (Eq. 4) to the TFA surface and resulting:

$$e = 37.42\ \left(\mathrm{kWh/m^2\ year}\right) \tag{8}$$

The energetic classes of building for every form of energy, expressed by the four statistic indicators, are shown in Table 9.

The evolution in time for each month of the specific energy, namely specific monthly consumption, is done with statistical indicators heating degree day and cooling degree day. Degree day method determines the heat and cold monthly consumption, per unit of TFA of building (relations 51–54 in Chap. "Structural Design of the mCCHP-RES System"). By processing the values of the specific energy consumption given in the Table 9, was obtained the specific monthly distribution of consumption, given in Table 10.

Table 10 The monthly specific consumption of residence

Month	Heating degrees days °C-d	Cooling degrees days °C-d	Heat q (kWh/m^2)	Cold q_c (kWh/m^2)	DHW q_{DHW} (kWh/m^2)	Electricity e (kWh/m^2)
Jan	611	0	17.46	0.00	1.18	3.12
Feb	521	0	14.89	0.00	1.18	3.12
Mar	403	0	11.51	0.00	1.18	3.12
Apr	207	33	5.91	0.94	1.18	3.12
May	40	208	1.14	5.94	1.18	3.12
Jun	0	300	0.00	8.56	1.18	3.12
Jul	0	363	0.00	10.36	1.18	3.12
Aug	0	363	0.00	10.36	1.18	3.12
Sep	6	234	0.17	6.68	1.18	3.12
Oct	195	53	5.57	1.51	1.18	3.12
Nov	390	0	11.14	0.00	1.18	3.12
Dec	539	0	15.40	0.00	1.18	3.12
Annually	**2,912**	**1,554**	**83.20**	**44.35**	**14.20**	**37.42**

5.5 Consumption Aggregating

The residence functional needs are met by incorporating into the system of several consumers as home appliance, with TCC, radiator, heater, and domestic facility. The structure of a mCCHP system with a thermally compression chiller is presented in Fig. 24 in Chap. "Structural Design of the mCCHP-RES System". The heat load of the mCCHP system must cover the heat consumption of the residence, as well as the heat consumption for cooling. The heat consumption for cooling depends on the coefficient of performance, COP_a, of the thermally compression chiller. The COP_a of cooling devices is considered equal to 0.8. The monthly power consumption of the residence is considered constant, the entire year. The diagrams of the power and heat consumption of the mCCHP system in the case of cooling with TCCs are presented in Fig. 10. These values are obtained by applying the relations 57 and 58

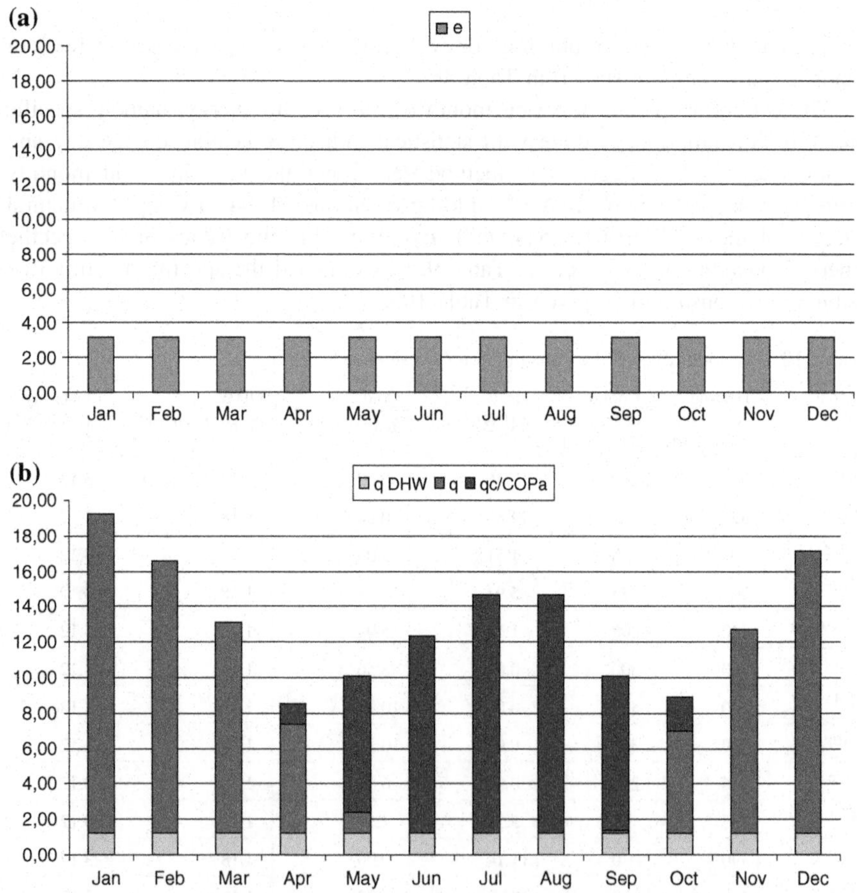

Fig. 10 Diagrams of the heat and power consumption of the mCCHP system. **a** Diagram of the system's power consumption. **b** Diagram of the system's heat consumption

Table 11 Monthly energy specific consumption and aggregated heat and power consumption

Month	Heat q (kWh/m^2)	Cold q_c (kWh/m^2)	DHW q_{DHW} (kWh/m^2)	Electricity e (kWh/m^2)	Aggregating energy consumption at the system level	
					e_{sys} (kWh/m^2)	q_{sys} (kWh/m^2)
Jan	17.46	0.00	1.18	3.12	3.12	18.64
Feb	14.89	0.00	1.18	3.12	3.12	16.07
Mar	11.51	0.00	1.18	3.12	3.12	12.70
Apr	5.91	0.94	1.18	3.12	3.12	8.27
May	1.14	5.94	1.18	3.12	3.12	9.75
Jun	0.00	8.56	1.18	3.12	3.12	11.89
Jul	0.00	10.36	1.18	3.12	3.12	14.13
Aug	0.00	10.36	1.18	3.12	3.12	14.13
Sep	0.17	6.68	1.18	3.12	3.12	9.70
Oct	5.57	1.51	1.18	3.12	3.12	8.65
Nov	11.14	0.00	1.18	3.12	3.12	12.33
Dec	15.40	0.00	1.18	3.12	3.12	16.58
Total	**83.20**	**44.35**	**14.20**	**37.42**	**37.42**	**152.84**

in Chap. "Structural Design of the mCCHP-RES System" to the data given in Table 10. The heat consumption of the mCCHP system for certain specific climate zones can be balanced between cool and warm season.

Aggregated heat and power consumption for a mCCHP system with TCC are recorded in the Table 11.

6 Load Estimation

Nicolae Badea

The system load is defined as the load necessary to fill the storage. It is estimated in two situations, namely during the system operation and during the system design. During system design is estimated the monthly load, namely that corresponding to each month of the year. No matter how big is, the estimation is carried out on the base of the monthly consumption, which was obtained by applying the estimation methods presented in the Sect. "Consumption estimation" (and selected in the Sect. 5.5). During system design, the ultimate goals of the load estimation are: (a) sizing the suppliers, so that they can support the estimated load, and (b) assess the performance indicators for structural model.

Table 12 Load sharing of the mCCHP system with thermally compression chiller

Month	System load		PV and ST panels load		CHP unit load		Boiler load
	e_{sys} (kWh/m^2)	q_{sys} (kWh/m^2)	e_{PV} (kWh/m^2)	q_{TP} (kWh/m^2)	e_{cg} (kWh/m^2)	q_{cg} (kWh/m^2)	q_{ad} (kWh/m^2)
Jan	3.12	18.64	0.23	1.53	2.89	8.67	8.44
Feb	3.12	16.07	0.33	2.17	2.79	8.38	5.52
Mar	3.12	12.70	0.54	3.58	2.58	7.74	1.37
Apr	3.12	8.27	0.75	4.98	2.37	7.11	0.00
May	3.12	9.75	0.93	6.19	2.19	6.57	0.00
Jun	3.12	11.89	1.01	6.74	2.11	6.32	0.00
Jul	3.12	14.13	1.02	6.83	2.09	6.28	1.02
Aug	3.12	14.13	0.90	5.99	2.22	6.66	1.48
Sep	3.12	9.70	0.66	4.43	2.45	7.36	0.00
Oct	3.12	8.65	0.46	3.09	2.65	7.96	0.00
Nov	3.12	12.33	0.25	1.65	2.87	8.61	2.06
Dec	3.12	16.58	0.19	1.24	2.93	8.80	6.55
Total	**37.42**	**152.84**	**7.26**	**48.43**	**30.16**	**90.47**	**26.45**

6.1 Load Sharing

The system load should be split between suppliers, to thus find the load for each supplier. This problem is called load sharing and the solution of the problem depends on the system structure. Using data from Table 11 for monthly energy specific consumption of the mCCHP system in the case of cooling with TCC and from Table 6 for monthly specific power and heat production of the panels and building, we can determine the load sharing of the CHP unit and back-up boiler. To determine the energy produced by the cogeneration unit and back-up boiler were used the Eqs. 59–61 in Chap. "Structural Design of the mCCHP-RES System". The load sharing of the mCCHP system with TCC is shown in Table 12.

7 Evaluation and Improving of the Performance

Nicolae Badea

7.1 Performance Indicators at the System Level

The first indicator is *Percentage of Energy Saving* or *Primary Energy Save* (PES) that refers basically to the percentage of fuel saved from the energy production of the CCHP system compared to the same energy produced by the reference system.

Table 13 Fuel consumption and performance indicators

Month	System load		PV and ST panels load		Fuel	PES (%)	EFF (%)
	e_{sys} (kWh/m^2)	q_{sys} (kWh/m^2)	e_{PV} (kWh/m^2)	q_{TP} (kWh/m^2)	q_{fuel} (kWh/m^2)		
Jan	3.12	18.64	0.23	1.53	23.82	27.25	85.04
Feb	3.12	16.07	0.33	2.17	20.09	31.97	84.93
Mar	3.12	12.70	0.54	3.58	14.43	43.02	85.26
Apr	3.12	8.27	0.75	4.98	11.86	40.10	64.79
May	3.12	9.75	0.93	6.19	10.95	49.38	71.21
Jun	3.12	11.89	1.01	6.74	10.53	56.66	82.04
Jul	3.12	14.13	1.02	6.83	11.60	57.22	88.66
Aug	3.12	14.13	0.90	5.99	12.75	52.98	87.85
Sep	3.12	9.70	0.66	4.43	12.27	43.12	73.85
Oct	3.12	8.65	0.46	3.09	13.27	34.48	69.90
Nov	3.12	12.33	0.25	1.65	16.64	33.05	83.29
Dec	3.12	16.58	0.19	1.24	21.94	27.31	84.33
Total	**37.42**	**152.84**	**7.26**	**48.43**	**180.16**	**41.38**	**80.10**

Conventionally, the reference system is the one, which produces heat and power with efficiency reference values η_{Href} and η_{Eref}, respectively. PES is calculated with relation 62 in Chap. "Structural Design of the mCCHP-RES System" where were considered the efficiency reference value for heat production $\eta_{Href} = 0.8$ and the efficiency reference value for power production $\eta_{Eref} = 0.33$.

The second indicator namely EFF sets the correlation between the primary and useful energy and shows how much of the primary energy is found to be useful energy and is given by the relation 63 in Chap. "Structural Design of the mCCHP-RES System".

To calculate the two indicators PES and EFF, we need to determine the specific fuel consumption of CCHP system. Specific fuel consumption is sum of fuel input for cogeneration unit and additional boiler. Its determination involves knowledge of the cogeneration unit and boiler energy production, as well as of the cogeneration unit and boiler efficiency.

If the overall efficiency of CHP unit is 80 % and the efficiency of additional boiler is 90 %, and using the monthly data from Table 12, the results are those recorded in Table 13 for the whole CCHP system.

7.2 Performance Improving

Improving the performance indicators of the system is achieved by analyzing the specific energy production system components and through actions to reduce losses but especially by diminishing the excess energy produced. The off-grid system

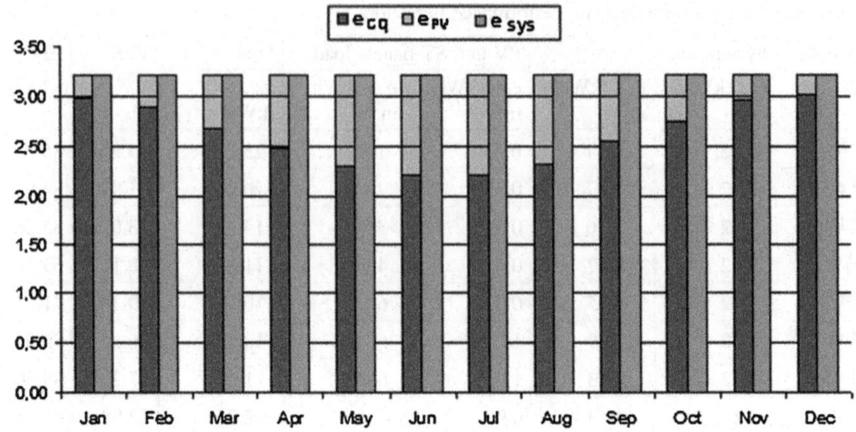

Fig. 11 Balance of electrical energy production and consumption

supplies the thermal and electrical energy for residential consumption only. During operation, the whole CCHP system control is based on balancing the electrical energy production and consumption. Control of this balance consists in appropriate modulation of the electrical energy obtained from CHP unit, so that, together with the current PV panel production, to cover the electrical energy current demands. Contributions to the energy production of the CHP unit and PV panels are shown in Fig. 11.

The balance of the thermal energy indicates excess energy in the spring and autumn seasons (Fig. 12).

To eliminate the excess of the thermal production should change the ST panel surface. In this case, the utilization factor of ST panel could be recalculated using the relation 64 in Chap. "Structural Design of the mCCHP-RES System"

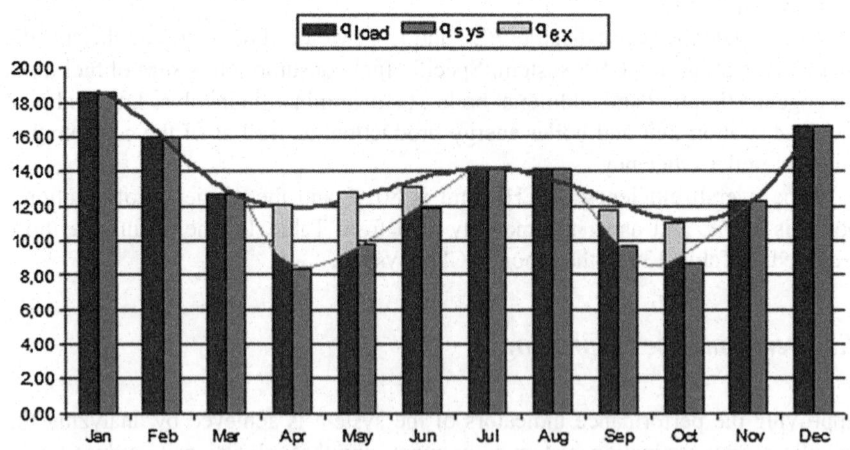

Fig. 12 Heat excess

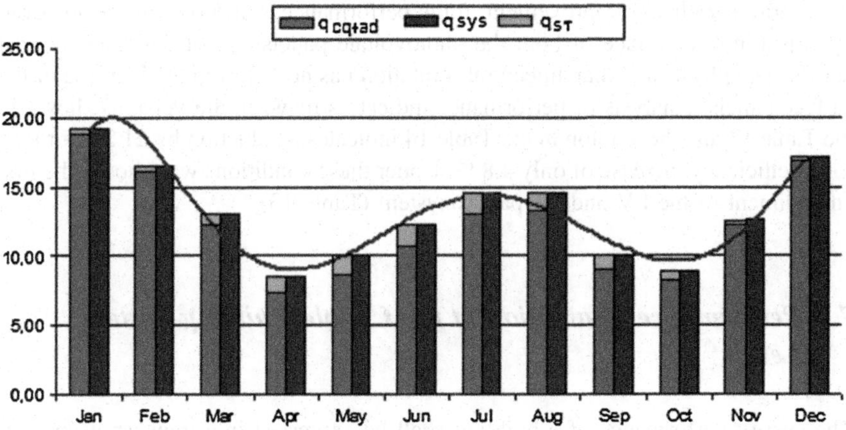

Fig. 13 Balance of thermal energy production and consumption

considering that in the month with maximum excess it is removed only by adjusting the ST panel production ($q_{ST} = q_{ST\ initial} - q_{excess}$). It results $A^r_{ST} = 4.18\,m^2$.

As consequence of changing the ST panel surface, the heat excess was eliminated as is shown in Fig. 13, where in yellow is the ST panels' contribution.

Eliminated heat excess leads to increase the system efficiency but also increase the fuel consumption of the mCCHP system and therefore reduce the PES. These observations are drawn from the comparison of data in Tables 13 and 14.

Table 14 Improving efficiency of the system

Month	System load		PV and ST panels load		Fuel	PES (%)	EFF (%)
	e_{sys} (kWh/m^2)	q_{sys} (kWh/m^2)	e_{PV} (kWh/m^2)	q_{TP} (kWh/m^2)	q_{fuel} (kWh/m^2)		
Jan	3.12	18.64	0.23	0.36	25.13	23.27	84.61
Feb	3.12	16.07	0.33	0.51	21.94	25.70	84.24
Mar	3.12	12.70	0.54	0.84	17.48	30.96	83.88
Apr	3.12	8.27	0.75	1.16	11.86	40.10	82.77
May	3.12	9.75	0.93	1.44	12.88	40.48	84.37
Jun	3.12	11.89	1.01	1.57	14.97	38.41	85.47
Jul	3.12	14.13	1.02	1.59	17.42	35.75	86.08
Aug	3.12	14.13	0.90	1.40	17.85	34.17	85.63
Sep	3.12	9.70	0.66	1.03	13.72	36.40	83.14
Oct	3.12	8.65	0.46	0.72	13.27	34.48	81.37
Nov	3.12	12.33	0.25	0.39	18.05	27.38	82.66
Dec	3.12	16.58	0.19	0.29	22.99	23.81	83.95
Total	**37.42**	**152.84**	**7.26**	**11.30**	**207.57**	**32.58**	**84.02**

A more significant improvement in the performance indicators can be achieved by adjusting the surface of both the photovoltaic panels and of the solar thermal panels. The physical and architectural limitation has not allowed the increase of the surface panels. Analysis of performance indicators between the versions shown in the Table 13 and the version in the Table 14 indicates a reduction by 21.2 % of PES for an efficiency increase of only 4.8 %. Under these conditions was chosen the first embodiment of the PV and ST panels system (Table 13).

7.3 Performance Evaluation at the Couple Building-System Level

The energy performance of a building shall be expressed in a transparent manner and shall include an energy performance indicator and a numeric indicator of primary energy use based on primary energy factors per energy carrier. The new numeric indicators of primary energy use refer to specific value for on-site production and the share of renewable energy.

Taking into account the data given in the Table 40 in Chap. "Structural Design of the mCCHP-RES System" and Table 12 in this chapter, and applying the relations from 69 to 71 in Chap. "Structural Design of the mCCHP-RES System" for the couple building-system, were obtained the values for the integrated energy performance $E_{p.tot}$ = **217.48** kWh/m^2 year and the share of renewable **RER = 25. 61** % (Table 15).

Table 15 Integrated energy performance of building

Energy carrier	Energy used (kWh/m^2 y)	Delivered energy (kWh/m^2 y)	fp. ren	Ep.ren (kWh/m^2 y)	fp. nren	Ep.nren (kWh/m^2 y)	fp. tot	Ep.tot (kWh/m^{2}y)
Electrical	37.42							
PV		7.26	1.00	7.26	0.00	0.00	1.00	7.26
CHP unit		30.16	0.00	0.00	1.10	33.18	1.10	33.18
Thermal	152.84							
ST		48.43	1.00	48.43	0.00	0.00	1.00	48.43
CHP		90.47	0.00	0.00	1.10	99.52	1.10	99.52
Back-up		26.45	0.00	0.00	1.10	29.10	1.10	29.10
Exported energy	0.00	On-site	0.20	0.00	2.30	0.00	2.50	0.00
Energy performance indicator of building			**Ep. ren**	55.69	**Ep. nren**	161.79	**Ep. tot**	**217.48**
Renewable energy ratio			**RER** (%)					**25.61**

8 Building the Functional Schemes

Nicolae Badea

For the structural model with thermally activated chiller was adopted, at the global level, the functional scheme presented in Fig. 1 in Chap. "Functional Design of the mCCHP-RES System".

8.1 Sizing of the System Components

The system components sizing consist in establishing the nominal values for the required characteristics of each component, the choice of those products which meets these characteristics of the component, and the evaluation of the daily operation time, t_{oper}, for each supplier, and the daily use time, t_{use}, for each consumer, taking into account:

- the monthly quantities of energy resulting from any kind of evaluation, whether it refers to consumption or load, and whether it refers to the main components or to the complementary, and
- the nominal technical characteristics of the components available in the market.

8.1.1 Electric Subsystem Components

In the house the electric installations have installed power $P_n = 12$ kW. Because the Romanian standard provides for power greater than 5 kW three phase distributions, it was chosen three phase and three mono-phase inverters. The electric subsystem scheme is similar to that shown in the Fig. 3 in Chap. "Functional Design of the mCCHP-RES System" while the micro CCHP system works in standalone mode (off-grid type).

A. *Photovoltaic panel*

The starting data for sizing the PV panel include the specific PV production in the month with the biggest PV production (July) and the medium sunshine time at the nominal power of the PV panel, corresponding to the same month.

The PV panel voltage, $V_{CC} = 48$ V, is chosen according to the recommendation given in Table 3 in Chap. "Functional Design of the mCCHP-RES System", the inverters input/output capabilities, and the consumers' voltage.

Using data from the Tables 6 and 12 and the Eqs. 1 to 3 in Chap. "Functional Design of the mCCHP-RES System", results for PV panels the values given in Table 16, where reduction factor which accounts for cloudy days of the month, k = 0.8, has been adopted.

Then, from the available products in the market [8] are chosen a solar module with $P_{SMn} = 190$ W nominal value of the maximum power and having $V_{SMn} = 36.6$ V nominal value of the voltage and operating current $I_{np} = 5.2$ A. The topology of

Table 16 Calculated values for sizing the PVs

July horizontal solar irradiation (Wh/m^2/d)	Sunshine time $t_{sunlight}$ (h)	Specific PV production $e_{PV}\left(kWh/m^2_{TFA}\right)$	Monthly energy produced by the PV panels $E_{PV_{max}}(kWh)$	Daily produced by the PV panels $E_{PV_{day}}(kWh)$	Power of the PV panels $P_{PV_{panel}}(kW)$
6,560	6.5	1.02	304.98	12.29	1.89

photovoltaic panel is established according to the power and voltage of both panel and its modules. The number of modules installed, $n_{SM} = 10$, has to be the whole even number, superior to the number resulted from the relation 5 in Chap. "Functional Design of the mCCHP-RES System" and the number of modules connected in series, is given by relation 6 in Chap. "Functional Design of the mCCHP-RES System". It results $n_{SMs} = 2$. Using the relation 7 in Chap. "Functional Design of the mCCHP-RES System" results the number of rows of modules, $n_{SMp} = 5$ connected in parallel. In this way the topology of the PV panels is established.

B. *Batteries capacity*

For dimensioning the batteries is necessary to determine the energy stored in the batteries (E_{need}), based on the condition imposed, namely that in the month with the highest consumption of electricity (February) when neither working the CHP unit and nor the PV panel. Choosing 1 day (24 h) as the $N_{day\ need}$ and using data from Tables 12 and Eq. 8 in Chap. "Functional Design of the mCCHP-RES System", it result $E_{need} = 30$ kWh.

Battery voltage is imposed to voltage DC network $V_{bateries} = V_{cc}$ and choosing the battery voltage $V_{battery} = 12$ V with $C_{battery} = 250$ Ah [9], results the number of the batteries connected in series $n_1 = 4$.

To determine the capacity of the batteries' system is used the Eqs. 10 and 11 in Chap. "Functional Design of the mCCHP-RES System" where ND = 0.7 and RT = 1 was used. Thus, we obtain $C_{batteries} = 625$ Ah and $C_{20} = 892$ Ah.

Were used the Eq. 12 in Chap. "Functional Design of the mCCHP-RES System" to determine the numbers of battery rows n_2 and Eq. 13 in Chap. "Functional Design of the mCCHP-RES System" to determine the number of batteries, resulting $n_2 = 4$ and $n_{batteries} = n_1 \cdot n_2 = 16$ where $\frac{C_{20}}{C_{battery}} = 3.57$. In this way results the total capacity of the batteries' rows of 1,000 Ah.

C. *Solar charge controller*

The PV panel is connected to batteries with the solar charge controller. The solar charge controller regulates battery voltage and output current based on the power available from the PV panel and the state of charge of the battery. The solar charge controller automatically charges the batteries in an optimal way with all the available solar power using its algorithm for Maximum Power Point Tracking (MPPT). The MPPT continuously seeks out the solar generator's optimal voltage to retrieve the maximum available energy. This operating point varies depending on

Table 17 Solar charge controller

Battery voltage (V)	Maximum power of the PVs (W)	Maximum voltage of the PVs (V_{DC})	Max. charging current to the batteries (A)
12	1,000	80	65
24	2,000	150	
48	4,000	150	

outdoor conditions (sunlight, temperature, etc.) to which it must adapt. According to the battery voltage $V_{cc} = 48$ V, its selection involves choosing:

- input voltage $V > 48$ V,
- input power in solar charge controller higher than maximum power of the PV panel at maximum voltage,
- maximum charging current to the batteries $I = 65$ A.

Then, from the available products in the market (Table 17) are chosen the solar charge controller [10] with 65 A and 150 V_{DC}.

D. *Cogeneration unit*

For sizing the CHP unit, using the Eq. 14 in Chap. "Functional Design of the mCCHP-RES System" and taking from the Table 12, the value of the specific energy in the month with the biggest load of the CHP unit was obtained $P_{CHP} = 5.65$ kW, where: daily time of operation was chosen $t_{oper} = 5$ h in correlation with the estimated time for battery recharging (Table 18 in Chap. "Structural Design of the mCCHP-RES System"). After that, was chosen from the market a CHP unit with nominal power $P_{CHP_n} = 9$ kW that satisfies the relation: $P_{CHP_n} \geq P_{CHP}$. Combined heat and power unit with Stirling engine chosen [11] has the following characteristics:

- V-stirling engine, alpha: 2-cylinder;
- Displacement: 160 cm ^{3};
- Operating gas: Helium;
- Maximum operating pressure: 150 bar;
- Work rated speed: 1,500 rpm;
- Produced electrical power: 2–9 kW (3 × 380/50 Hz);
- Produced thermal power: 8–26 kW;
- Overall efficiency: 92–96 % for 50–100 % load;
- Electrical efficiency: 22–24.5 % for 50–100 % load;
- Power to heat ratio: 0.34;
- Cooling water temperature 65 °C;
- Flow rate in the outer system 1–2 m^3/h.

E. *Inverters*

Inverters connect the batteries to the electrical appliances and convert DC voltage to AC voltage at 50 Hz. Also, they should allow the CHP unit to supply residential consumers and also to load the batteries.

Table 18 Inverter technical data

Output voltage	Battery voltage	Charge current	Maximum current	Power
230 V_{AC}/50 Hz	48 V_{DC}	50 A_{DC}	50 A_{AC}	4,000 VA

For sizing the inverters the following conditions were satisfied:

- Rated output capacity of inverter $P_{output} > P_{CHP}/3$, resulting $P_{output} = 3$ kW;
- Input voltage inverter $V_{input} = V_{CC} = 48$ V (equal with the rated batteries voltage);
- Rated output voltage $V_{output} = 220 \sim 230$, V_{AC} 50 ± 0.05 Hz, THD < 4 %.

The inverters chosen have perfect sine wave output [12] with technical data shown in Table 18 and the schematic diagram presented in the Fig. 14.

In addition to the output wave type, the main features of the inverter are:

- Inverters have a higher efficiency near the rated power so that the operating recommendation is close to the nominal load. Usually, the manufacturer gives the inverters performance to a 70 % load of rated power;

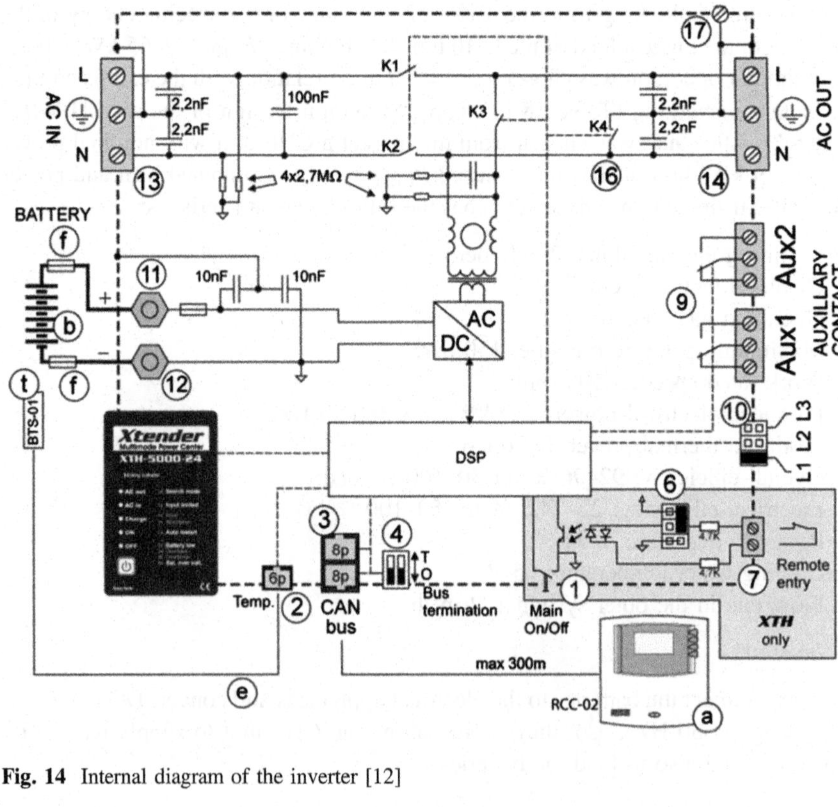

Fig. 14 Internal diagram of the inverter [12]

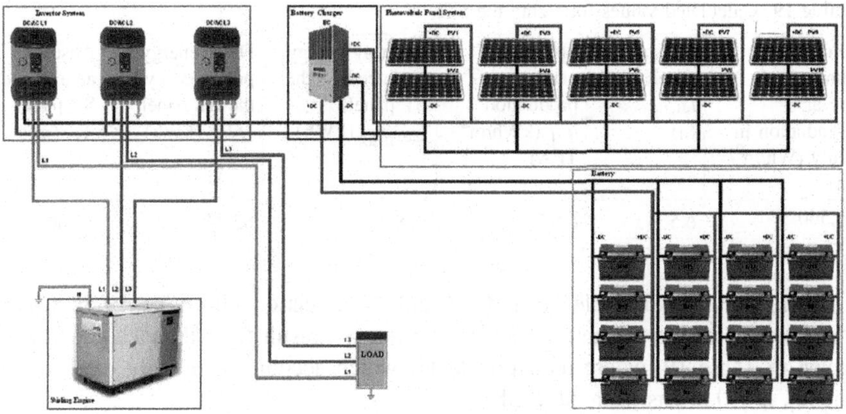

Fig. 15 Functional electric diagram of the experimental mCCHP system [17]

- Possibility of charging the batteries from several alternative current sources (network, generator, beside PV panel) by automatic switches between these different sources;
- Reliability in the overvoltage and over current presence (overvoltage and over current protections);
- Allows parallel connection for extend to three phase.

The functional electrical diagram of the experimental mCCHP system is shown in Fig. 15.

8.1.2 Thermal Subsystem Components

First, the ST panel and back-up boiler must be dimensioned. After that, the hydraulics and the pumps will be dimensioned. The third source is the Stirling engine which was dimensioned in electrical subsystem. The Stirling engine has an internal cooling system with a pump and expansion vessel, and a plate heat exchanger. The second circuit which makes connection between the plate heat exchanger and the storage tank (outer system) must be dimensioned. The flow rate in the outer system (aided by the pump which runs continuously when the CHP is operating) is 1–2 m^3/h, resulting in a spread of 20–10 °K at maximum load. Finally, will be dimensioned the heat storage tank and the expansion tank. All hydraulic circuits must be protected at over temperature and pressure.

A. *Solar thermal panel*

The starting data for sizing the ST panel include the specific ST production in the month with the biggest PV production (July) and the medium sunshine time corresponding to the same month.

Table 19 Calculated values for sizing the STs

July horizontal solar irradiation in July (Wh/m^2/d)	Sunshine time, $t_{sunlight}$ (h)	Specific ST panel production, q_{ST} (kWh/m^2 TFA)	Monthly energy produced by the ST panel, $Q_{ST_{max}}$ (kWh)	Daily energy produced by the ST panel, $Q_{ST_{day}}$ (kWh)	Absorber area of the ST panels, A_{ST} (m^2)
6,560	6.5	6.83	2,042.1	82.34	12.66

Using data from Tables 6 and 12 and the equations from 16–19 in Chap. "Functional Design of the mCCHP-RES System", result for ST panels the values given in Table 19, where reduction factor which account for cloudy days of the month, k = 0.8, has been adopted.

From the product data sheet [13] was chosen direct flow solar collectors with 3.02 m^2 absorber area and 0.77 optical efficiency (solar collector contains 30 tubes number). Applying the Eq. 20 in Chap. "Functional Design of the mCCHP-RES System", with $\frac{A_{ST}}{\eta_{ST_o} \cdot A_{ST_n}} = 5.44$, was determined the number of the ST collectors and were chosen $n_{ST} = 6$. The heat flow $\dot{Q}_{day}$ was calculated with relation 18 in Chap. "Functional Design of the mCCHP-RES System" and results $\dot{Q}_{day} = 12.66$ kWt.

B. *Back-up boiler* (*additional boiler*)

For sizing the back-up boiler, using the Eqs. 22–24 in Chap. "Functional Design of the mCCHP-RES System" and the value of the specific energy in the month with the biggest load of the boiler (given in the Table 12) was obtained $\dot{Q}_{ad} = 20.35\,\text{kW}$, where daily time of operation was chosen $t_{oper} = 4\,\text{h}$. After that was chosen from the market a pellet boiler [14] with heat output range $\dot{Q}_{ad} = 14 - 49$ kWt, (that satisfies the relation $\dot{Q}_{ad_n} > \dot{Q}_{ad}$) and fully automated pellet feed system. The pellet supply system transports the pellets from the storage facility to the boiler hopper by suction system. The technical data of the boiler are shown in Table 20.

C. *Chiller*

The amount of heat required to power the chiller is calculated by subtracting the DHW consumption from the total consumption of the hottest month ($q_c = q_{sys} - q_{DHW}$).

For sizing the chiller, have been used the Eqs. 26–28 in Chap. "Functional Design of the mCCHP-RES System", where the value of the specific energy in the month with the biggest load of the chiller (July) is that given in the Table 11, the

Table 20 Technical data of the boiler

Heat output range (kW)	Boiler efficiency (%)	Operating/test pressure (bar)	Water flow at, $\Delta T = 10$ °K (m^3/h)	Maximum boiler flow temperature (°C)	Minimum boiler operating temperature (°C)
14–49	>90	3.0/4.5	4.2	80	60

Table 21 Technical data of chiller sizing

Specific heat need for input chiller (kWh/m^2 month)	Monthly energy amount (kWh/month)	Daily energy amount (kWh/d)	Power of the chiller (kW)
12.95	3,872.05	156.13	15.61

daily time of operation was chosen of 10 h and the reduction factor for cloudy days of the month is $k = 0.8$. The daily operation time of the chiller at the nominal value of the cooling power in the month with the biggest load of the chiller (July) was determinate as the ratio between the average monthly sunshine duration and number of days (ds = 307 h/month from Table 4). It results for chiller the values given in the Table 21.

After that was chosen from the market an adsorption chiller [15] with cooling heat output capacity $\dot{Q}_{\text{cooling}_n} = 15\,\text{kW}$ cooling.

Absorption chiller unit [15] has the following features:

- Cooling nominal capacity: 15 kW;
- Cooling max.nominal capacity: 23 kW;
- Nominal coefficient of performance COP: 0.6;
- Work agent: water-silicagel;
- Hot water I/O temperature: 72/65 °C;
- Volume flow: 3.2 m^3/h;
- Cold water out temperature: 6–15 °C;
- Provided flow: 4.0 m^3/h.

For cooling storage was chosen a buffer tank with 500 l capacity.

D. *Heat storage*

The starting data for sizing the heat storage tank include the specific residence consumption in the month with the biggest system demand. For residence location, the maximum heat consumption is in January and the maximum cold demand is in July.

The minimum possible temperature of the water in tank is a decisive variable for determining the required tank volume. The value adopted was 10 °C. The maximum temperature was adopted as 90 °C. In these conditions, using data from Table 12 and the equations from 30 to 33 in Chap. "Functional Design of the mCCHP-RES System", it results for heat storage tanks the values given in the Table 22. From the product data sheet [16] was chosen the hot water tank with 2,000 l of water volume of storage with sanitary tank inside.

E. *The hydraulic circuits*

The sizing of a hydraulic circuit consists first in placement of all elements which must contain pumps, valves, protection elements, connection elements.

Then all circuits are connected in the hydraulic scheme (Fig. 16) and are positioned within the building.

Table 22 Technical data of heat storage

	Specific system consumption, q_{sys} (kWh/m^{2}TFA)	Monthly energy consumption, $Q_{demand_{max}}$ (kWh)	Daily energy consumption, $Q_{demand.day}$ (kWh)	Mass of the tank fluid (kg)	Volume of the storage tank (liter)
Heating	18.64	5,573	179	1,932	2,000

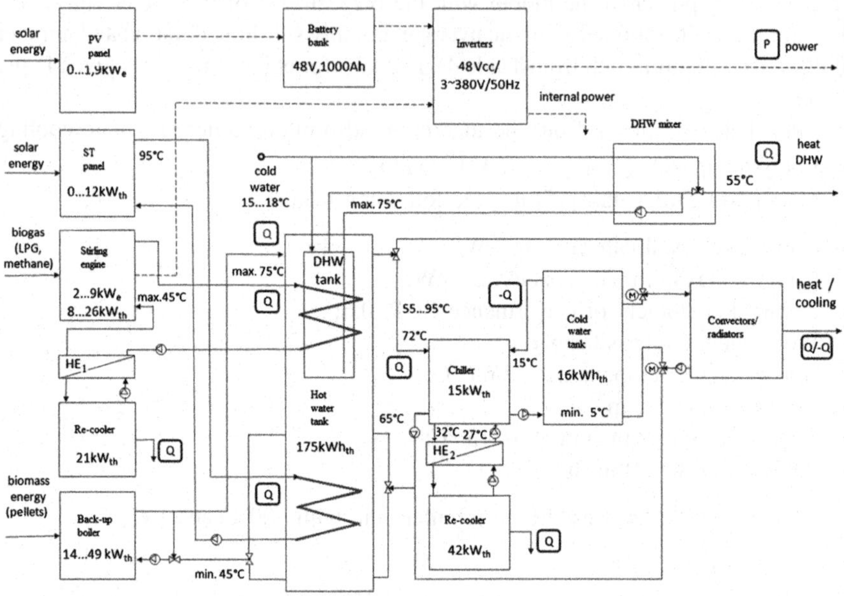

Fig. 16 Hydraulic scheme

In this way we can identify the number of bends, fitting T-junctions between pipes, and real lengths of pipelines for each hydraulic circuit to determine the pipe diameter and pressure loss.

The main components of the circuits are:

- separation cocks, which ensure the separation of the active elements to allow local intervention to the device without affecting the other equipment;
- recirculation pump;
- mixing valve, used to control the start temperature and to reduce the duration of the transition mode;
- division valve, through which the work area in the hot water tank is chosen, depending on the temperature of the water at the bottom of the tank;
- separation valve with the role of select hot or cold water to supply the coil convectors.

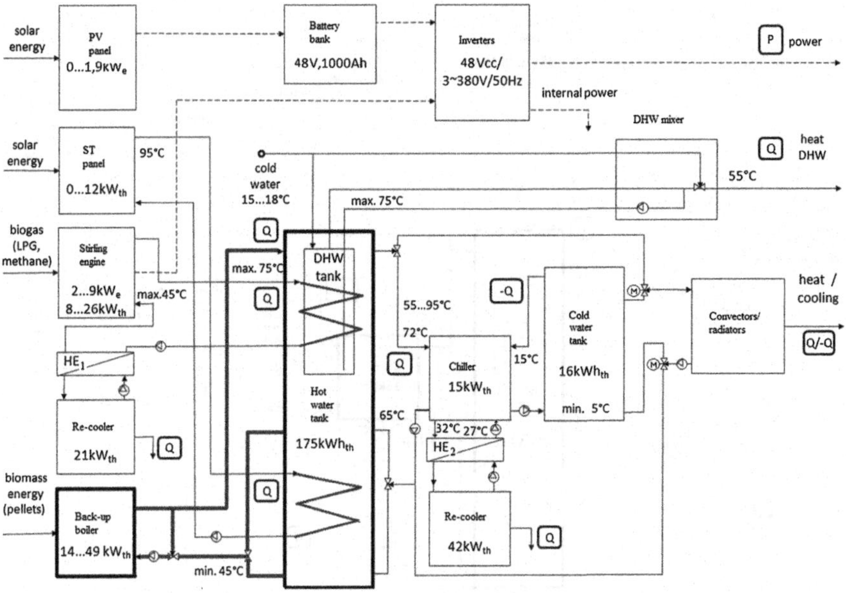

Fig. 17 Hydraulic circuit of pellet boiler (CH1)

To protect from overheating the Stirling engine and to operate the thermal chiller two additional cooling circuits have been used in the hydraulic scheme. They are connected to the equipment through heat exchangers. All circuits have safety valve (not shown in Fig. 16), with the role of protection from the increase of pressure in the circuit over the maximum calibrated value. Thus was obtained the 12 hydraulic circuits, which will be described below.

Circuit of hot water from pellet boiler to hot water tank (CH1)

This hydraulic circuit connects the pellet boiler to the hot water tank (Fig. 17).

The heat resulting from the burning of the pellets is taken up with the aid of an internal heat exchanger (inside of the boiler) and sent to the tank. The thermal agent used is pressurized water nonharsh.

In this hydraulic circuit besides the recirculation pump there are a mixing valve, used to control the start temperature of the tank (to reduce the duration of the transition mode and thus reduce the quantity of ash and smoke sent toward the chimney) and a three-way valve, through which the work area in the hot water tank is chosen, depending on the temperature of the water at the bottom of the tank (at different stratified temperatures values of the tank).

Circuit of hot water from heat storage to the chiller (CH2)

The chiller has three outlets in its inner structure, all of which are coupled thermally:

- High temperature input (HT),
- Medium temperature rejection output (MT),
- Low temperature output (LT).

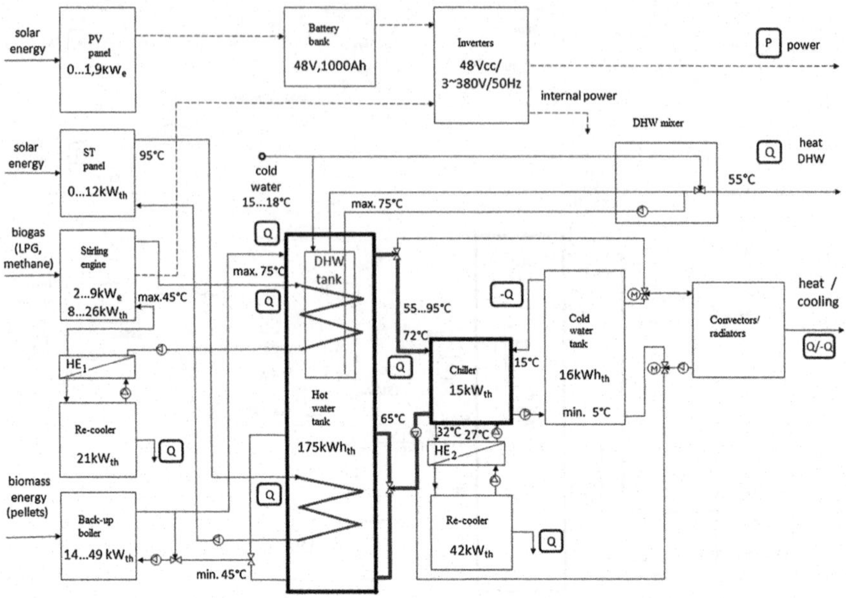

Fig. 18 Hydraulic circuit to supply chiller (CH2)

The hydraulic circuit CH2 connects the hot water tank with the HT input of the chiller, which is of the adsorption type.

Figure 18 presents the position of the CH2 circuit in the functional scheme of the system.

The chiller is coupled with the CH3 circuit of heat discharge through the re-cooler and with the CH4 circuit for cold storage. At the chiller entrance the range of the temperature is 60–95°C.

Circuits of the chiller for heat discharge (*CH3 and CH11*)

The structure of the circuits is presented in Fig. 19. The heat taken up from the spaces of the residence, multiplied by $1/COP_a$, must be discharged outside.

To this end, the chiller is connected to the medium temperature (MT) outlets with the aid of a hydraulic circuit (CH3), which contains a pump and a safety system. Since the exterior radiator is outside the building, an exterior CH11 circuit was introduced, through which water with antifreeze circulates to prevent freezing. The CH3 and CH11 circuits are thermally coupled through a heat exchanger with plates (HE2).The CH3 hydraulic circuit achieves the actual connections between the MT outlet and the HE2 heat exchanger.

Circuit of cold water from the chiller to the cold water tank (*CH4*)

The chiller is the equipment which generates cold water with a minimum temperature of 5 °C, stored in the cold water tank. The CH4 hydraulic circuit makes the connections between the LT outlet of the chiller and the cold water tank. Passive circuit elements used in CH4 are: separation valves, filter for impurities, temperature indicator, pressure indicator, one-way valve, airing valve, and safety valve. The structure of the circuit is presented in Fig. 20.

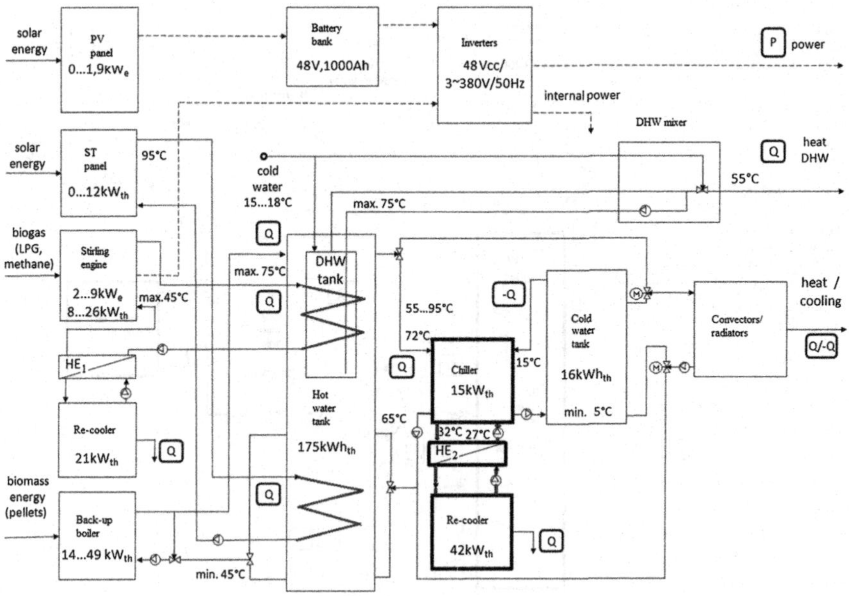

Fig. 19 Hydraulic circuits for discharge the heat of chiller

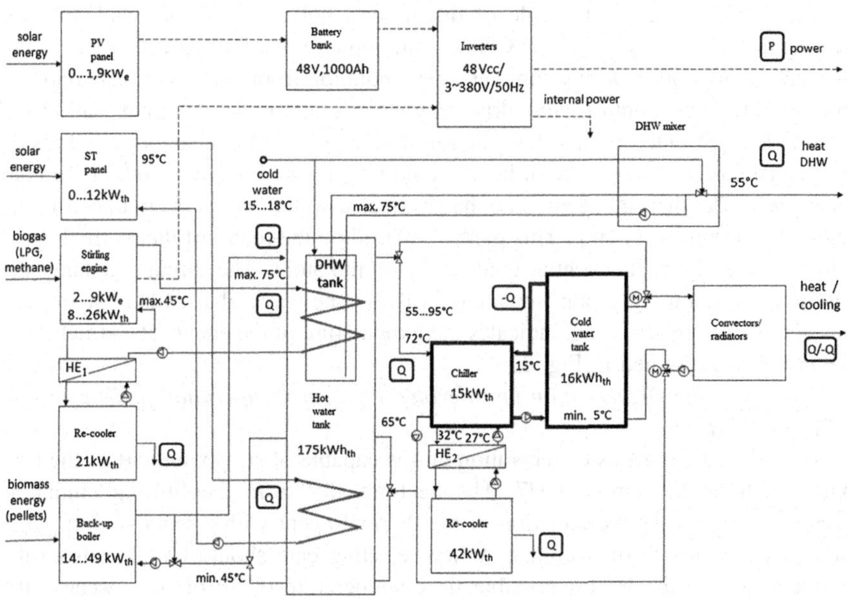

Fig. 20 Low temperature hydraulic circuits of the chiller (CH4)

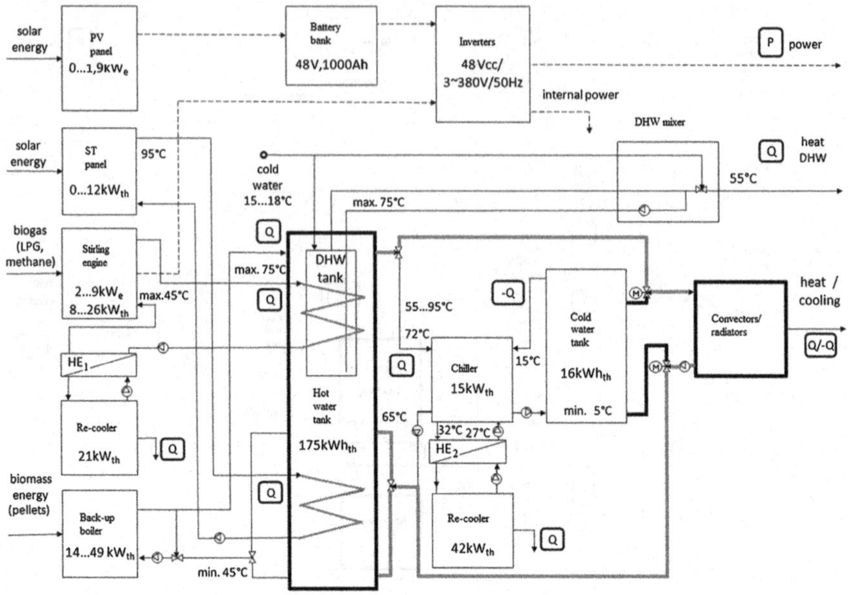

Fig. 21 Thermal agent distributions in residence

Circuit of hot/cold water distribution to the residence (*CH5/CH6*)

The CH5 circuit has the role of distributing hot water to the residence (for heating during winter), while the CH6 circuit—that of distributing cold water to the residence (for cooling during summer). Commutation from hot/cold water supply is through three-way control valve driven by electric motor. The thermal load of the circuits is represented by a system of ventilo-convectors and heaters placed inside the heated spaces. The source of heat or cold may be the hot water tank or the cold water tank. The thermal agent is nonharsh, treated water, which is circulated in the hydraulic circuit by a pump. This pump is placed on the return of the thermal agent which comes from the ventilo-convectors or radiators. The circuit contains the following: a pumping group, separation valves, one-way valves, filter for impurities, and thermometer which indicates the temperature of the agent. The structure of the circuit is presented in Fig. 21.

Circuits of the cogeneration unit devoted for heat recuperation and protection (*CH7 and CH8*)

The Stirling engine as cogeneration unit is capable of supplying part of the heat with the hydraulic circuit CH7. The heat transfer from the Stirling engine is achieved through the recirculation of the thermal agent with a recirculation pump located inside the Stirling engine. When selecting care should be taken from the outset to ensure the lowest possible flow temperature (CHP outlet), because the electrical efficiency depends on the cooling water temperature. This should not exceed 65 °C. For this reason the system must automatically vent to the outside the heat. The CH8 circuit has the role of rejecting the heat taken up from the CH7 and

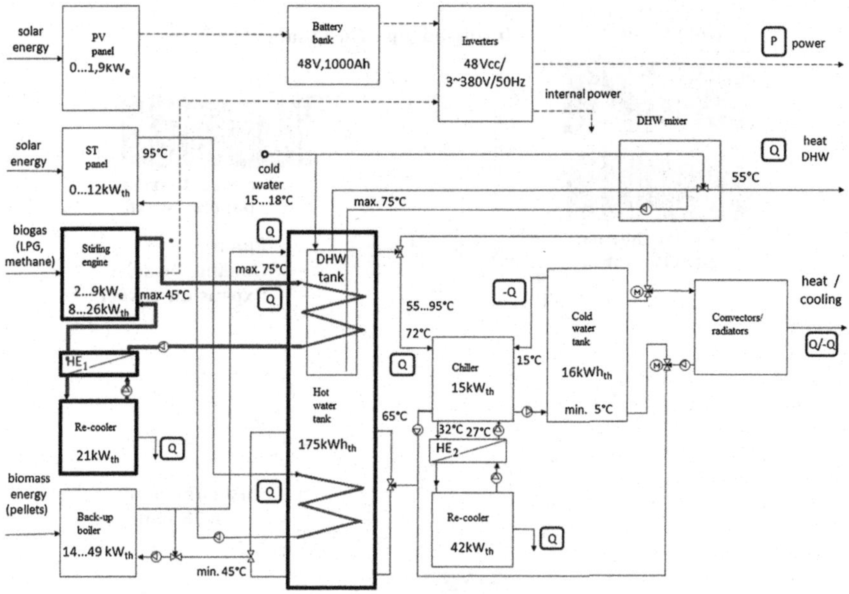

Fig. 22 Stirling heat recuperation and protection hydraulic circuit

of thus maintaining the efficiency of the Stirling engine. Circuit CH7 is placed inside of the residence. Circuit CH8 has heat exchanger (HE_1) located inside of the residence while the re-cooler is located outside. The thermal load is represented by a heat sink, fitted outside the system, which has the role of reducing the temperature of the thermal agent to 45 °C. The heat sink is cooled through forced convection with a ventilator, so as to dissipate the heat flux. Figure 22 presents the position of the CH7 and CH8 circuits in the functional scheme of the system.

Circuits of hot water from ST panel to heat storage (*CH9.1 and CH9.2*)

Hydraulic circuit for obtaining heat from solar energy and storing it in the hot water tank (CH9) has six solar panels of the direct flow type (Fig. 23).

Placing the solar panels on the roof of the building was done according to the available space; that is why two groups were formed, one of four panels and one of two panels. Each group has its own pumping group. The CH9.1 and CH9.2 circuits are united into a single circuit which goes to the inferior coil of the hot water tank. The hydraulic circuit is characterized by the use of a soft water solution and mono-ethylene glycol as thermal agent, so that the normal liquid water use limits increase in the range of negative temperatures (during winter), but also in the range of temperatures exceeding 100 °C (during summer).

Circuit of DHW (*CH10*)

The DHW circuit is fuelled with energy from the DHW tank. The amount of water is sufficient for the daily consumption of the residence.

This water is heated through conduction from the water of the hot water tank. The continuous refreshing of the water in the DHW tank presupposes resolving the

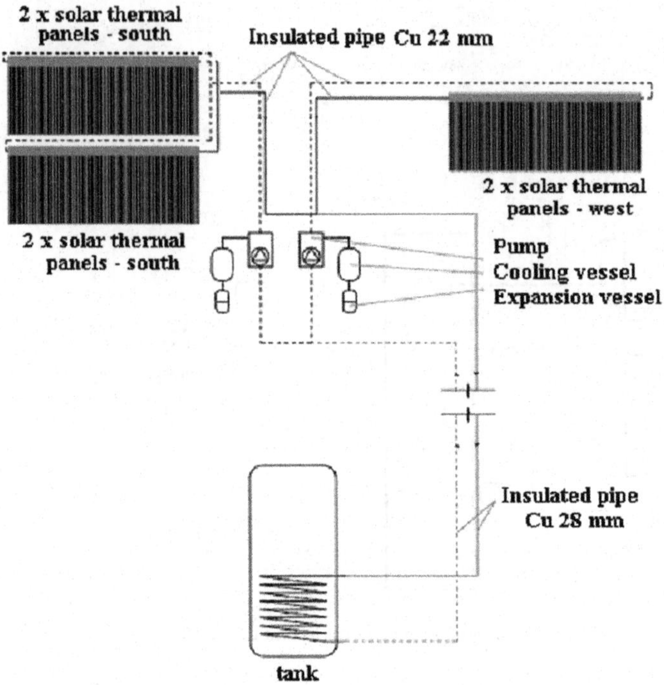

Fig. 23 Solar thermal panels connecting scheme

problem of calcareous residues with negative effects on the heat transfer. In the case of this tank, the method of the soluble magnesium anode is used.

The thermal load of the circuit is represented by DHW consumption in the adequate spaces (kitchen, baths). The consumption is variable and requires an adequate replacement of the consumed water volume with fresh tap water. The CH10 circuit (Fig. 24) is made up of two parts. One part is included in the residence and complies with the construction and exploitation rules of the domestic hot and cold water distribution systems. Another part is included in the boiler appliance, located in the basement.

The recirculation pump has the role of maintaining constant temperature in the pipes, so that the time interval between opening a consumption valve and reaching the DHW optimal temperature is the lowest possible. The circuit is equipped with separation valves in order to allow easy interventions, and elements belonging to the pumping and safety groups: filter, indicating manometer, indicating thermometer, one- way valve, ventilation valve, safety valve.

The following variables represent the operating parameters of each hydraulic circuit when it is intended to carry thermal energy:

$\dot{Q}$—Heat flow (kW);
$\dot{V}$—Volume flow (m^3/h);

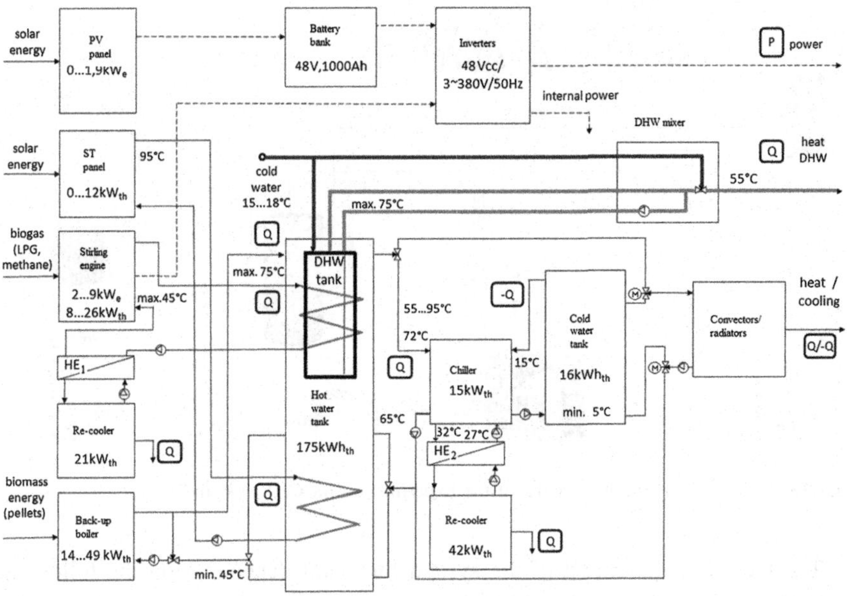

Fig. 24 DHW hydraulic circuit scheme (CH10)

$\Delta\vartheta$—Temperature difference (°C);
ρ—Density of thermal fluid (kg/m^3);
Ho—Pressure of the thermal fluid (mbar);
c—Specific heat capacity of the thermal fluid (kWh/kg °C);
v—fluid velocity (m/s).

The results obtained by applying the relations from 38 to 45 in Chap. "Functional Design of the mCCHP-RES System" for hydraulic circuits sizing, are given in the Table 23.

The connection of all circuits in a hydraulic scheme is shown in Fig. 25.

Table 23 Hydraulic circuits features

Circuits	CH1	CH2	CH3, CH11	CH4	CH5, CH6	CH7, CH8	CH9	CH10
DN (mm)	32	30	40	40	40	25	25	20
Volume flow (m^3/h)	4.2	3.2	7.0	4.0	6.0	4.0	1.5	1
Delivery head (m)	25	12	12	12	12 (2 rotors)	12	12	6
Power consumption (W)	170	20–70	25–345	20–190	25–450	190	90	70

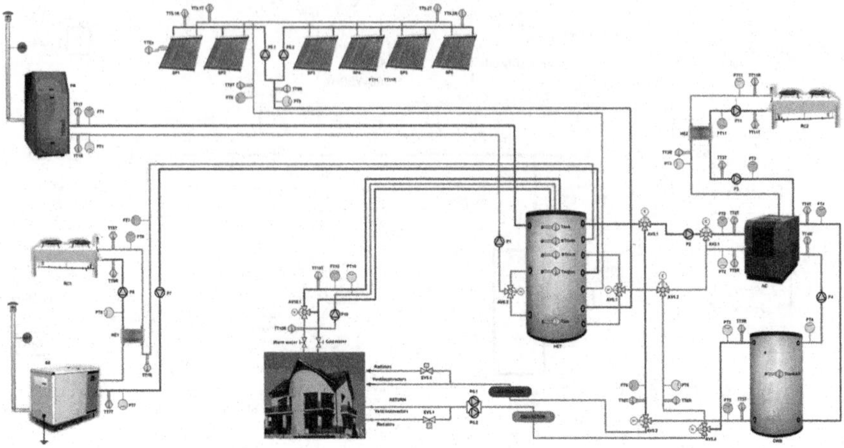

Fig. 25 Functional hydraulic circuits of the experimental mCCHP system

The thermal requirement of the system fluctuates depending on the following factors: user behaviors and comfort, extraneous heat influence, influence of hydraulic control devices (three-way valve of the ventilo-convectors).

Any reduction of volume flow in convectors or convector closed leads to higher return temperature. The volume flow conveyed through a circulating pump is depending on the thermal output/cooling output requirement of the system being supplied. For this reason was necessary the control of the volume flow conveyed through circulating pumps. The controls consist in to adapt the pump performance (and thus the power consumption) continuously to the actual requirement/demand. The control mode adopted was $\Delta p - T$, where the power electronics circuitry varies the set point differential pressure value, to be maintained by the pump as a function of the measured fluid temperature.

9 System Operating and Control

Nicolae Badea

System operating for mCCHP system is off-grid type. In this case the Stirling engine is in electricity-driven operating mode. The principle applied in the control strategy consists in using the voltage of the electrical energy accumulator and the temperature of the heat accumulation tank, as values sensitive to the misbalance between the produced power and the consumed one. In the case of both accumulations, if the power produced is less than the consumed one, then the electrical/thermal potential decreases and vice versa. The control system must maintain at constant (nominal) values the capacities through which the electrical/thermal potential (voltage and temperature) are evaluated, by adjusting the produced power.

Thus, the equilibrium between production and consumption is achieved. It has to be mentioned the fact that the Stirling engine is driven by the control loop of battery voltage and the fact that the back-up boiler is driven by the control loop of temperature tank. Two types of control methods could be adopted:

(a) using a PI continuous controller with PWM control. It ensures a smooth variation of the system variables;
(b) control on/off that has the advantage of simplicity in implementation.

Further the first control solution has been taken into consideration in both operating regimes (winter and summer).

10 System Dynamics Analysis

Marian Barbu

10.1 System Dynamics Analysis in Winter Regime

For simulation of the whole system, the scheme presented in Fig. 41 in Chap. "Functional Design of the mCCHP-RES System" was modified by taking into account the equipment characteristics mentioned above. The system simulation was done over a 3-days interval.

The operating conditions are given by the following mandatory requirements:

- the thermal power consumed in the domestic water circuit is that given by Fig. 26;
- the consumed electrical power (electrical load) is that shown in Fig. 27. The evolution of this power is the result of the useful consumed electrical power over a 24-h interval, to which, in permanent regime, the following consumptions were added: the consumption of the pump for the hot water circulation from the hot water tank to ventilo-convectors, the consumption of the Stirling engine heat exchanger, and the consumption of the circulation pump of the thermal agent from Stirling engine to the hot water tank. The consumption of the pellet boiler,

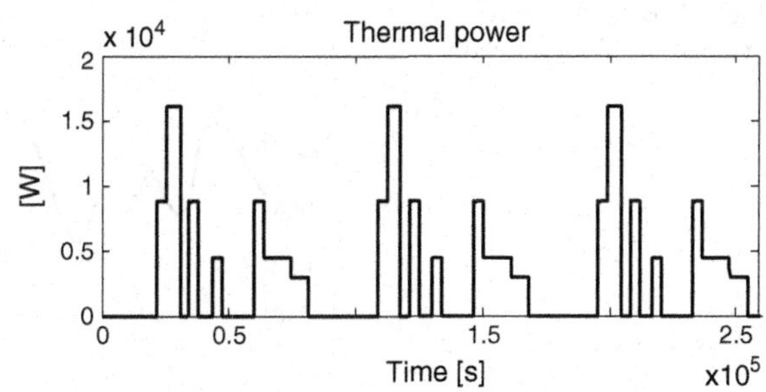

Fig. 26 Thermal power consumed in the domestic water circuit

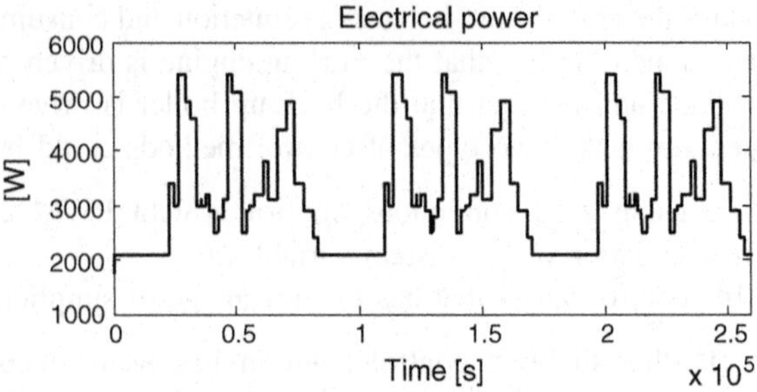

Fig. 27 Total consumed electrical power in Winter Regime

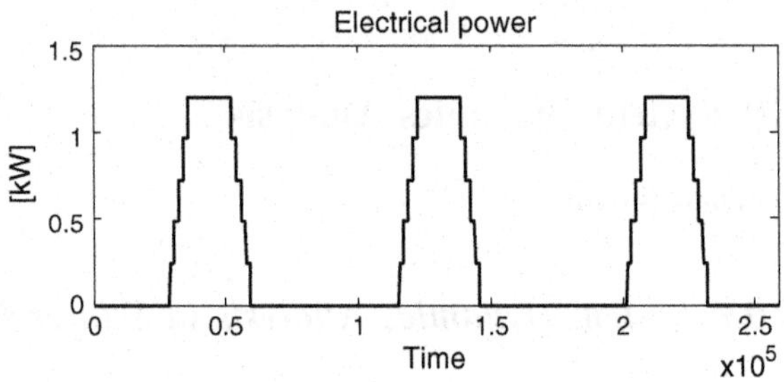

Fig. 28 Profile of the power generated by PV panel

the consumption of the circulation pump of the thermal agent from the pellet boiler to the hot water tank, and the consumption of the circulation pump for the DHW are taken into account only when they start to operate;

- the electric power of the PV source is that presented in Fig. 28.

Figure 29 presents the evolution of the battery voltage while the Fig. 30 presents the energy accumulated in the battery. It can be noticed that, after a transient regime, these evolutions practically indicate a permanent regime. That means the electrical energy sources cover the consumption if there is (enough) energy accumulated in the battery. Figure 31 shows the variation of the electrical power generated by Stirling engine. It can be noticed its continuous operating regime according to the producer specifications (no more than one stop per day is recommended). The ratio between the thermal and the electrical power of Stirling engine is 3:1.

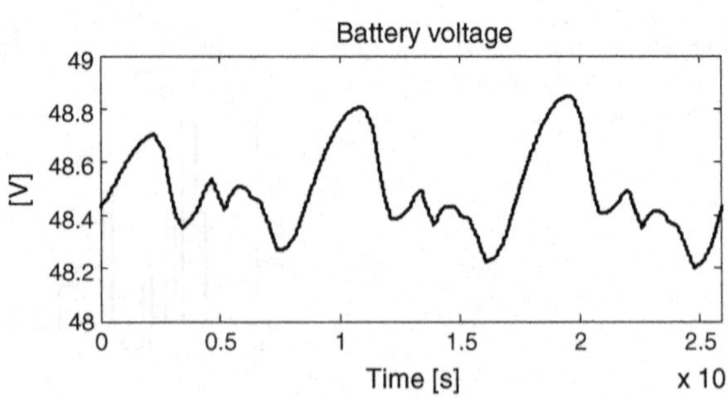

Fig. 29 Battery voltage evolution in Winter Regime

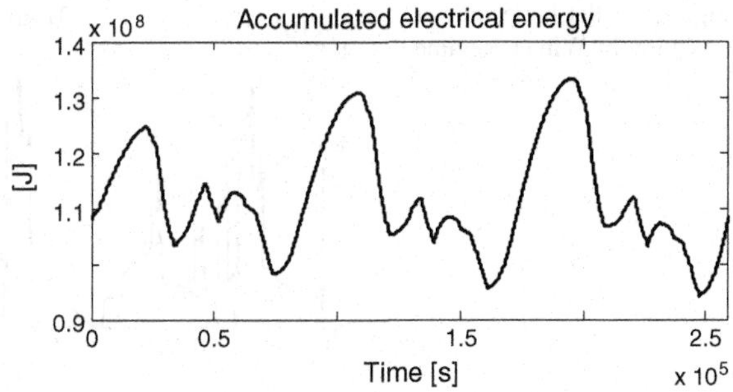

Fig. 30 Accumulated electrical energy evolution in Winter Regime

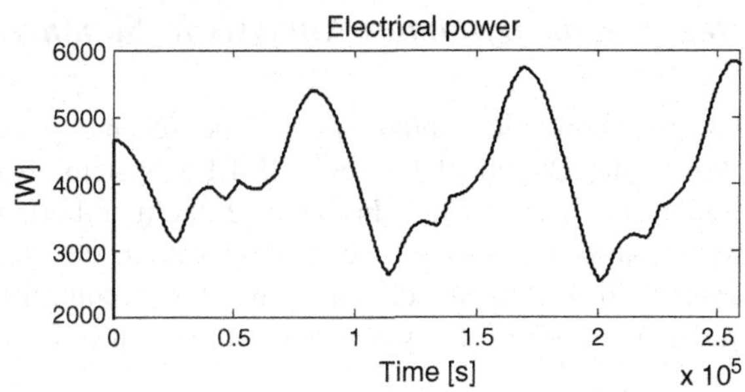

Fig. 31 Evolution of the electrical power produced by Stirling engine

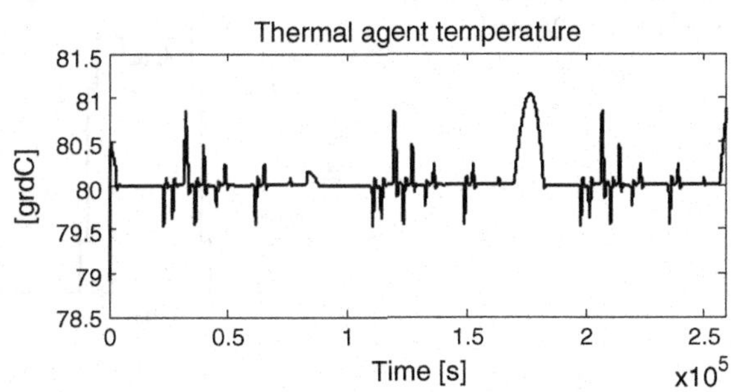

Fig. 32 Thermal agent temperature evolution in the hot water tank in Winter Regime

Figure 32 presents the performances of the temperature control system of the thermal agent in the hot water tank (the set point is equal to 80 °C). The figure shows a very good behavior of the temperature control loop, the variations from the set point being smaller than 1 °C. The temperature variations are determined by the following: the variation of the power in the domestic water circuit and the variations of thermal power produced by Stirling engine. The temperature control of the thermal agent is achieved through the control of the pellet boiler (Fig. 33).

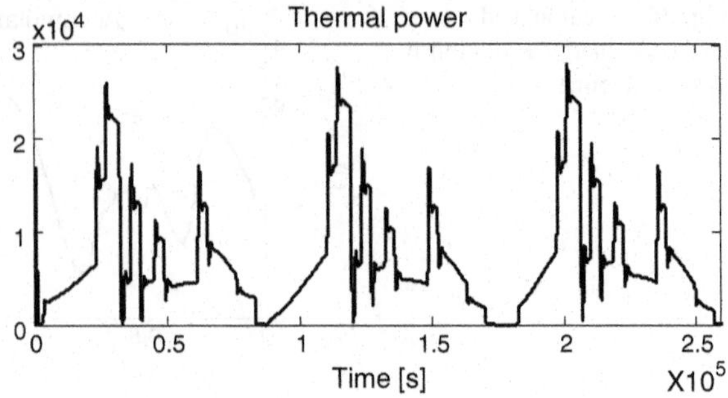

Fig. 33 Pellet boiler power evolution in Winter Regime

10.2 System Dynamics Analysis in Summer Regime

For simulating the whole system, the scheme presented in Fig. 53 in Chap. "Functional Design of the mCCHP-RES System" was modified considering the equipment characteristics. In this case, the time horizon for the system simulation was 3 days. The sources which discharge in the electric subsystem are the following: Stirling engine, driven by the voltage controller, and PV panel. The power variation graphs of the two sources are shown in Figs. 34 and 35.

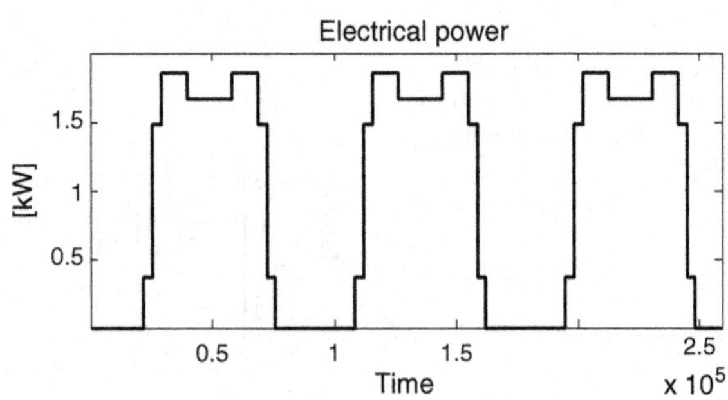

Fig. 34 Profile of the power generated by PV panel

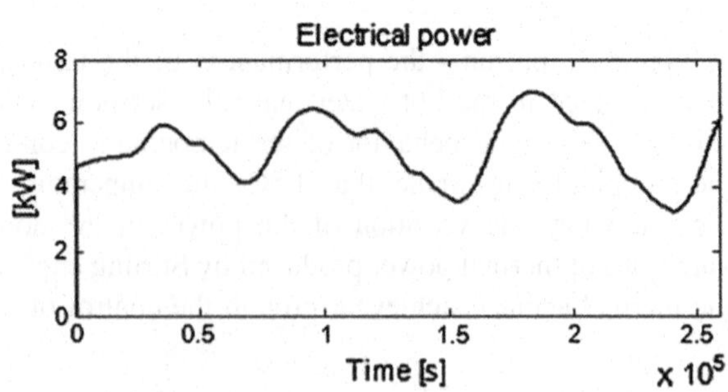

Fig. 35 Evolution of the electrical power produced by the Stirling engine

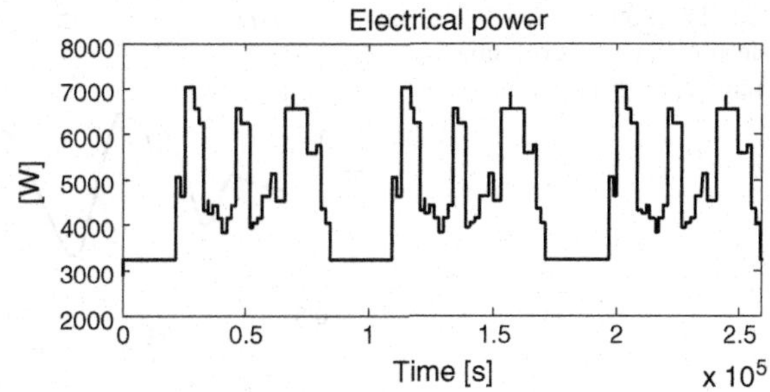

Fig. 36 Total consumed electrical power

The electric load has the following components:

- hydraulic station pumps for the adsorption chiller;
- circulation pump of the cold water toward the ventilo-convectors;
- circulation pump of the thermal agent from the pellet boiler to the hot water tank;
- circulation pump of the thermal agent from Stirling engine to the hot water tank;
- circulation pump of the solar thermal panel.

Figure 36 presents the graph of the total consumed electrical power. This graph was established taking into consideration a series of consumptions that occur only while the equipments operate (pellet boiler, DHW) and it was considered that the cooling equipment, together its pumps, operates permanently.

In the given conditions regarding the evolutions of the source power and of the load, the battery voltage and the energy accumulated in the battery are presented in Figs. 37 and 38. It can be noticed that a permanent regime is obtained. Therefore, the operation in dynamic regime of the electrical subsystem takes place accordingly to the requirement of the battery voltage control to the set point (48.5 V).

In the thermal subsystem there are three sources of energy: Stirling engine, the ST panel, and the pellet boiler, which are controlled by the temperature controller of the thermal agent in the hot water tank. The evolutions of the powers of the ST panel and the pellet boiler are presented in Figs. 39 and 40. The evolution of the thermal power of Stirling engine is the one from the graph of its electrical power, presented in Fig. 35, multiplied by three.

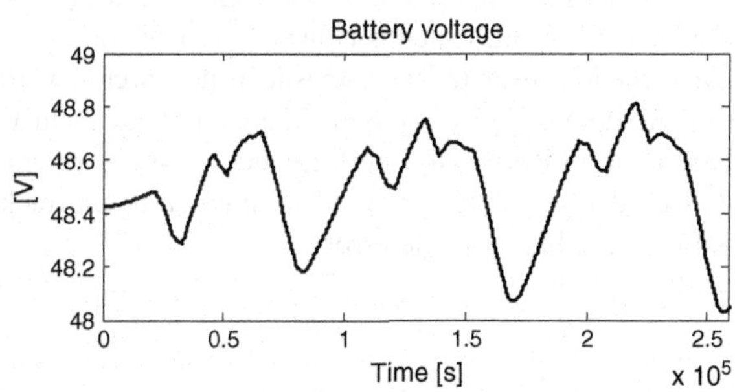

Fig. 37 Battery voltage evolution in Summer Regime

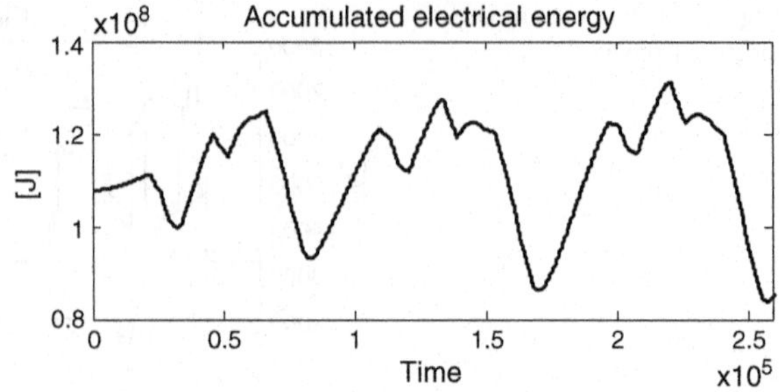

Fig. 38 Accumulated electrical energy evolution in Summer Regime

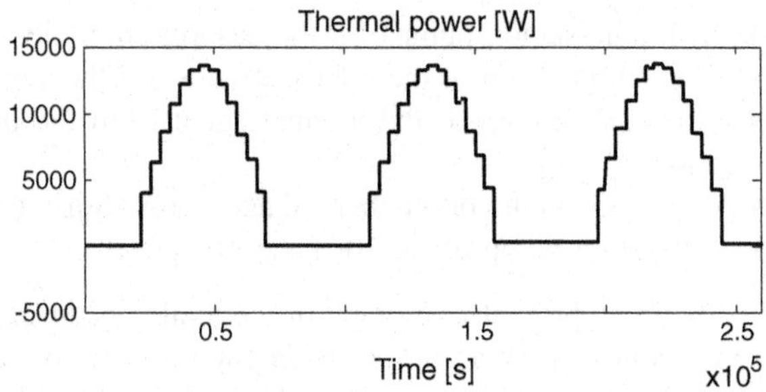

Fig. 39 Profile of the power generated by the ST panel

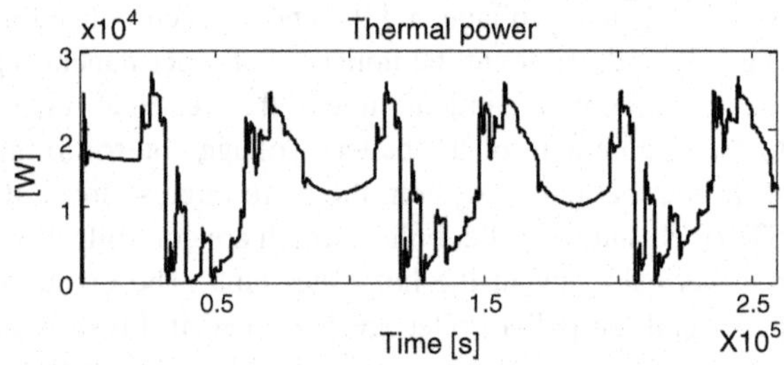

Fig. 40 Pellet boiler power evolution in Summer Regime

The thermal load contains two components: the power absorbed by the conditioning equipment, which is considered to be constant at approximately 30 kW, and the variable power in the domestic water circuit, which was presented in Fig. 26 (for the domestic water circuit, the same load as in winter regime is considered). Figure 41 illustrates the performance of the temperature control loop. It can be noticed that the temperature of the thermal agent tracks the setpoint value (80 °C) with admissible dynamic errors.

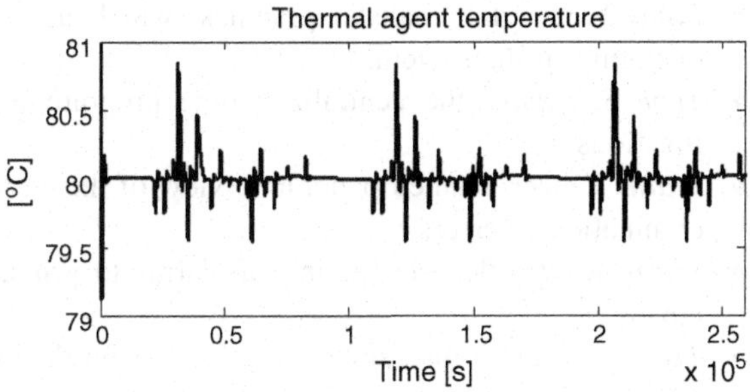

Fig. 41 Thermal agent temperature evolution in the hot water tank in Summer Regime

11 Design of the Control Subsystem

Nicolae Badea and Marian Barbu

One of the basic functions of the automation and control system of the building is a timely control of the procedures and processes that provide an efficient energy operation.

The control programs make sure that the lighting and heating are not automatically shut down at the end of the day, that the building temperature is reduced during the night and that the mCCHP installation does not function more than it is necessary. The operations of switch on time provided by the automation and control system are implemented through the BACnet standard, using the BACnet objects "Schedule" and "Calendar". This function allows a flexible management of the building. From ordinary programs to unusual exceptions, these functions make the monitoring more flexible. By using the BACnet standard functions, the time programs BACnet can be operated from the entire system from the touch panel unit type, as well as from the management station. For the exchange of information between its own components, three standard protocols are used, namely BACnet, LONWORKS, and Konnex (KNX) S-mode (EIB). The BACnet communication protocol is used for the exchange of information between the distributed automation equipment and among the touch panel operating units, using LonTalk or point-to-point (PTP, modem or null modem) as transport medium. Inside the room automation, the communication is done through the LONWORKS and KNX S-mode (EIB) standard.

The clearly structured, modular software, oriented on objects of the management station is based on the latest standard Windows technology. The functional purpose and the ease in using the software reduces the operating costs and those of getting familiar with it, while the operational level is maintained. The main applications transmitted through web are described as follows:

- *Plant Viewer* the realistic graphics, which allows the rapid monitoring and operating of the system.
- *Time Scheduler* the centralized time programming of the building's service functions.
- *Alarm Viewer* detailed panoramic view of the alarms for rapid localization and elimination of errors.
- *Alarm Router* flexible routing of alarms to printers, faxes, mobile phones, or emails.
- *Trend Viewer* Comfortable analysis of trend data for the optimization of operations and the increase of EFF.
- *Report Viewer* questions for the meeting of the clients' necessities and showing them in reports. The reports provide information on the analysis on the functioning of the installation as well as on the evaluation and documentation.
- *Object Viewer* an efficient tool for the navigation through the hierarchic structure of all the data points of the system. These points can be read or manipulated, depending on the user's access rights.
- *Log Viewer* The alarms, errors, and user's activities are registered in chronological order and may be displayed for a subsequent evaluation, depending on the necessity.
- *Database Audit Viewer* The registration of the unauthorized modifications in the data base which guarantees the highest integrity possible of the data.
- *System Configurator* It is used for the general configuration of the management station and of the associated applications.
- *Graphics Builder* The efficient creation of the installation graphics.
- *BACnet, OPC, KNX S-Mode (EIB), LONWORKS drivers* For the direct integration of various interfaces in the management station.
- The online tool that operates on the existent system.

The user's access in the system is simplified through the specific start-up sequences of the preselected programs and of the installation.

The *Plant Viewer* application shows the areas of the building of the installation associated in graphic form. The user works interactively with these images for the monitoring and control of the data points in the building. The values may be modified and the alarms noticed. The measured values, the set points, the functioning modes, and the alarms are displayed on the screen in real time and continuously actualized. The display form is determined during the visualization. The modifications either are indicated by a symbol, for example an animation or change in forms and colors, or by the movement, the change in color or in the text of the affected value.

The *Time Scheduler* application is used for the centralized programming of all controlled functions depending on the time, of the building's installations, including the control system for individual rooms.

The *Alarm Viewer* application shows the alarms depending on their type and provides helpful information depending on the action needed by the system.

Together with the extensive filter and the search functions, Alarm Viewer facilitates rapid and clear access to the necessary information.

The *Trend Viewer* application is used for the reviewing of the process's actual data in real time (online) and of the subsequent data (off-line) for a certain period. Trend Viewer is user friendly, being used to optimize the operation of the installation and to reduce the consumed energy. A few of the Trend Viewer's facilities are presented below:

- The registration of the process value and of the measured values for a period of time and the minimal and maximum values registered in the graphics according to a period time;
- The monitoring of the present conditions of the installation;
- The optimization of the installation and its adjustment;
- Response times for the data base support.

12 System Interface

Nicolae Badea

The experimental data processing and interpretation is possible with SCADA interface. We have 148 variables monitored for experimental data processing and interpretation. The monitoring, control, and protection system through a WEB terminal with real time access from the residential building, equipped with a mCCHP system, is divided in two categories:

a. *The monitoring and control system of the residential house*

In the monitoring, control, and protection system of the residential house, the monitored parameters are taken directly from the inside of the house with the use of the dedicated equipments and introduced in the database. By using the software, the technical parameters from the data base, setting the conditions necessary for a good interior comfort, the access in each room are processed and displayed on a SCADA interface (Fig. 42).

b. *The monitoring, control, and protection system of the mCCHP installation.*

In the monitoring, control, and protection system of the mCCHP installation, the monitored parameters are taken directly from the process and introduced in the database. By using the software, the technical parameters from the database are processed and displayed on a SCADA interface for thermal circuits (see Fig. 43) and electrical circuits (Fig. 44). On the same SCADA interface, one can see the parameters of the mCCHP system as well as the conditions inside the house with the navigation menu.

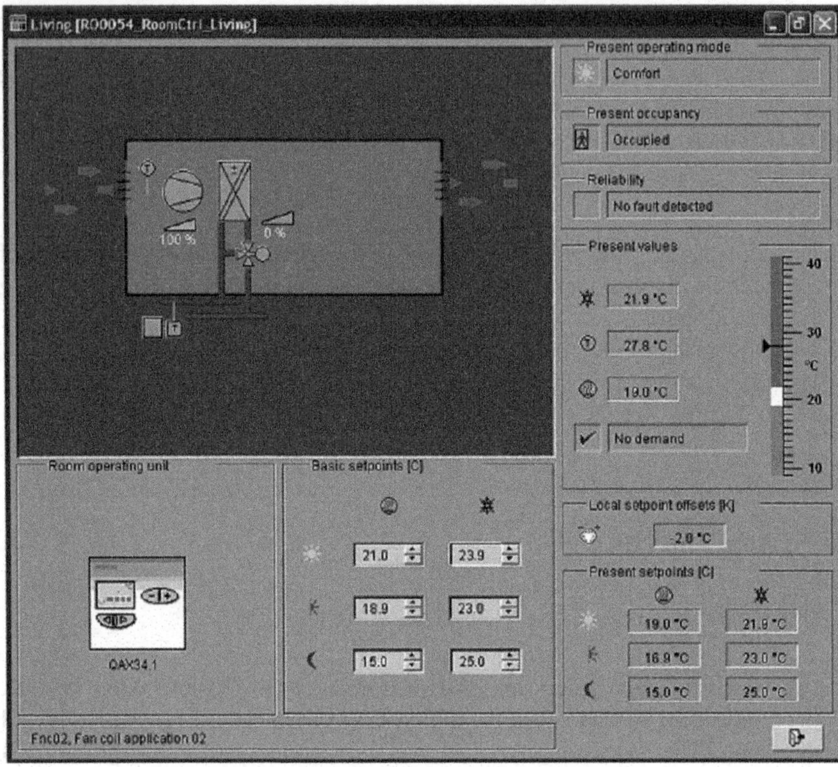

Fig. 42 Menu for house condition

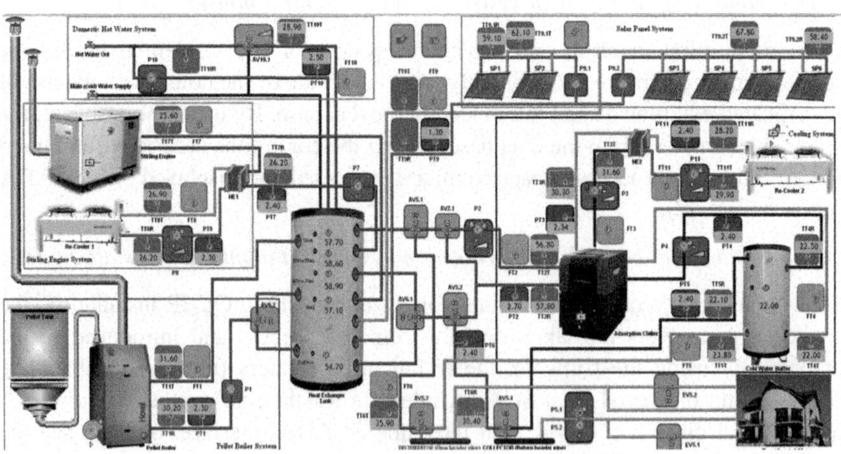

Fig. 43 The thermal monitoring of the experimental mCCHP installation—SCADA interface [17]

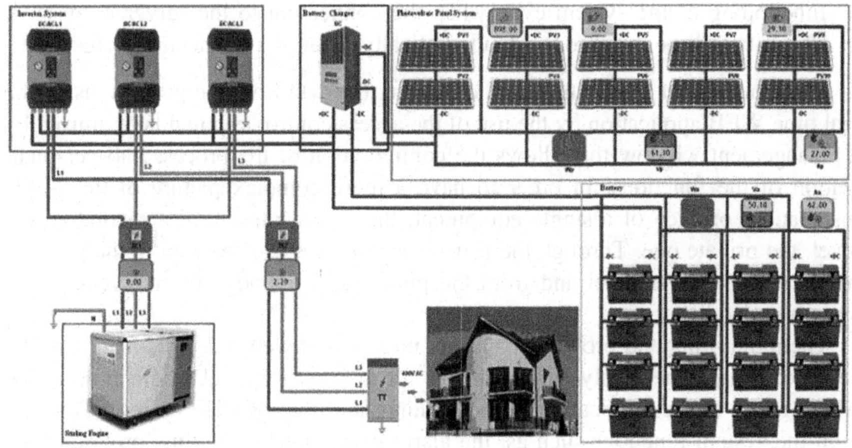

Fig. 44 The electrical monitoring of the experimental mCCHP installation—SCADA interface [17]

13 The mCCHP-SE-RES System Data

Nicolae Badea

The solution of the access in real time by the use of the web in order to process data allows the mCCHP installation to be monitored, unifying in this way the functionality of the remote control with the applicability of the WEB system. The developed application is easy to be configured and easily accessed; it can be visualized and managed to a certain extent in real time by the use of the Internet browser, like any other WEB application, with any type of connection and from any location. The use of the application, the control of the data and of the installation monitoring, allows for data to be stored in a database on a dedicated server, having the ability to connect through an ID and a password.

The software application together with the touch screen terminal from the residence is able to collect, validate, and monitor the data transfer in real time, depending on the rules and access rights configured in the data base.

Ways of management and monitoring:

- *Online*, when all the monitoring operations managed by the program and applied to the open "web terminals" are validated in real time through the control of the access politics registered in the database.
- *Off-line*, when the access politics created in the program are automatically transmitted to the operator display and to the management station inside the house. In this case, the connection to the application is no longer possible through a web browser, as it is made directly from the house.
- *Mixed*, when the system normally works online, but, due to the transfer flexibility of the data through the application's access politics, it allows the correct

functioning of the system even when the connection to the server is missing. The data will be synchronized automatically when the connection is restored.

The functions of transmitting and collecting the data from the process, using the real time WEB application by the use of the access control terminal is optimized by a management window that allows the configuration of the process and personalization of the functions. In order to have a more complex picture of the whole installation, or even of a single equipment, there was introduced a second access level, the private one. Through the private access, one can see data in the process, but from this access level and from the public one, nobody can intervene in the system or modify data.

The monitoring and protection system, through the web access, is made up of a series of equipments freely programmable for management and automation for the whole range of the application in the building and the mCCHP system. Together with the system functions such as: the alarm management, the time programmers, and the trend registration, combined with the sophisticated command functions, the monitoring and protection system on WEB access represents a versatile asset of the building. The innovative web technology, the large databases as well as the open communication make this to be an advanced application.

Appendix 1: Experimental System Pictures

Picture 1 Adsorption chiller and pellets boiler

Picture 2 Chiler pumps station

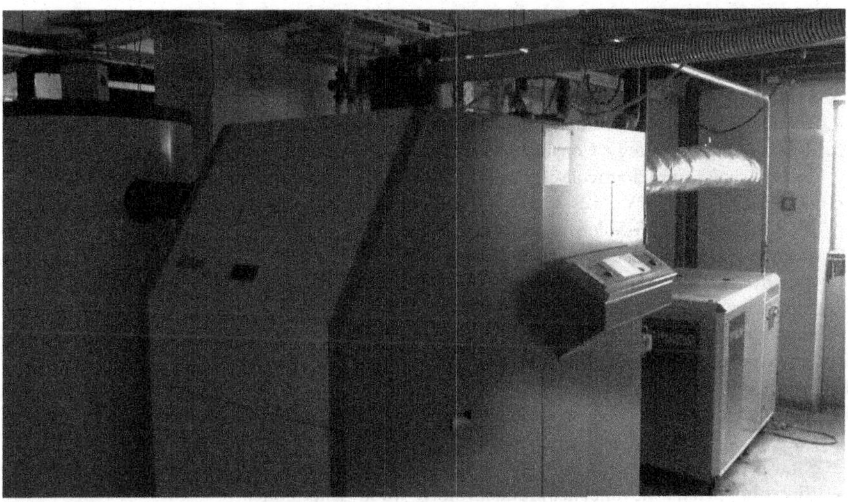

Picture 3 Pellet boiler and Stirling engine

Picture 4 Experimental house

References

1. https://www.google.ro/maps/preview?source=newuser-ws
2. SR 4839:1997 Heating system—Number DD
3. STAS 6648/2—Ventilation and air conditioning. External climatic parameters (in Romanian)
4. SR 1907-1-1997. Calculation prescriptions. (in Romanian)
5. Roland Berger Strategy Consultants SRL (2010) Green Energy in Romania. http://oldrbd.doingbusiness.ro/ro/5/articole-recente/1/373/green-energy-in-romania
6. EU Commission, PV GIS—2011 GeoModelSolar. http://solargis.info/doc/_pics/freemaps/1000px/ghi/SolarGIS-Solar-map-Romania-en.png
7. http://re.jrc.ec.europa.eu/pvgis/apps4/pvest.php
8. http://www.solarshop-europe.net/product_info.php?products_id=1381
9. http://www.solarlinerenovables.com/gb/batteries/671-bateria-monoblock-solar-12v-fs.html#/modelo-fs_250
10. http://www.solar-electric.com/xaxwmp60amps.html
11. Stirling V161. www.cleanergyindustries.com
12. Studer Innotec from Xtender series 4000-48 XTM model. http://www.studer-inno.com/?cat=sine_wave_inverter-chargers&id=432
13. Kindspan Solar data sheet http://www.makethesunwork.com/
14. Biolyt http://www.hoval.co.uk/products/biolyt-wood-pellet-boiler/
15. Sortec http://www.sortech.de/en/adsorption-chiller-aggregates/
16. http://www.cordivari.it/product.aspx?id=2&gid=136&prd=254&lng=0
17. http://www.mcchp.ugal.ro/

Printed in the United States
By Bookmasters